Toward Sustainable Development

An
Ecological Economics
Approach

Toward Sustainable Development

An Ecological Economics Approach

by

Philip A. Lawn

CRC Press
Taylor & Francis Group
Boca Raton London New York

CRC Press is an imprint of the
Taylor & Francis Group, an **informa** business

ISEE
International
Society for
Ecological
Economics

CRC Press
Taylor & Francis Group
6000 Broken Sound Parkway NW, Suite 300
Boca Raton, FL 33487-2742

First issued in paperback 2020

ISBN 13: 978-0-367-57895-4 (pbk)
ISBN 13: 978-1-56670-411-3 (hbk)

Visit the Taylor & Francis Web site at
http://www.taylorandfrancis.com

and the CRC Press Web site at
http://www.crcpress.com

Library of Congress Cataloging-in-Publication Data

Lawn, Philip A.
    Toward sustainable development : an ecological economics approach / Philip A. Lawn.
        p.   cm. (Ecological economics series)
    Includes bibliographical references and index.
    ISBN 0-56670-411-1 (alk. paper)
    1. Sustainable development. 2. Environmental economics. I. Title.
    II. Ecological economics series (International Society for Ecological Economics)
HD75.6 .L38   2000
338.9'27—dc21   00-039082

Library of Congress Card Number 00-039082

Toward Sustainable Development

An Ecological Economics Approach

by

Philip A. Lawn

CRC Press
Taylor & Francis Group
Boca Raton London New York

CRC Press is an imprint of the
Taylor & Francis Group, an **informa** business

ISEE
International
Society for
Ecological
Economics

CRC Press
Taylor & Francis Group
6000 Broken Sound Parkway NW, Suite 300
Boca Raton, FL 33487-2742

First issued in paperback 2020

ISBN 13: 978-0-367-57895-4 (pbk)
ISBN 13: 978-1-56670-411-3 (hbk)

Visit the Taylor & Francis Web site at
http://www.taylorandfrancis.com

and the CRC Press Web site at
http://www.crcpress.com

Library of Congress Cataloging-in-Publication Data

Lawn, Philip A.
    Toward sustainable development : an ecological economics approach / Philip A. Lawn.
        p. cm. (Ecological economics series)
    Includes bibliographical references and index.
    ISBN 0-56670-411-1 (alk. paper)
    1. Sustainable development. 2. Environmental economics. I. Title.
    II. Ecological economics series (International Society for Ecological Economics)
HD75.6 .L38   2000
338.9'27—dc21   00-039082

Library of Congress Card Number 00-039082

Author

Philip A. Lawn received his Bachelor of Economics in 1991 from the Flinders University of South Australia. Dr. Lawn worked as a tutor at Flinders before embarking on postgraduate studies at Griffith University in Brisbane, Australia. In 1998, he received his Ph.D. in ecological economics and returned to Flinders University to take on a position as a lecturer in environmental and ecological economics.

Dr. Lawn's research interests include the valuation of environmental services, green national accounting, the development of alternative sustainability and welfare indicators, modelling the interaction between economic and environmental systems, and the design of policy instruments and institutional mechanisms to facilitate a transition toward sustainable development. In recent years, he has published a number of articles in the field of ecological economics. He is a member of the International Society for Ecological Economics as well as its Australian and New Zealand branches.

Contents

Tables

Figures

Acknowledgments

This book is the product of a long and continuing intellectual journey that began around 15 years ago. Rarely is a journey undertaken alone or without some guidance and assistance. This journey is no exception. Since this book is a rework of my Ph.D. dissertation, I would like to begin by thanking my Ph.D. supervisors, Aynsley Kellow and John Tisdell, for their constructive comments and advice during my time at Griffith University. I would also like to thank the academic and administrative staff at Griffith University's Faculty of Environmental Sciences and the Office for Research. A Ph.D. scholarship from the Australian Federal Government is also greatly appreciated. Special mention must also be made of the contribution made by Richard Sanders, a fellow Ph.D. student at Griffith University. Not only did Richard help me gather the data I needed for this book, but the book benefited enormously from the many discussions we had over a cup of tea in the Common Room. I also owe a considerable intellectual debt to all my past teachers and the ecological economists who have both educated and inspired me.

Of my past teachers, I, particularly, would like to thank Stewart Fraser, Richard Damania, and Dick Blandy for their contribution during my time as an undergraduate and Honours student at the Flinders University of South Australia. As for the ecological economists from whom I have learnt so much, the contribution from two of them is worthy of special acknowledgment. The first is that of Robert Costanza. Costanza not only offered a number of useful suggestions to incorporate into this book but was wholly responsible for initiating the contact between Lewis Publishers and myself. The second is the contribution from Herman Daly whom I feel I owe my largest intellectual debt. As anyone familiar with Daly's work will soon appreciate, the impact of Daly on my thinking and this book is significant. I would also like to thank Mick Young and Richard Norgaard for making some helpful comments and suggestions on an earlier draft of this book. However, my greatest debt is owed to my friends and family, in particular my father and mother, Graham and Lesley Lawn. Without their love and support, this book would never have been possible.

"The significant problems we face cannot be solved at the same level of thinking we were at when we created them."

— *Albert Einstein*

chapter one

Introduction

1.1 The aims and objectives of the book

As the title of this book suggests, the principal aim is to adopt an ecological economic approach to sustainable development (SD) in order to suggest the ways and means of moving toward a just, sustainable, and efficient society. It will endeavour to achieve this by challenging the conventional understanding of SD and the neoclassical models adopted by most economists to address this critical issue. While the rationale for adopting an ecological economic approach should become self-evident as the book unfolds, it stems from the following broadly conceived research questions. First, can SD be achieved by employing conventional and subjectivist philosophies as well as reductionist and atomistic-based models? Second, because we aim to show that it cannot, what philosophical and theoretical approaches provide a better understanding of the SD concept itself and a more insightful conceptual framework from which to draw appropriate policy conclusions?

To achieve its principal aim, this book is set out in five main parts. Parts I, II, and III constitute its theoretical foundation. It is within these three parts that the guidelines, tools, and the conceptual framework necessary to achieve SD are established. Part I begins with a review and assessment of existing SD definitions and concepts. After showing why most are devoid of operational value, an ends-means spectrum is introduced as a theoretical foundation upon which to establish an ecological economic-based SD concept. At the conclusion of Part I, an SD concept is put forward to serve as a benchmark for the remainder of the book.

Part II deals with the macro- and microeconomics of SD. In the macroeconomics chapter, a distinction is made between "economic" and "uneconomic" growth. By demonstrating why growth is eventually uneconomic, the theoretical notion of an *optimal macroeconomic scale* is introduced. An optimal macroeconomic scale is one where the physical scale of a nation's macroeconomy and the qualitative nature of the goods with which it is comprised maximises the sustainable economic welfare enjoyed by its citizens. The notion of an optimal macroeconomic scale is a crucial one because it enables one to understand how a nation can achieve SD without the perceived need for continued growth. In the remaining chapters of Part II, the three microeconomic-related aspects of economic efficiency, the market, and relative prices are considered closely with the aim of showing how each can assist in the attainment of an optimal macroeconomic scale. It is through a better understanding of the macro- and microeconomics of SD that the respective policy goals and instruments required to achieve SD are revealed.

Part III is devoted to the establishment of a more appropriate conceptual framework from which to draw SD-based policies. It begins with an assessment of the neoclassical economic paradigm and aims to demonstrate that it cannot be usefully applied to prescribe SD policies, even in a modified form, because it is methodologically wedded to the principles underlying a Newtonian-like, atomistic-mechanistic world view. A Newtonian world view, as we will endeavour to show, is incommensurate with concrete reality. In response, a *co-evolutionary-based* conceptual framework is developed on the understanding that a co-evolutionary world view more closely reflects the evolutionary and interdependent nature of the real world. Because a co-evolutionary-based conceptual framework recognises such real-world phenomena as novelty, surprise, human ignorance, and the path-dependent nature of evolutionary systems, it is shown to have important institutional and policy implications that would otherwise go unnoticed if a neoclassical approach was employed. A co-evolutionary-based conceptual framework also reveals an extraordinary feature of the market — namely, that it is a co-evolutionary feedback mechanism that may or may not facilitate the movement toward SD. To what extent it does depends very much on the market's price-determining parameters.

With the theoretical foundation of the book out of the way, Part IV focuses on how to move toward SD. It does this by putting forward a number of national accounting reforms to qualitatively measure SD. It then outlines the type of policies deemed necessary for SD to be achieved. While the aim of the prescribed policies is to facilitate more appropriate forms of human behaviour, the preaching of required policy changes does not guarantee their implementation. For this reason, Part IV concludes with a chapter dealing with the various sources of human power that facilitate or impede the movement towards SD, in particular, the relationship between human morality and humankind's ability to initiate appropriate action.

The final part of the book, Part V, includes four appendices outlining the methods and the data sources used to calculate the items that make up the national accounts revealed in Part IV. To assist the reader, each appendix includes a set of tables to show how the individual items have been calculated.

1.2 What is ecological economics?

Given that the aim of this book is to adopt an ecological economic approach to SD, the remainder of this chapter is devoted to explaining what ecological economics is and how it differs from conventional economic paradigms.* The official journal of the International Society for Ecological Economics describes ecological economics as a transdisciplinary paradigm that seeks to integrate the study and management of nature's household (the ecosphere) and humankind's household (the macroeconomy). Ecological economics has evolved largely out of a general failure of the standard neoclassical paradigm to deal satisfactorily with the *interdependence* of social, economic, and ecological systems. Thus, in many ways, the development of an ecological economic paradigm has been a concerted attempt to overhaul the standard neoclassical approach by bringing the many false "pre-analytical visions" underpinning its assumptions into line with biophysical and existential realities. Having said this, ecological economics does not abandon conventional economics altogether. To the contrary, ecological economics remains firmly wedded to a number of its fundamental underlying principles.

While the preceding definition provides a clear picture of what ecological economics seeks to achieve, its weakness lies in the fact that it does not indicate how ecological economics differs from the conventional economics paradigm. The best way to explain the difference between the two paradigms is to highlight the shortcomings of mainstream economics. In a nutshell, the problem with mainstream economics is that it is has been too physical in relation to human welfare but not physical enough in terms of the economic process and what is required to sustain it. Economics has been too physical in relation to human welfare because many economists wrongly believe that human welfare is a direct function of the quantity of goods produced and consumed. However, as Georgescu-Roegen (1971) pointed out, human welfare depends, not on the rate of a physical flow, but on a *psychic flux* — the psychic enjoyment of life. While a psychic flux cannot be experienced without the existence of physical goods, it is not the rate at which physical goods are produced and consumed that determines the intensity of humankind's psychic enjoyment of life. It is determined primarily by the quantity of the stock of human-made capital (at least up to a certain amount), the quality of the stock, and its ownership distribution.

Paradoxically, economics has not been physical enough in the sense that it has long overlooked the entropic connection between the economic process and the ecosphere and the fact that the former, to be sustained, requires the continual absorption of low entropy resources and the excretion of high

* For a more thorough explanation of what ecological economics is and what constitutes some of its core principles, see Lawn (1999a).

entropy wastes.* This connection has considerable implications for sustaining the economic process because the only source of low entropy resources and the ultimate repository of all high entropy wastes is natural capital (i.e., forests, soil, rivers, the seas and oceans, mines, wetlands, and the earth's atmosphere). In addition, natural capital is the sole provider of life-support services. Ecological economists are at pains to emphasise three of these implications. First, natural and human-made capital are complementary forms of capital. Second, natural capital stock maintenance is necessary to sustain the economic process. Third, markets are limited in their ability to achieve desirable and ecologically sustainable human goals.

Many people are familiar with the arguments in support of the first two implications, even if they disagree with them. The argument supporting the third implication is less obvious and therefore requires further explanation. Consider the following. Economics is essentially the study of how to best allocate scarce means to satisfy competing ends. Insofar as exchange mechanisms are very effective at facilitating the efficient allocation of scarce resources, economics and markets have a very important role to play in the movement toward SD. However, market prices only provide information regarding the scarcity of one thing relative to another (e.g., the scarcity of oil relative to coal). It is because the market is very effective at revealing *relative* scarcities that it constitutes an effective allocative mechanism. But ecological sustainability is a question of the *absolute* scarcity of the very nonsubstitutable stuff that sustains the economic process — namely, natural capital and the low entropy matter-energy it provides — and no amount of

---

* To understand what is meant by low and high entropy matter–energy, the importance of the first and second laws of thermodynamics requires explanation. The first law of thermodynamics is the *law of conservation of energy and matter*. It declares that energy and matter can never be created nor destroyed and thus imposes a condition of finitude. The second law is the so-called *entropy law*. It declares that whenever energy is used in physical transformation processes, the amount of usable or "available" energy always declines. While the first law ensures the maintenance of a given quantity of energy and matter, the entropy law determines that which is usable. This is critical because, from a physical viewpoint, it is not the total quantity of matter–energy that is of primary concern, but the amount that exists in a readily available form.

The best way to illustrate the relevance of these two laws is to provide a simple example. Consider a piece of coal. When it is burned, the matter–energy embodied within the coal is transformed into heat and ash. While the first law ensures the total amount of matter–energy in the heat and ashes equals that previously embodied in the piece of coal, the second law ensures the usable quantity of matter–energy does not. In other words, the dispersed heat and ashes can no longer be used in a way similar to the original piece of coal. To make matters worse, any attempt to reconcentrate the dispersed matter–energy, which requires the input of additional energy, results in more usable energy being expended than that reconcentrated. Hence, all physical transformation processes involve an irrevocable loss of available energy or what is sometimes referred to as a "net entropy deficit." This enables one to understand the use of the term *low entropy* and to distinguish it from *high entropy*. Low entropy refers to a highly ordered physical structure embodying energy and matter in a readily available form such as a piece of coal. Conversely, high entropy refers to a highly disordered and degraded physical structure embodying energy and matter that is, by itself, in an unusable or unavailable form such as heat and ash. By definition, matter–energy used in economic processes can be considered low entropy resources whereas unusable by-products can be considered high entropy wastes.

relative scarcity information can render the market effective at ensuring the sustainable use of natural resources. Similarly, distributional equity requires a fair and just distribution of income and wealth, yet the market has no organs to sense or taste fairness and justice. Hence, while markets are the most useful mechanisms at facilitating the efficient allocation of scarce resources, they are woefully inadequate at ensuring an ecologically sustainable rate of resource use and an equitable distribution of income and wealth (to be fully explained in Part II of this book). It is this recognition of the limits to both economics and the useful application of economic instruments, such as markets, that is probably the single most important contribution of ecological economics.

There are, of course, many other features of the ecological economics paradigm that distinguish it from standard economics. Unlike standard economists who depict the macroeconomy as an isolated system whereby the output of the economic process is considered circular, self-feeding, and self-renewing, ecological economists view the macroeconomy as an open subsystem of the larger ecosphere. By viewing the macroeconomy this way, ecological economists are forced to consider two important questions. First, how big can macroeconomic systems get before they are no longer ecologically sustainable? Second, and perhaps more importantly, how big can macroeconomic systems get before the additional benefits of growth — the increase in the psychic enjoyment of life — are exceeded by the additional cost in terms of the reduced ability of the ecosphere to provide its source, sink, and life-support services? Clearly, by recognising macroeconomic systems as subsystems of the larger ecosphere, ecological economists are forced to consider the possibility of "uneconomic" growth. Such a consideration is not immediately obvious for the mainstream economist who views the macroeconomy as an isolated system. Indeed, for some, it is never an issue to begin with. In all, whereas a maximised rate of growth constitutes the primary macroeconomic objective of mainstream economists, for ecological economists it is the attainment of an optimal macroeconomic scale.

Although ecological economics recognises the need to maintain intact the ecosphere's source, sink, and life-support services, it remains an anthropocentric field of study in the sense that it is primarily concerned with the long-run survival and welfare of human beings. However, it differs to the anthropocentric nature of conventional economics insofar as it is wedded to principles underlying a *biocentric* rather than *egocentric* world view. That is, as much as ecological economics is concerned with human-related issues, the ecosphere is valued not just because it provides a range of instrumental services for humankind, but because it exists in its own right and is home for many sentient, nonhuman creatures. Consequently, ecological economics is concerned with the intrinsic as well as the instrumental values of natural capital.

The importance of the ecosphere's intrinsic value leads to one of the more crucial elements of the ecological economic paradigm, namely, the moral and existential dimensions of human endeavour.* Their significance rests on the basis that existential scarcity and the scarcity of moral capital itself is probably as important as ecological scarcity (Hirsch, 1976). Ecological economics recognises, for instance, that shared social values play an indispensable role in the successful operation of an individualistic-based, contractual economy and that self-motivated behaviour can only operate effectively in tandem with supporting social and moral capital. In addition, ecological economics understands that the moral and existential dimensions of human endeavour are necessary for economic activity to be in any way purposeful. Finally, and as will be explained in Part IV, the growth of moral capital is necessary to initiate the movement toward an SD future (Boulding, 1991).

In another critical departure from standard economic paradigms, ecological economics rejects the atomistic-mechanistic (A-M) foundations of neoclassical models. Ecological economics attempts to conform more closely with reality by basing the conceptual framework upon which it draws conclusions and policy prescriptions on a *co-evolutionary* world view (Norgaard, 1985, 1988, and 1992). Co-evolution involves the reciprocal responses of any two or more closely interdependent and evolving subsystems. In a co-evolutionary world, the evolution of individual subsystems — whether they be a small ecosystem or a large macroeconomy — is characterised by two opposing tendencies. The first is a self-assertive tendency of systems to preserve their individual autonomy. The second is an integrative tendency of systems to function as part of a larger whole. Just in what direction or in what form all subsystems evolve, adapt, and mutate depends very much on how they respond to changes occurring within the broader global ensemble. However, unlike the A-M epistemological foundations of the neoclassical paradigm where the future states of all subsystems are considered both predictable and reversible, the same cannot be said to exist in a co-evolutionary world. Novelty, surprise, and human ignorance abound in a co-evolutionary world because the underlying parameters of all interacting systems are incessantly changing. Consequently, it is often only possible to make predictions in very broad rather than specific terms (Faber and Proops, 1990). It is the generation of uncertainty and ignorance that is of great interest to ecological economics because it has a marked impact on the policy setting process and the design of appropriate institutional mechanisms. Categorically speaking, uncertainty and ignorance affect policy setting in the following three ways:

---

* It is an appropriate time to explain what we mean by "moral" in the context of this book. By moral, we mean virtuous conduct according to civilised standards of right and wrong. This having been said, at no stage will we specifically define what constitutes a morally acceptable form of conduct. Nor will we be suggesting that a particular subgroup's notion of right and wrong should be imposed on the rest of society. Indeed, in a world where most contemporary societies are characteristically pluralistic, it is our belief that a growing acceptance and strengthening of civilised standards of right and wrong — what might be deemed "moral growth" — is best achieved via an envisioning process involving each and every member of society.

1. Policies must be based on a prior recognition that humankind's teleological freedom and ability to achieve worthwhile goals depends upon a recognition that it is a slave to the "rules" governing evolutionary processes, as opposed to being a slave to the evolutionary process itself, which it clearly is not,
2. Policies must incorporate the *precautionary principle* by making allowances for the emergence of unforseen circumstances (e.g., Ecological damage from economic activities that cannot be predicted *a priori*).
3. Policies must be based on human management rather than planetary management.

In view of what has been briefly outlined in this chapter, what is ecological economics? As we see it, ecological economics can be considered a form of economics that takes account of the interdependent and co-evolutionary relationships that exist between the various economic, social, and ecological systems that make up the global ensemble. In doing so, ecological economics recognises economic instruments to be useful solely as a means to achieve an efficient allocation of scarce resources. Because allocative efficiency does not guarantee ecological sustainability or distributional equity, ecological economics recognises the limits to how far economic instruments can foster humankind's sustainable development interests. From a macroeconomic perspective, ecological economics emphasises the desirability of growth insofar as it increases a nation's sustainable economic welfare. Growth of a nation's macroeconomy should cease once sustainable economic welfare begins to decline. A nation's macroeconomy should never exceed its maximum sustainable scale.

1.3   *Epilogue*

This chapter has set out to achieve three things. First, it has highlighted the aims and objectives of the book. Second, it has provided a clear indication of both the approach being adopted and the manner in which the aims and objectives of the book will be achieved. Third, it has attempted to familiarise the reader with the basic elements of the ecological economic paradigm. In doing this, the chapter has laid the basic foundations upon which to begin the process of articulating how humankind can, as the title of book indicates, move toward sustainable development.

I

An ecological economic view of the sustainable development concept

chapter two

An overview of the sustainable development concept

2.1 An examination of a range of sustainable development concepts

Because the movement toward SD constitutes the central theme of this book, the initial aim is to establish a rigorous and credible SD concept to serve as a measuring stick for the remainder of the book. This will enable the theoretical and policy implications arising from its analysis to be considered and discussed later. According to Pearce et al. (1990), little headway has been made in terms of establishing a rigorous SD concept. This, they believe, has resulted in few and generally unpersuasive attempts to operationalise SD and incorporate it into the decision-making process. Indeed, without intellectual clarity and rigour, the concept of SD is in danger of becoming what Lele (1991, page 607) has described as "another cliche-like appropriate technology — a fashionable phrase that everyone pays homage to but nobody cares to define." Given the vagueness of the SD concept, there has been a growing tendency to label anything deemed subjectively and arbitrarily "good" as an example of SD. This has not only relieved decision makers from having to demonstrate whether a proposed policy is genuinely sustainable and desirable, it has allowed many ill-conceived policies to parade as SD-based initiatives in order to gain broad acceptance and credibility.

As a forerunner to the establishment of a credible SD concept, it is my belief that SD stands as an ethical guiding principle. Clearly, SD is not a futuristic state that can be visualised and aimed for because guiding principles provide rules for appropriate action, not specific or concrete targets. This has three major implications. First, it is not possible to design an exact blueprint for achieving SD. Second, for an SD concept to be sufficiently credible, it must incorporate parameters to act as rules or precepts for appropriate action. Finally, any precepts established must be erected from a sound theoretical foundation. This third implication is of particular importance given that the widespread failure to undertake such a task has left the concept of SD without a "coherent theoretical core" (Adams quoted in Mitlin, 1992, page 113). Not only has this deprived SD of considerable operational value, it has led to severe criticisms of the SD concept itself (e.g., Beckerman, 1992). More importantly, it has increased the difficulty of ascertaining how SD can be properly institutionalised and by what indicators it can best be measured.

2.1.1   A brief look at the process that led to the establishment of the sustainable development concept

To understand why the SD concept lacks a coherent theoretical core, one needs to carefully consider the historical process which culminated in the publication of the Brundtland Report by the World Commission on Environment and Development (WCED, 1987). It was from this Report that SD became firmly established as a new catch cry broadly embraced by governments and development agencies alike. An examination of this type demonstrates the extent to which the biases and prejudices embodied within the concept are a legacy of the very process which led to its initial conception and ensuing development.

Exactly when the historical process began is highly debatable and is reflected by the many wide and varied opinions that exist within the SD literature. For instance, while Barbier (Mitlin, 1992) believes the SD concept has its origin in the 1970s, Hundloe (1991) suggests the evolution of the SD concept can be traced as far back as the origin of the economics discipline itself. Hundloe bases his argument on a strong belief that the concept has logical links with the founders of economics and others who have since assisted in the development of the discipline to its present state. Some commentators have argued that rather than having been born out of any single historical event, SD has evolved from the foundations laid by critical actions and events in the past. An example of this view is one expressed by van den Bergh and van der Straaten (1994). They insist that SD emerged as a result of the following events

- The Stockholm Conference on the Human Environment in 1972 and the subsequent establishment of the United Nations Environment Programme (UNEP).
- The Limits to Growth Report (Meadows et al., 1972).
- The World Conservation Strategy (WCN/IUCN, 1980).
- The *U.S. Global 2000 Report to the President* (Barney, 1980), and its response, *The Resourceful Earth* (Simon and Kahn, 1984).
- The IIASA report, *Sustainable Development of the Biosphere* (Clark and Munn, 1986).
- And the previously mentioned United Nations report, *Our Common Future* (WCED, 1987).

Without underestimating the significance of any of the preceding events, it is my belief that the SD concept emerged largely from the forces generated by the "limits-to-growth" debate of the early 1970s. The limits-to-growth debate was ignited by the findings of the so-called Club of Rome Report (Meadows et al., 1972). This controversial and widely publicised report involved an in-depth study of the environmental and resource limits to growth and the potential role of technological change in removing such limits. The results it generated indicated the rapid depletion of many key resources and the fast approaching limits to the growth in resource use and the growth of macroeconomic systems. In keeping with its findings, the Club of Rome condemned what it considered to be the reckless growth orientation of the many policies and programmes implemented by both governments and international development agencies.

Given the nature of the Club of Rome Report, it became uncomfortably clear in the early 1970s that any broad acceptance of its findings would require radical policy changes directed toward the possibility of low- or no-growth economies. As a consequence, a range of socially and politically unpalatable questions in relation to economic growth quickly emerged and were hotly debated. To avoid these questions and their obvious ramifications, a palatable change in the course of the limits-to-growth debate ensued — a process that can be likened to Ruttan's (1994) description of "establishment appropriation." Establishment appropriation occurs when dominant institutions subvert legitimate external challenges by appropriating or embracing the symbols promoted by opposition forces. By playing a significant role in the evolution of these emerging symbols, dominant institutions are able to demobilise external challenges and ensure their own longevity and continued dominance. As an upshot of the process of establishment appropriation that followed the publication of the Club of Rome Report, it was the SD concept that eventually emerged as the central focus or symbol of the newly directed debate. Yet, because of the influence of dominant institutions such as the World Bank and the Organisation for Economic Co-operation and Development, the SD debate proceeded in a manner that deliberately avoided critical issues concerning the desirability of economic growth (e.g., what is the optimal physical scale of the macroeconomy?). As a consequence, dominant institutions were able to legitimate and justify the status quo. Furthermore, by ensuring that the SD concept remained vague and obscure, dominant institutions succeeded in engendering a widespread belief that growth-based policies warranted little more than cosmetic changes (Lele, 1991; Ekins, 1993).

Perhaps the clearest evidence to support the fundamental change in emphasis from the limits-to-growth debate to the SD debate comes from none other than the Brundtland Report itself. By backing away from any attempt to distinguish quantitative growth from qualitative development, the WCED avoided having to consider whether continued growth was genuinely desirable and ecologically sustainable. It simply assumed that growth was a laudable macroeconomic objective by advocating a five- to tenfold expansion of the global economy (WCED, 1987, page 47). Despite occasional references relating to the need for a qualitative change in the nature of all future economic growth, none of which directly addressed the issue of growth as opposed to development, the WCED successfully evaded many of the concerns initially raised in the Club of Rome Report.

Overall, the general vagueness and lack of intellectual rigour and clarity of the SD concept can be attributed to a number of related factors, most of which essentially emerged from the limits-to-growth debate. First, as a recently conceived concept, SD was considerably influenced by the prevailing philosophical and conceptual research paradigms and assumptions of the time (e.g., continuous growth is a SD prerequisite). Second, it was politically expedient to avoid unpalatable questions initially raised by the limits-to-growth debate. Finally, because of the lack of any genuine attempt to erect the SD concept from a sound theoretical foundation, the concept emerged devoid of a coherent theoretical core. Thus, by collectively attenuating the emergence of a key set of parameters to act as precepts for appropriate future action, business-as-usual was tacitly if not explicitly sanctioned.

2.1.2   An examination of a range of sustainable development definitions

Over the past decade, a plethora of SD definitions has emerged of which only some have been genuine attempts at establishing an intellectually rigorous and operational SD concept. The aim here is to briefly examine a number of these definitions. This is necessary, not only to build on their strengths, but to reveal why their many weaknesses are a legacy of the failure to establish an SD concept from a sound theoretical foundation.

Perhaps the most appropriate place to begin is with the definition given by the WCED itself. They concluded that SD is:

> "...development that meets the needs of the present without compromising the ability of future generations to meet their own needs" (WCED, 1987, page 43).

A few pages on in the report they argue:

> "Sustainable development is a process of change in which the exploitation of resources, the direction of investments, the orientation of technological development, and institutional change are all in harmony and enhance the current and future potential to meet needs and aspirations" (*Ibid.*, page 46).

There is little doubt that both definitions have self-evident appeal. The widespread reference to them in the SD literature is testimony to that.

However, because both definitions are devoid of specific parameters, guide-lines, or minimum positions, one must question whether they leave us any more advanced in terms of what truly constitutes SD. For example, if meeting needs and aspirations constitutes development, then

- Exactly what needs and what aspirations are the WCED referring to?
- Is the WCED referring only to basic physiological needs or the full spectrum of human needs, including "higher-order" needs?
- Does development include wants as well as needs?
- Is development achieved through the attainment of any needs, wants, or aspirations irrespective of their moral content, that is, irrespective of whether they are virtuous according to civilised standards of right and wrong?
- Who determines the needs or wants that, if attained, constitute de-velopment?
- Is development, to some degree, dependent upon how well or effec-tively these needs and wants are being met?
- Is the continued growth of macroeconomic systems necessary for these needs and wants to be attained?

And if not compromising the potential of future generations to meet their own needs is an example of sustainability, then

- What are the necessary or minimum conditions required to ensure such potential is not compromised?
- Does it mean having to bequeath posterity a stock of suitable assets?
- If so, what type of assets — human-made, naturally occuring, or both, and in what combinations?

The openness of these questions reflects the extent to which the SD concept put forward in the Brundtland Report is devoid of operational value. For instance, if sustainability means guaranteeing that future needs are never compromised, what is required to achieve sustainability will differ enor-mously between a limits-to-growth advocate and a technological optimist. While a limits-to-growth advocate would consider the conservative use of natural resources an SD priority, a technological optimist, as someone who believes technology can overcome any impending resource constraint, would not.*

The apparent shortcomings of the previously mentioned definitions can be equally extended to the majority of SD definitions. Consider the following definitions which represent the broad range of SD concepts to be found.

---

* For example, Feige and Blau (1980) suggest the substitution possibilities induced by changes in relative resource prices are so great that physical resource scarcity has no obvious economic consequences.

1. SD involves "...a pattern of social and structural economic transformations which optimises the economic and other social benefits available in the present, without jeopardising the likely potential for similar benefits in the future [.....] A primary goal of sustainable development is to achieve a reasonable and equitably distributed level of economic well-being that can be perpetuated continually for many human generations" (Goodland and Ledec, 1987, pages 35–36).
2. SD is a concept which involves "...satisfying the multiple criteria of sustainable growth, poverty alleviation, and sound environmental management" (World Bank, 1987, page 10).
3. SD is "...development that is likely to achieve lasting satisfaction of human needs and improvement of the quality of life" (Allen, 1980, page 23).
4. SD involves "...learning how long-term and large-scale interactions between environment and development can be better managed to increase the prospects for ecologically sustainable improvements in human well-being" (Clark and Munn, 1986, page 5).

Not unlike the SD concept put forward by the WCED, one could easily ask the following different but totally related questions concerning the aforementioned definitions. In terms of development considerations:

- What is meant by economic well-being and how is it related to development?
- How does one determine a reasonable and equitable distribution of economic well-being — that is, to what extent must economic well-being be redistributed to constitute development?
- Is sustainable growth possible and, if so, does it necessarily equate to development?
- In what way does the satisfaction of human needs and the improvement in the quality of life constitute development? What needs? What about wants?

In terms of the sustainability considerations implied by the above definitions and concepts,

- What are the minimum conditions required to prevent the jeopardisation of economic and social benefits in the future — that is, how can a flow of economic and social benefits be perpetuated for many generations?
- Because many generations does not imply all future generations, how many is many?
- If sustainability is conditional upon ensuring future generations are not made "worse off," what is meant by worse off? Moreover, once known, what does its prevention entail?

- How and by what means does sound environmental management constitute sustainability — in other words, could human management be more important than environmental management?
- What are the minimum requirements necessary to ensure human-kind's ecologically sustainable interaction with the environment?

2.1.3   The perception of continuous growth as a sustainable development prerequisite

Because the SD concept is vague and considerably open to individual inter-pretation, it is often taken for granted that continuous growth is a necessary condition for the achievement of SD. This, of course, is not unexpected. Apart from the broad acceptance of economic growth as a desirable macroeconomic objective, one of the principal aims of the SD debate was, through a process of establishment appropriation, to eliminate the perceived need for any radical policy adjustments in line with the limits-to-growth conclusions. Thus, the sheer absence among the majority of SD proponents to call for a halt to economic growth is simply a reflection of how successful the process of establishment appropriation has been. Evidence of the extent to which growth is regarded as an SD prerequisite can be gauged by the following three definitions gathered by Daly (1991a, pages 196–197).

1. SD is "development that maintains the highest rate of economic growth without fuelling inflation" (OECD, 1990, page 5).
2. SD is "a dynamic concept taking into consideration the expanding needs of a growing world population, implying by this a steady rate of growth" (Sachs, 1989, page 5).
3. SD requires "...a sustainable growth in the rate of increase in economic activity" [quote appears in Daly (1991a, page 196)].

By their very nature, the preceding definitions of SD suggest, firstly, that economic growth constitutes development in its own right and, secondly, that economic growth is unquestionably sustainable. Yet, all three definitions fail to show in what way growth is directly related to human well-being. Furthermore, neither definition makes an explicit reference to the natural resource base that sustains economic activity. One can only conclude, there-fore, that all three definitions are predicated on pure articles of faith. Hence, not unlike the preceding definitions, they demonstrably lack intellectual rigour as well as any vestige of operational value.

In one of the few rigorous examinations of the SD concept, Pearce et al. (1990) point out that the explicit use of the term "development" rather than economic growth implies a broad acceptance that SD embraces quality of life concerns which are not necessarily reflected by annual estimates of production and consumption (e.g., educational attainment, nutritional sta-tus, access to basic freedoms, and spiritual welfare). Hence, they believe development acknowledges the limitations inherent in the utilisation of the popular measure of economic growth — Gross Domestic Product (GDP) — as an indicator of the well-being and development status of nations. Moreover, Pearce et al. suggest the elements contained within a given development "vector" are open to ethical debate. In other words, what constitutes development is an ethical consideration and will differ between countries or even within countries. Therefore, one may be forced to recognise that GDP and many other quantitative-based indicators are unlikely to serve as very good indicators of development, let alone SD.

2.1.4   Hicksian income and a constant stock of capital as a sustainable development condition

Pearce et al. believe SD to be a situation in which a development vector — comprised of ethically desirable social objectives — is sustained indefinitely. They have also argued that the use of the term "sustainability" suggests the existence of a minimum condition for development to be sustainable, irrespective of how development might be perceived. According to Pearce and his colleagues, such a condition necessitates a nondeclining stock of natural capital over time, where natural capital includes resources such as "soil and soil quality, ground and surface water quantity and quality, land and water biomass, and the waste assimilation capacity of receiving environments" (Pearce et al., 1988, page 6).

The minimum or constant natural capital stock condition posited by Pearce et al. has been employed elsewhere, albeit in different terms, and is essentially derived from the so-called Hicksian interpretation of income — defined as the maximum monetary value of final goods and services that can be consumed over a given period without reducing the capacity to sustain the same consumption stream over time. It is the only true measure of income because, as Hicks emphasised, the central criterion for defining the concept of income is "to give people an indication of the amount which they can consume without impoverishing themselves" (Hicks, 1946, page 172). The Hicksian definition of income subsumes the notion of sustainability by guaranteeing that any individual, entity, or nation is at least as well off at the end of a given accounting period as it was at its commencement.

Given the need to maintain a constant stock of capital assets over time, the most important implication of the Hicksian interpretation of income is its potential to serve as an operational rule-of-thumb. However, what remains problematic is knowing what type of assets or what combinations of assets are necessary to, sustain at the very least, a constant consumption stream over time. The variation in opinion with regard to this matter has led to the emergence of three distinct Hicksian income-based sustainability measures. They are

1. A *strong sustainability* measure of income: This is a measure of in-come based on the principle that the quantities of both natural capital

(resource assets, ecological system services, etc.) and human-made capital (producer goods such as plant, machinery, and equipment) must be kept intact. The need to maintain both types of capital assets intact stems from a belief that both assets are perfect or near-perfect complements and, as such, the productivity of one is dependent upon the availability of the other (except in the case of the natural ecosystem services provided by the stock of natural capital). This is a critical assumption because it implies that any augmentation of the quantity of human-made capital is unable to offset the depletion of natural capital sufficiently to maintain a constant stream of income, or consumption, over time. Only through improvements in the efficiency with which human-made capital converts natural capital into final goods and services can a constant consumption stream be sustained. But, even then, there are thermodynamic limits to such improvements, meaning inevitable limits on the capacity of human-made capital to overcome the diminution of natural capital stocks.

2. A *weak sustainability* measure of income: This is a measure of income based on the premise that a combination of both natural capital and human-made capital assets needs to be kept intact, rather than each variety of assets being individually maintained. It is based on the belief that both categories of assets are perfect or near-perfect substitutes. As such, it is assumed that the depletion of the natural capital stock will not undermine the capacity to maintain a constant income stream over time so long as the stock of human-made capital is augmented sufficiently to compensate. The weak sustainability measure of income is more commonly referred to in the economics literature as the *Hartwick Rule* for sustainability (Hartwick, 1977 and 1978). The debate surrounding the Hartwick Rule and the relationship between natural and human-made capital is dealt with in Chapter 14.

3. A *very weak sustainability* measure of income: This is a more recent measure of sustainability defined as "nondeclining utility over time" (Pezzey, 1989). It is based on the capacity of a combined stock of capital (irrespective of its makeup) to sustain a given level of utility rather than a particular stream of income. It, like the weak sustainability measure, is also based on the assumption of perfect substitutability between natural and human-made capital.

This latter neoclassical derivation of the Hicksian concept of income has a number of inherent weaknesses. First, by referring to utility instead of income, the concept is largely deprived of operational value (a key normative value of the Hicksian concept to begin with). Second, and depending on the nature of the assumptions associated with the utility and production functions employed, there is the potential for the proponents of a very weak sustainability measure to design and promote a physically infeasible set of resource use rules (Daly, 1991a). The potential for this problem to emerge arises because, as Pearce and Turner (1990, page 37) point out:

> Simply saying that the end purpose of the economy is
> to create utility, and to organise the economy accord-
> ingly, is to ignore the fact that, ultimately, a closed
> system (such as the ecosphere within which the econ-
> omy is embedded) sets limits, or boundaries, to what
> can be done by way of achieving that utility (paren-
> theses added).

Hence, unless the limits to the benefits provided by various forms and combinations of capital are properly recognised, specifically targeted utility or welfare levels could well prove ecologically unsustainable simply because the subsequent organisation of economic activity leads to the excessive depletion of natural capital.

The first two measures of sustainability have not avoided their fair share of criticism either. Perhaps the most crucial form of criticism involves the supposed correlation between income and development, a point already mentioned in connection with the concerns of Pearce et al. There is good reason to doubt the applicability of income as an indicator of development and, therefore, SD. Because, in many instances, critical development factors do not obtain a market price, many noneconomic factors are excluded from conventional measures of GDP. Moreover, included in the measurement of GDP are defensive and rehabilitation expenditures that are, in essence, a reflection of the opportunity cost of economic activity. Of course, the dilemma concerning the inability to capture and reflect the full spectrum of sustainability and development factors is not confined to measurements of GDP. It is a problem consistent with all indicators and arises largely because of the difficulty associated with quantifying many of the quality of life dimensions of SD (Holmberg and Sandbrook, 1992). However, despite the complexities involved, some attempts have already begun in terms of estab-lishing quantitative indices to reflect qualitative SD factors. A number of these are revealed and discussed in Chapter 14.

2.1.5   Conclusion

Overall, the widespread attempt at establishing a meaningful and opera-tional SD concept has, in nearly all cases, been brief, shallow, and devoid of intellectual rigour. There has been a general failure or reluctance to develop and establish the SD concept from any sound theoretical foundation. Instead, the SD concept is largely the product of conclusions drawn from a mish-mash of prior assumptions and, in most instances, falsely held pre-analytic visions. Only with respect to the sustainability side of the SD coin has there been any attempt to establish operational guidelines (e.g., the minimum capital stock rule). A considerable problem still remains, however, in terms of the development side of the SD coin. Consequently, the aim of the remain-der of this chapter is to outline the theoretical foundation upon which a rigorous SD concept can be developed in Chapters 3, 4, and 5.

2.2 An ends-means spectrum as the theoretical foundation for a sustainable development concept

The notion of SD is an implicit recognition of desirable ends and limited means. Development, for instance, recognises a society's desire to attain a set of desirable goals and objectives. Sustainability acknowledges that, in striving to attain such goals, society is constrained by the existence of limited means. Hence, in many ways, the SD concept is an encapsulation of the "economic problem" — a problem broadly regarded as how to best utilise scarce means to satisfy competing ends. For instance, how and at what rate scarce means are presently exploited determines the extent to which scarce means are available for future exploitation. Limited means, on the other hand, compel humankind to make choices which, in turn, demands human valuation. Exactly what valuation is placed on limited means, what criteria are employed to rank and prioritise ends, and what choices are made in allocating scarce means determines the extent to which humankind progresses or develops in the broadest sense.

Figure 2.1 is an illustration of Daly's (1973) ends-means spectrum showing its extreme poles as well as the spectrum's intermediate levels and categories. It is our belief that the majority of SD definitions so far established have dealt almost exclusively with the intermediate segment of the ends-means spectrum. By overlooking biophysical and existential realities, the extreme and ultimate poles of the spectrum have largely been ignored. From an ecological economic perspective, the entire spectrum of ends and means warrants consideration because, in a very real sense, achieving SD involves the prudent and sustainable use of the ultimate means in servicing the ultimate end — what, for the purposes of this book, is referred to as the *ecological economic problem*. The entire spectrum of ends and means will therefore serve as the theoretical foundation upon which will be established an ecological economic-based SD concept.

The following is now a brief explanation of the ends-means spectrum.* There are many reasons why human beings establish and maintain social and economic systems. None is perhaps greater than the desire to satisfy human needs and wants. Of course, social and economic systems do not satisfy all needs and wants because many are directly served by the ecosphere or natural capital, particularly those related to the life-support services it provides. Like the supporting ecosphere, social and economic systems are open thermodynamic systems. As such, maintaining the human-made capital of which both are comprised requires the continued input of low entropy matter–energy. Although the variety of low entropy enables one form to be substituted for another (e.g., coal for oil, and timber for stone), there is, unfortunately, no substitute for low entropy itself. That is, low entropy matter–energy constitutes the fundamental stuff of the universe and,

* Much of the following on the ends-means spectrum has been drawn and adapted from Daly (1973, 1979, 1980, and 1991b).

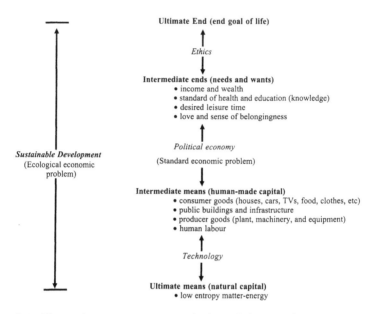

Figure 2.1 The ends-means spectrum. (Adapted from Daly, 1973, page 8. With permission.)

without low entropy, human beings cannot produce and maintain the human-made capital (intermediate means) required to satisfy human needs and wants (intermediate ends). For this reason, low entropy matter–energy constitutes the *ultimate means* of all economic activity and is represented at the base of the ends-means spectrum.

Given the fundamental importance of low entropy matter–energy, one of humankind's obvious concerns is its continued availability and the prospect of its eventual exhaustion. Yet, such a prospect is of little long-run concern if low entropy can be humanly created to replenish depleted stocks. Unfortunately, the direct creation of low entropy is impossible because it is only through natural biogeochemical processes that low entropy generation takes place. Of course, it is true that human beings are able to facilitate the generation of low entropy through cultivation practices such as agriculture, aquiculture, and plantation forestry. But this, in itself, does not guarantee the replenishment of low entropy resource stocks. Complete replenishment only takes place if the low entropy being generated is no less than the entropic dissipation of matter–energy arising from all physical transformation processes, including human induced processes. This leaves three important things to consider, all of which are readily overlooked. First, in view of the ever increasing entropic dissipation of matter–energy caused by a burgeoning rate of global economic activity, it is doubtful whether the low entropy generated by human cultivation practices and natural regeneration processes is increasing sufficiently enough to keep low entropy stocks intact (e.g., consider the natural capital stock account for Australia in Table 14.5 of Chapter 14). Second, humankind's current dependence on low entropy is biased toward nonrenewable natural capital which, by definition, cannot be replenished. Finally, because the creation of low entropy occurs almost entirely from the product of photosynthesis, the current rate of global deforestation is diminishing the planet's capacity to replenish low entropy stocks.

At the top of the ends-means spectrum is the *ultimate end* which Daly (1980, page 9) sees as "that which is intrinsically good and does not derive its goodness from any instrumental relation to some higher good." Despite the likelihood of being unable to precisely define the ultimate end, societies are compelled to recognise its existence if only because they must prioritise their goals and objectives which, in turn, necessitates the prior existence of an ordering or ethical guiding principle. Indeed, the failure to recognise the need for an ultimate end, and the inadequate nature of the ultimate ends of many societies that do, stands as the major obstacle in the movement toward SD. More on this will be discussed in later chapters.

Between the ultimate means and the ultimate end of the ends-means spectrum are the two intermediate categories. Both constitute ends with respect to a lower order category and means with respect to a higher order category of the spectrum. For instance, a hierarchy of *intermediate ends* exists below the ultimate end in recognition of the fact that

a. A dogmatic belief in objective value is required to rank human needs and wants from most to least important.
b. The satisfaction of needs and wants in line with their hierarchical ordering is instrumental to approaching the ultimate.*

As the entire array of human needs and wants, intermediate ends include the desire for such things as wealth, income, and comfort; health and education (knowledge); leisure time, love, and a sense of belongingness, just to name a few.

Below the intermediate ends are the *intermediate means* which are the prerequisites for the attainment of one or more intermediate ends. While intermediate means are means with respect to intermediate ends (i.e., without intermediate means, needs and wants cannot be satisfied), they are

---

* Because of the importance of the ultimate end and its relationship with objective values, it is necessary to explain what is meant here by *objective value*. To do so, one needs to acknowledge, as Williams (1970) has demonstrated, the distinction between value in the sense of an evaluation of an object relative to another, and value in the sense of a set of standards by which an evaluation is made. Because standards of any worth constitute the "desirable" objective elements of all human perception, objective values are the standards upon which something is adjudged as either "good" or "bad." Subjective values, on the other hand, reflect the tastes and preferences of individuals. As such, they do not necessarily derive their goodness from any instrumental relation to some higher good — such as a set of objective value-based standards and principles. Hence, overall, the difference between objective value and its subjective-based equivalent is akin to Kluckholm's (1964) distinction between "that which is intrinsically desirable" (objective value) and "that which is conveniently desired" (subjective value).

alternatively ends directly served by the entropic throughput of mat-ter–energy (i.e., without the input of the ultimate means, intermediate means cannot exist). The stock of intermediate means is comprised of such things as residential dwellings; public buildings and infrastructure; consumer dura-bles (cars, refrigerators, TVs); producer goods such as plant, machinery, and equipment; and all forms of human labour. In this sense, intermediate means constitute the entire stock of human-made capital as defined by Fisher (1906).*

Figure 2.1 indicates two vastly different relationships linking the inter-mediate and ultimate categories of the ends-means spectrum. Because inter-mediate ends are ranked in reference to a dogmatic belief in objective value, the relationship between the ultimate and intermediate ends is essentially an ethical one. Conversely, because the production of human-made capital involves the initial extraction and subsequent transformation of low entropy resources, the relationship between the ultimate means and intermediate means is a purely technological one. There are two types of technology relevant to this relationship. The first is the "natural technology" embodied in natural capital such as forests, estuarine environments, rivers, deserts, and the seas. Natural technology involves the biogeochemical processes that lead to the regeneration of renewable resources, the absorption of high entropy wastes, and the maintenance of the ecosphere's life-support system (Daly, 1980). The second category of technology is the "human technology" embod-ied in human-made capital. Improvements in human technology lead to better methods of resource exploitation, more efficient resource use, and an increase in the service-yielding qualities of all newly produced human-made capital.

The central or intermediate segment of the ends-means spectrum involves not so much a relationship, but the standard economic problem of valuing, organising, and allocating intermediate means in ways that best serve the hierarchy of intermediate ends. Exactly how well the resolution of the standard economic problem contributes to the movement toward SD depends critically on key aspects related to the extreme poles of the ends-means spectrum. These key elements bring to bear the following questions.

- Are the intermediate ends appropriately ranked with respect to the ultimate end?
- How appropriate is the ultimate end in the first place?
- Are human beings appropriating the ultimate means in ways that ensure the entropic dissipation of matter–energy is no greater than the capacity of the ecosphere to provide a continuous source of low entropy and to assimilate high entropy wastes? In other words, is

* In the Fisher (1906) sense, capital is regarded as any physical object that is subject to human ownership and capable of directly or indirectly satisfying human needs and wants. Hence, human-made capital refers to all producer and consumer goods. Although not subject to ownership, other than by the individual who possesses productive knowledge and skills, labour can also be included as part of the stock of human-made capital.

the appropriation and utilisation of the ultimate means ecologically sustainable?

2.2.1 *The movement toward sustainable development as the ecological economic problem*

The preceding questions are important because the successful resolution of the standard economic problem is not enough to guarantee the movement toward SD. It is entirely conceivable to solve the standard economic problem but move progressively away from the SD goal because

a. Intermediate ends are inappropriately ranked or ranked according to an inadequate ultimate end.
b. The ultimate means are being exploited in an ecologically unsustainable manner.

Why might this come about? There is a very logical reason. It was earlier pointed out that the primary SD concern is the prudent use of the ultimate means in servicing the ultimate end — what was deemed to be the ecological economic problem. According to Daly (1980), referring to the problem in this manner emphasises both its wholeness and the necessity of breaking the problem into more manageable subproblems. While the latter underscores the importance of addressing the ecological economic problem one step at a time, the former serves as a reminder that tackling any one step will lead to worthless conclusions if the remaining subproblems are ignored or overlooked. In essence, the various parts of the ecological economic problem are highly interrelated and cannot be dealt with independently of each other.

Consider, therefore, the consequences of addressing the standard economic problem in isolation. Because it deals exclusively with how well intermediate means serve the hierarchy of intermediate ends, it is but one of many subproblems subsumed by the larger ecological economic problem. To focus solely on the standard economic problem and resolve it at the expense of other subproblems means, inevitably, that the ecological economic problem will go unresolved. Clearly, the resolution of the standard economic problem is a necessary but insufficient condition for achieving SD. It is the failure of many policy makers to recognise this fact that has led to the establishment of an inadequate institutional framework and the implementation of many ill-conceived policies. This will become more obvious in Parts II and IV of this book.

The failure to address subproblems in light of their relationship with the ecological economic problem is also evidenced by the propensity for academic disciplines to go beyond their legitimate areas of study to encompass the entire ends-means spectrum. The economics profession, for instance, by preoccupying itself with the standard economic problem, has erroneously assumed "middle-range pluralities, relativities and substitutabilities among competing ends and scarce means to be representative of the whole spectrum" (Daly, 1980, page 10). This has given rise to an approach labelled *economic imperialism* — a chauvanistic attempt at reducing everything to the level of intermediates and relativities in order for everything to obtain a relative price. However, economics is not the only guilty profession. Narrowly defined exercises by ecologists and physicists have given rise to the "entropy" or "energy theory of value" and its close relative, the "maximum power principle." Considered an example of *ecological reductionism*, these principles rule out human purpose and choice and ignore the role played by exchange in the generation of value.

2.2.2 *The interdependent relationship between sustainability and development*

Implicit in the recognition that SD involves the prudent utilisation of ultimate means in servicing the ultimate end is the strict interdependency between the two conditions of sustainability and development. In many respects, one begets the other. For instance, concerns for the moral issues of distributive justice and intergenerational equity, both of which are important elements of the human development process, are unlikely to emerge if questions regarding the biophysical limits to growth are ignored. Similarly, an initial lack of concern for critical moral issues invariably leads to a lack of concern for biophysical constraints (Daly, 1980). For this reason, a credible SD concept must be predicated on biophysical realities and moral concerns. The physical laws of thermodynamics and ecology provide the biophysical anchor of any SD concept. Concerns for posterity, intragenerational inequities and injustices, sentient nonhuman beings, and a deep respect for and the continuation of the evolutionary process provide the minimum conditions for a corresponding moral anchor.

The interdependency between the two conditions of sustainability and development does not, however, preclude their independent assessment. Indeed, and as will soon be demonstrated, it is highly expedient to address each condition separately. After all, how is a society to move toward SD if there is no clear conception of what it means to operate sustainably or to experience true human progress and development? What ultimately matters is that, in assessing both conditions separately, the guidelines established to achieve development are internally consistent with the guidelines to achieve ecological sustainability. To ensure consistency, a number of questions related to both ends-based and means-based arguments need to be seriously considered and addressed. These questions form the basis of Chapters 3 and 4. From an ends-based and, more particularly, a development perspective, it is necessary to consider the following.

- How and in what way do the ultimate end and the associated hierarchical ordering of intermediate ends relate to human development?

- What constitutes the ultimate end — that is, by what criteria should intermediate ends be ranked?
- Is the growing accumulation of human-made capital always desirable or does it eventually render the ultimate end a disservice?

Conversely, from a means-based or sustainability perspective, the following questions need to be considered.

- Given that low entropy matter–energy constitutes the ultimate means, in what way and to what degree are they limited?
- Are ultimate means limited in ways that technological innovation cannot overcome?
- If there are insurmountable limits to technological innovation, then what is required to ensure sustainable outcomes, and does such a requirement impose biophysical limitations on the maximum sustainable scale of macroeconomic systems?

chapter three

What is development?

3.1 Development: A qualitative rather than quantitative phenomenon

Development of any kind is an evolutionary process involving the improvement of a particular state or condition over time. As such, human development can best be described as an evolutionary process toward human betterment (Boulding, 1990). Of course, the potential for betterment is not always guaranteed. If deterioration in the conditions required to sustain the well-being of humankind is imminent, development becomes a measure of how successfully human beings minimise regressive change.

Assuming the potential for human betterment will continue for some time, development is likely to be evidenced by the progression of a society toward a predetermined set of desirable social goals and objectives. Eventually, it is likely to be characterised by the progressive attainment of more desirable social goals and objectives as each preceding set is subsequently satisfied. This suggests, first, that development will be difficult to achieve without a clear perception of what human betterment involves, and, second, without a "rule of right action" to guide the human development process.

Before an ordering principle is established to serve as a development-based rule of right action, the following needs emphasising. From an ecological economic perspective, there is a fundamental difference between qualitative- and quantitative-based notions of development. The distinction is best described by Daly (1990, page 1):

> To grow means to increase naturally in size through the addition of material through assimilation or accretion. To develop means to expand or realise the potentialities of; bringing gradually to a fuller, greater, or better state. In short, growth is the quantitative increase in physical scale while development is qualitative improvement or the unfolding of potentiality. An

> economy can grow without developing, or develop
> without growing, or do both, or neither.

Owing to the past and present emphasis on the importance of growth, development has long been associated with the expansion of the physical quantity and technological capabilities of human-made capital. This explains why the difference in GDP, as a supposed indicator of the growth of macroeconomies, is used to compare the development status of all nations. Indeed, nations with per capita GDPs at the lower end of the poor–rich scale have long been considered "underdeveloped" while, more recently, some at the upper end have been outrageously described as "overdeveloped."

Because ecological economists focus on the qualitative nature of the human condition, human-made capital is only considered useful if it serves positively in the human development process. Of course, development will always be a considerable function of human-made capital because, up to a point, the size of the stock and its technological capabilities are principal factors in expanding the human potential to experience a fuller and more rewarding sense of well-being (Coombs, 1977). Nevertheless, the departure from quantitative growth suggests an accumulation of human-made capital beyond a certain physical quantity need not always better the human condition. Unless it can be shown that a quantitative expansion in economic activity genuinely contributes to the betterment of a nation's citizens, it cannot be assumed that an increase in per capita GDP elevates a nation's development status. Indeed, as Chapter 14 will demonstrate, it appears that growth is rendering some countries poorer rather than richer.

3.1.1   Development requires intermediate ends to be appropriately ranked and the ultimate end to be appropriately conceived

Because development involves the qualitative improvement in human well-being over time, two things need to be carefully considered. First, if the qualitative nature of the human condition is dependent on the ranking of intermediate ends and their subsequent attainment, how is human betterment to be recognised and achieved if an ordering or ethical guiding principle — an ultimate end — does not exist? To consciously deny the existence of an ordering principle is clearly an error in fundamentals. Moreover, to go a step further and consider the indiscriminate attainment of intermediate ends as equivalent to having served the ultimate end is to confuse ends with means. To wit,

> humankind will seek ends, but only right ends (an
> appropriate hierarchy of intermediate ends) are des-
> tined to give true satisfaction in fruition (Hawtrey,
> 1946, page 364; parenthesis added).

And, as Hobson (1929, page 328) once explained:

> It is only the extent to which we can identify the de-
> sired from the desirable, that the evolution of custom-
> ary standards of life becomes a sound human art.

Second, as critical as it is to rank intermediate ends in accordance with how well they serve the ultimate end, so is the appropriateness of the hierarchy itself. Development will not take place if, to begin with, the hier-archical ordering of intermediate ends is erroneously conceived. As remarked by Knight (1933, page 4),

> Living intelligently includes more than the intelligent
> use of means in realising ends; it is as fully as impor-
> tant to select the ends intelligently.

Regrettably, there has been a growing tendency to blur the distinction between the ultimate and intermediate ends (Hobson, 1929; Horkheimer, 1947; and Daly, 1973). This has allowed both to become conflated and, in so doing, has freed the process of capital accumulation from an objective assess-ment criteria that would otherwise allow desirable and undesirable forms of human endeavour to be easily distinguished. As a consequence, the accu-mulation of human-made capital has become an end in itself where, as Colvin (1977, page 109) claims, one rarely considers, "Doing what for what?" Even less, it seems, does one explore the consequences of, "Doing what for what with what?" Instead, it is simply assumed that an ever expanding stock of human-made capital corresponds to a logical improvement in human well-being and that, by implication, growth is always desirable. Thus, as far as Weisskopf is concerned, it is the conflation of intermediate ends and the ultimate end that has permitted "rational means to justify irrational and, in many instances, bad ends" (Weisskopf, 1971, page 92).

According to Horkheimer (1947), the conflation of intermediate ends and the ultimate end began as the philosophers of the Enlightenment, following the Reformation of the sixteenth century, attacked religion in the name of reason. What was set in motion was "not a process which saw the death of the church, but one that saw the death of metaphysics and the objective concept of reason itself" (Weisskopf, 1971, pages 17–18). Whereas objective reason had previously emphasised ends rather than means and had served as a measuring rod to guide individual thoughts and actions, it was even-tually "liquidated as an agency of ethical, moral, and religious insight" (Weisskopf, 1971, page 4). As far as Daly (1979) is concerned, this indicates the "temper of the modern age" — a progressive banishment of once dom-inant concepts as teleology and purpose in light of a deterministic, materi-alistic, and reductionist world view conceived during the Enlightenment itself. Broadly referred to as "scientism," it has all but destroyed any notion of good and evil by reducing ethics to the level of individual tastes and preferences (Daly, 1987). From a developmental perspective, the growth in scientism and the subsequent confusion of means with ends has rendered all genuine attempts to perceive, assess, and measure qualitative change impossibly difficult. Development has literally become a term devoid of substantive meaning and purpose. Virtually anything vaguely desirable is now able to constitute and be promoted as a form of development.

3.1.2   What is the ultimate end?

Given the need for an ordering or ethical guiding principle, the following needs to be asked: What is the ultimate end? In an absolute sense, the ultimate end cannot be precisely defined. Although this presents an imme-diate obstacle, recognising the need for an ethical guiding principle is of greater importance because, as Tawney once remarked, "ideals of religion, art, and understanding are not hard to find for those who actively seek them, or those who seek them first. But if they are sought second they are never found at all" (Alonzo Smith, 1980, page 231). Broadly speaking, the ultimate end can be considered the end goal of life and reflects, to a large extent, the objective values pertaining to a society's belief system — a system of inter-dependently arranged and interacting images, beliefs, and values (Boulding, 1956; 1970a; and 1985). These so-called objective values constitute and deter-mine, either correctly or incorrectly, the true meaning and purpose in life and human development generally. As such, they form the basis upon which intermediate ends are hierarchically ranked and which, ultimately, human-made capital ought to be produced to serve.

Despite its objective value basis, the ultimate end must exhibit some degree of subjectivity if only because a society is left to collectively deter-mine, whether on religious grounds or otherwise, what it believes to be an appropriate ordering of intermediate ends. This subjectivity, of sorts, is evi-denced not only by the immense variation in religious and cultural beliefs, but also by the manner in which these very beliefs have considerably evolved over time. Ultimately, however, a society must accept some form of objec-tivity. It simply must, at a particular point in time, hold some things to be true and others false to in any way establish and maintain civilised standards of right and wrong. If not, there is nothing substantive from which a con-sensus might be formed in relation to what constitutes the true meaning of life and development generally.

It is important to recognise that an adherence to objective value need not deprive individuals from making choices for themselves. Even if a per-son's ranking of intermediate ends largely is determined socially, which it must be to some degree, they are still essentially free to choose the means to their satisfaction. Hence, an adherence to objective value need not deter-mine whether a person prefers a chocolate bar to a cream cake because both are merely alternative means to satisfying a particular intermediate end — in this case, the desire for flavoursome food. But, should a person rank his desire for good health above his desire for something flavoursome because of a widely held belief that individuals are morally obligated to maintain a reasonable state of health, his adherence to objective value could see him preferring an apple to a chocolate bar, though not necessarily an apple over a banana. Of course, the choices between different foods will also depend on what a person has recently eaten. Should it be that this person has eaten plenty of bland, nutritious food, he may opt for a chocolate bar over an apple if the amount of recently eaten healthy foods has been sufficient to ensure the more important of the two intermediate ends — a desire for a healthy diet — has been fully satisfied. In any case, objective forms of reason have, on the whole, never sought to deprive individual subjectivity outright. By merely demanding a specific mode of human behaviour, objective forms of reason have always permitted the latter to exist as an opportunity for individual expression of "a universal rationality from which the criteria for all things and beings are derived" (Horkheimer, 1947, page 4). Consequently, one should never view the adherence to objective value as a call for authoritarian interference. Nor should it be considered synonymous with "egotism and intolerance" (Daly, 1980, page 352). Instead, the individual subordination of subjective desires to a dogmatic belief in objective value should be viewed as a reflection of one wanting to belong to a "community" because, as Swaney (1981, page 619) has argued, "the free individual and the community grow together." Moreover, one's freedom invariably increases in line with one's ongoing involvement in the community because, as the individual and the community grow together, so the individual "grows deep" (Swaney, 1981). While a commitment to a community is likely to restrict one's ability to rank intermediate ends according to individual tastes and preferences (as opposed to choosing the means to satisfying them), what is invariably overlooked is that a commitment of this kind is a prerequisite to gaining an individual sense of autonomy and personality. Consequently, it stands as a means to individual freedom and development, particularly so if the community is open to inquiry (Swaney, 1981). It is, unfortunately, a measure of precisely how far many have gone in abstracting away from our individual concrete experience of "persons in community" that many observers continue to label an adherence to objective value as "anti-individual" (Daly and Cobb, 1989).

Adhering to a shared sense of, and dogmatic belief in, objective value highlights a very important distinction between ecological and conventional economics — a distinction concerning the principle of *consumer sovereignty*. The conventional view of consumer sovereignty is based on privileging individual valuations and related preferences above any coercive authority. Society, as such, is merely perceived as a mechanism characterised by the interplay of individuals seeking their own self-interested and motivated ends. Ecological economics, on the other hand, espouses the importance of individual self-autonomy but recognises the importance of a coercive authority and the subsequent need for individual choices to remain commensurate with civilised standards of right and wrong. Similar restrictions on individual self-autonomy are also necessary to maintain the health of ecosystems which, after all, are the very foundation upon which an individual derives and sustains his or her development needs (Common and Perrings, 1992).

3.1.3   Development requires the ultimate end to be updated suitably

To some extent, a society's belief system will always contain erroneous beliefs, images, and values. Quite literally, some things held to be true will eventually prove to be false. As a consequence, one can never expect the corresponding hierarchy of intermediate ends, even if effectively satisfied, to maximise the betterment of a society's members. Nevertheless, one can expect some ultimate ends to advance the human development process more so than others.

Given the historical tendency for "fit" cultural traits to survive and multiply (Norgaard, 1992), the failure to update the ultimate end in light of conflicting messages and signals can lead to cultural implosion and ultimate societal extinction. Thus, as the growth in knowledge reveals the extent to which a society's belief system is replete with erroneous images, beliefs, and values, it is necessary, in the first instance, to question all dubiously held truths. This often requires questionable images and beliefs to undergo a testing process (Boulding, 1970a). Second, once conflicting messages persist long enough to reveal some established truths as falsehoods, all erroneous images and beliefs should be rejected. Third, as newly emerging truths become increasingly difficult to deny, a society must endeavour to incorporate them into its belief system. Finally, because the ultimate end reflects the objective values pertaining to a society's belief system, as the belief system is systematically revised, so the ultimate end must be appropriately updated.

Unfortunately, a revisional process of this nature is rarely a smooth one. It is customary for the modification of human belief systems to be painstakingly slow because they are characteristically resistant to change (Boulding, 1956). Indeed, the resistance of human belief systems to beneficial change can often hold the key to a society's development and, in some instances, its long-term survival. This having been said, the obstinate nature of human belief systems is not entirely undesirable. At least some degree of conservatism is necessary for belief systems to survive and for a shared sense of and dogmatic belief in objective value to persist for any length of time. Exactly how resistant a society's general belief system is to its beneficial revision often depends on the nature of its institutional structures and arrangements. Successful societies possess institutions that, in the Hegelian sense, are predominantly nondialectical in nature. Nondialectical institutions minimise factional conflict, intolerance, and ideological inertia, and allow questionable truths to be openly scrutinised. It is for this reason that nondialectical institutions have a meritorious record of facilitating the appropriate revision of belief systems and of allowing the ultimate end to be suitably updated. The importance of nondialectical institutions as a means to moving successfully toward SD is revisited in Chapter 16.

3.2 *Establishing a development-based rule of right action*

It is one thing to demonstrate the developmental importance of an ordering or ethical guiding principle. It is another entirely to establish an ordering principle to serve as a development-based rule of right action. Given the need to rank intermediate ends in accordance with the ultimate end, establishing a broadly based guiding principle is best achieved by searching for an objective conception of human need (Hodgson, 1988).

We shall begin our search for an objective conception of human need with a fundamental hierarchy of human needs first posited by Maslow (1954), better known as Maslow's *needs hierarchy*. Because the needs hierarchy is based on an empirical assessment of human motivation, it is important that a couple of points are first brought to bear. First, as instructive as Maslow's needs hierarchy might be as a starting point in the broader understanding of an objective conception of human need, it must never be taken as a concrete assessment of the hierarchy of human needs. Indeed, there is considerable debate surrounding Maslow's choice of the individual categories of human need and, in particular, their precise order. Nevertheless, such a debate does not override an important conclusion to be drawn from Maslow's psychological insights — namely, the indisputable existence of different categories of human need and an inextricable relationship between human welfare and the subsequent attainment of higher- as well as lower-order needs. Second, there is always cause for concern when an attempt is made to base an objective conception of human need from an empirical theory of human motivation. There is the potential to fall into the trap of deriving a set of "oughts" from an observation of what "is." While concerns of this nature are legitimate and cannot be ignored, all great religious and philosophical systems — as essentially objective conceptions of the "ought" — have almost always been based on some prior, albeit empirically derived, conception of what "is" or "has been." Hence, should objective value have already existed in a world out of which humankind has evolved, it is through a learning process, facilitated largely by an empirical analysis and interpretation of what best contributes to a balanced system of physiological and psychological need satisfaction, that such truths are best revealed to human beings.

Let us now return to Maslow's needs hierarchy. The hierarchy is classified next in accordance with Maslow's ranking of lower- to higher-order needs. Beginning with the lowest form of human needs, they are ranked as follows (Maslow, 1954).

- *Physiological needs* — such as food, clothing, shelter, etc.
- *Safety needs* — this includes the need for physical and mental security, including the requirement for stability, dependency, and protection; the freedom from fear, anxiety, and chaos; and the need for structure and order. It also includes the general requirement for a comprehensive "philosophy that organises the universe and the men (sic) within

it into a sort of satisfactory coherent meaningful whole." Satisfying safety needs necessitates such things as:

a. A minimum level of income and an appropriate welfare safety net — overall, a strict adherence to the principle of intragenerational equity and justice.

b. A strong sense of "community."

c. The establishment of institutions based around the need for social coherence and stability.

d. Ecological sustainability and the continuation of the evolutionary process to ensure physiological needs are safely sustained into the foreseeable future.

- *The need for belongingness and love* — this includes the need for affectionate relationships with people in general; the hunger for contact and intimacy; the desire for a sense of place in the group, family, and society; and the urgent need to overcome or avoid the pangs of loneliness, of ostracism, of rejection, and of rootlessness. A true and fully encompassing sense of belongingness and love also involves a sense of identity with posterity. Hence, satisfying the need for belongingness and love demands a corresponding adherence to the principle of intergenerational equity and justice.

- *The need for esteem* — this includes the need for a stable and high evaluation of oneself, for self-respect or self-esteem, and for the esteem of others. It essentially involves

a. The desire for strength, for achievement, for adequacy, for mastery and competence, and for independence and freedom.

b. The desire for recognition, attention, importance, dignity, and appreciation.

The need for esteem is satisfied by ensuring that in work, in leisure, and in play, one gains a sense of personal contribution to society at large.

- *Self-actualisation needs* — the need for self-actualisation concerns an individual's ultimate desire for self-fulfilment, that is, the desire to become fully actualised in what he or she is capable of becoming. At the pinnacle of the hierarchy of human needs, Maslow considers self-actualisation needs to be the most "creative and rewarding phase of the individual human development process." This is because, at the highest level of desire, there is, more than ever, a need for meaning and purpose in life.

By organising human needs into a hierarchy of relative prepotency, Maslow's needs hierarchy not only reflects the multidimensionality of the human existence, it reveals the human personality as an integrated whole in which every part, level, and dimension is interdependent. Most importantly, however, the needs hierarchy indicates that once basic physiological needs have been satisfied, intermediate ends originating from a higher level of human existence begin to emerge. As they do, one's behaviour is no longer dominated by the need for such things as food, clothing, and shelter, but by the need to satisfy emerging psychological needs. Consequently, to maintain a healthy human existence, it becomes necessary for the emerging higher-order needs to be satisfied along with basic physiological needs — what Weisskopf (1973) has referred to as a healthy *existential balance*.

Maslow's insights reveal three very important aspects concerning the human developmental process. First, need satisfaction that continuously increases the supply of means along one level and neglects needs on a different level is likely to disturb the balance of human existence. Second, because development itself will require a balanced system of need satisfaction, the process of human-made capital accumulation, which is aimed predominantly at meeting lower-order needs, must not occur at the expense of higher-order needs. Finally, given the nature of the higher-order needs propounded by Maslow, it would seem, at the very least, that human development will demand a deep respect for the continuation of the evolutionary process as well as a widespread concern for posterity and intragenerational inequities and injustices. In particular, development will necessitate the alleviation of absolute poverty. Not only does poverty alleviation ensure the satisfaction of basic physiological needs, it constitutes a prerequisite for the attainment of the higher-order needs necessary for a balanced and healthy human existence.

3.2.1   Human existential balance demands an optimal stock of human-made capital

Because an overemphasis on the accumulation of human-made capital can lead to higher-order needs being neglected, Maslow's needs hierarchy suggests a possible limit to the desirability of capital accumulation. If such a limit exists, as ecological economists believe it does, maintaining a healthy existential balance will require the accumulation of human-made capital to cease once a sufficient or optimal quantity is reached. For this reason, it can be said that a healthy existential balance required to achieve development demands the accumulation of a so-called "optimal stock" of human-made capital — essentially a stock of human-made capital that, when expanded to and maintained at a particular physical scale, best facilitates the human development process.

The notion of an optimal stock of human-made capital also conceivably emerges because, as Daly (1973) has pointed out, human beings presumably have a limited ability to experience and enjoy the services yielded by human-made capital. There simply must be, it seems, a physiological limit to personal satisfaction. Furthermore, when referring to the well-being of a society at large, Weisskopf (1973) has demonstrated that the law of diminishing marginal utility applies to the entire stock of human-made capital and not just individual commodities. That is, as the stock grows in physical size and number, the combined economic welfare it yields increases at a diminishing rate. Of course, the change in the stock of human-made capital is not confined to its growth.

It can also be qualitatively improved and the range of available items can be increased. However, progress of this type is never time and effort free, and is likely to become increasingly time-expensive the closer one's capacity to experience higher levels of personal satisfaction nears the physiological limit. Thus, even if technological progress continues to marginally augment the service-yielding qualities of human-made capital, the reduced time remaining for high-order pursuits indicates that any subsequent increase in lower-order need satisfaction will be achieved at the rising expense of psychological need satisfaction. Should the process continue, the marginal benefits of the former must eventually exceed the marginal cost of the latter, resulting, therefore, in an unhealthy existential imbalance. In sum, the capacity for a growing stock to better serve the ultimate end is likely to approach an asymptotic limit. Worse, still, the accumulation of human-made capital beyond a sufficient quantity is likely to render the ultimate end a disservice.

Despite the evidence now emerging to support the preceding claims (e.g., Easterlin, 1974; Daly and Cobb, 1989; Max-Neef, 1995; Jackson and Stymne, 1996; Castaneda, 1999, and Lawn and Sanders, 1999), just how well is a concept of a sufficient or optimal stock of human-made capital likely to be received? Given that growth is generally considered the only surefire means to development, not particularly well. Yet, as much as conventional economists, for example, are among those who are likely to be antithetic toward the concept, it is certainly nothing novel or foreign to them. Because the attainment of the optimum requires the marginal benefits generated by any addition to the overall stock to be equated with any marginal costs incurred, the notion of an optimal stock of human-made capital is entirely consistent with the most fundamental of all economic principles — the principle of *constrained optimisation*. Moreover, the concept of an optimal stock of human-made capital was long ago implied by Ruskin, a humanistic economist, when he sought to differentiate between wealth (objectively desirable forms of human-made capital) and undesirable forms of "illth" (capital that renders the ultimate end a disserve). To Ruskin, while some items of human-made capital could truly aid humanity and be construed as "true wealth," others contributed to the degradation of the human condition. It was the latter of these that Ruskin labelled illth because the more of them accumulated and acquired, the more they contributed to humanity's illness. According to Tawney (1921), Ruskin's idea of avoiding the production of anything that renders the ultimate end a disservice is an appropriate principle for any society to follow — a way of assigning economic activity its "proper place as the servant, not master, of society" (Tawney, 1921, page 183). Such a principle was regarded by Tawney as a *principle of limitation* because it implied a simple standard by which all economic activity could be judged.

In spite of the aforementioned, one is still left to consider the following questions. First, by what standards are the relative merits of human endeavour to be properly judged and assessed? Second, if there is a need to limit the accumulation of human-made capital, how is poverty to be alleviated?

To the first question, it has been the result of the growing tendency to avoid questions related to the ultimate end that the term "goods" in the general sense of human-made capital (producer and consumer goods) has replaced the objective notion of "good" (Daly, 1980, page 353). This led Bentham to devise the broadly embraced and adopted *maximum principle* — the "greatest good for the greatest number." Yet, it was largely because of the failure of Bentham to recognise the existence of an ultimate end that Tawney put forward the principle of limitation as an alternative rule of right action. Daly, too, points out a serious problem concerning the maximum principle — the impossibility of maximising more than one variable at a time. Somewhere, it seems, greatest must be replaced with *sufficient*. Daly suggests that sufficient should be substituted for "greatest" in relation to human-made capital while any reference to the "greatest number" is best understood in terms of the greatest number over time, not at any point in time. Should Daly's suggestion be embraced, the revised principle becomes a "sufficient per capita product for the greatest number over time" (Daly, 1980). When understood in these terms, Daly's sufficiency principle explicitly accepts, like Ruskin, Tawney, and Weisskopf before him, that an accumulation of human-made capital beyond the sufficient level, by rendering the ultimate end a disservice, is an impediment to the human development process.

To the second question, growth is not the only means to poverty alleviation and, in the case of wealthy nations, is probably the most undesirable way to overcome the problem because, apart from sustainability concerns, it comes at the expense of an increasing existential imbalance. The perceived need for continued growth is owing largely to the GDP-poverty link built into the institutional structures and arrangements of most nations. However, poverty can be alleviated and the GDP-poverty link severed by establishing alternative institutional structures that focus more heavily on the redistribution of income and wealth and a transition toward a stable population. Precisely how this can be achieved is revealed in Chapter 15.

3.2.2 A principle of limitation as a development-based rule of right action

Because Tawney's principle of limitation accords with Weisskopf's concept of human existential balance and Ruskin's differentiation between wealth and illth, it will henceforth serve as a development-based rule of right action. It is also consistent with Daly's sufficiency principle. Indeed, so much so, Daly's revision of Bentham's maximum principle — "the sufficient per capita product for the greatest number over time" — shall become the guide to appropriate forms of limiting behaviour. In adopting the principle of limitation as a development-based rule of right action, what constitutes the optimal stock of human-made capital — that is, what types and categories of human-made capital, and how much of each? There are two things that need to be considered here. First, and as previously pointed out, any perception of the optimal stock cannot be determined independently of the ultimate end because any attempt to do so makes it impossible to distinguish wealth from illth. With this in mind, the perception of an optimal stock of human-made capital will clearly differ between societies and, as such, will be more for some and less for others. For societies that have elevated the accumulation of human-made capital to the status of ultimate end goal, the optimal supply is effectively the maximum supply.

Second, how effectively human-made capital serves the ultimate end is, in many regards, conditional upon "human technology." By improving the quality of the stock, human technology increases the utility or service it yields. This implies, *ceteris paribus*, that the ultimate end can be served by fewer items. Hence, as the stock of human-made capital evolves, technological progress has the potential to reduce the physical scale of the optimum, or equivalently, increase the welfare that a given sized optimum can yield. Whichever it may be, technological progress need not, as is commonly believed, always lead to a progressively larger optimal stock of human-made capital.

Because the principle of limitation leaves one no closer to knowing exactly what is required to achieve human development, there will always be plenty of critics eager to proclaim the principle as too vague to be of any worthwhile value. Certainly it offers, as Daly points out, "no magic philosopher's stone for making difficult choices easy" (Daly, 1980, page 354). But if development is a condition rather than a specific or concrete target, not unlike the SD concept itself, the choices a society is required to make to facilitate development are always inordinately difficult. Yet, it is by recognising that human development is conditional upon the ranking of intermediate ends in reference to the ultimate end that the principle of limitation is far superior to its Benthamite counterpart. It not only forces one to consider the question of *purpose*, which the Benthamite rule transparently avoids, it implicitly recognises that a development concept devoid of an ultimate end is no meaningful development concept at all. While it will be a difficult and complicated exercise to re-establish a shared sense of and dogmatic belief in objective value, it will be considerably more perilous, as Daly (1980) points out, to continue operating on the erroneous principle of "there is no such thing as enough."

3.2.3   *Development requires the maintenance of moral capital*

Because of the importance of an ultimate end, development requires a minimum stock of what Hirsch (1976) has referred to as *moral capital*. Moral capital includes the social virtues of "truth, acceptance, restraint, and obligation," all of which are critical to the very idea of maintaining mutual standards of honesty and trust and important co-operative relationships. Social virtues can be collectively considered a form of capital because their key role in the successful functioning of an individualistic, contractual-based, market economy renders them necessary "agents" in the production, maintenance, and subsequent provision of an optimal stock of human-made capital.

Because the ultimate end reflects the objective values pertaining to a society's general belief system, moral capital is further required for two additional reasons. First, to ensure the ultimate end is properly conceived and, where necessary, appropriately revised. Second, to prevent the erosion of objective assessment criteria necessary to distinguish between wealth and illth. It is for this last reason, in particular, that Hirsch believes moral capital may be the most crucial form of capital, suggesting that if and when moral capital is depleted, a society must engage itself in a process of "moral growth" to ensure the stock of moral capital is kept at some necessary or minimum level. The need for moral growth and its relevance to SD is discussed in greater detail in Chapter 16.

3.3   What is development? A summary

Development is an evolutionary process involving the qualitative improvement in the human condition over time. Because human betterment is only achievable if the human condition can be evaluated in accordance with objective assessment criteria, development demands an ethical guiding principle — an ultimate end — to rank the intermediate ends that human-made capital ought to be produced to serve. Clearly, to achieve development, the ultimate end must be appropriately conceived. In addition, because the ultimate end reflects the objective values pertaining to a society's belief system and, as is always the case, belief systems contain a certain number of erroneous images and beliefs, development also requires a society to revise its belief system and to correspondingly update the ultimate end. Only then can it be certain that newly produced commodities constitute wealth rather than illth and that human-made capital is contributing positively to the human development process.

As for the accumulation of human-made capital itself, if it is continued beyond a certain quantity, it is likely to render the ultimate end a disservice. From a development perspective, this means two things. First, there is a limit to how much the quantitative growth of human-made capital can improve the human condition. Second, it is necessary to abandon the popular Benthamite principle of the "greatest good for the greatest number" and adhere to the revised principle of a "sufficient per capita product for the greatest number over time." Should this latter principle of limitation be followed, a nation should, in due course, move closer toward an optimal stock of human-made capital. Once reached, a society can direct its energies toward its qualitative improvement.

Last, but not least, it would seem the minimum conditions for achieving development are a deep respect for the continuation of the evolutionary process and a widespread concern for posterity and intragenerational inequities and injustices. In keeping with the latter, development necessitates the alleviation of absolute poverty. By guaranteeing the satisfaction of basic physiological needs, poverty alleviation provides the foundation required to attain higher-order needs and a healthy balanced system of human need satisfaction.

chapter four

What is sustainability?

4.1 Does sustainability matter?

At the end of Chapter 2, it was pointed out that sustainability considerations arise because the ultimate means of human activity, low entropy matter–energy, is absolutely scarce. However, the presence of limited ultimate means does not mean that ecological sustainability must be a desirable human goal. Human beings are at liberty to ignore biophysical constraints and opt for a profligate and short existence. Indeed, as Tisdell (1988) points out, there are some economists who believe the depletion, exhaustion, and extinction of inorganic and living resources is justifiable on so-called economic grounds.

Clearly, before considering what sustainability is and what is required to achieve it, an argument in support of ecological sustainability as a desirable human goal must first be sustained. To do this, the following question is considered: Does sustainability matter? To support sustainability as a desirable human goal, two brief arguments are now presented. While the first is based on a moral imperative to operate sustainably, the second directly concerns the immediate impact of operating unsustainably on human well-being and, therefore, human development.

4.1.1 Sustainability and the higher-order moral principle of intergenerational equity

Because sustainability requires a suitable stock of assets to be kept intact, operating sustainably implies having to invoke the principle of intergenerational equity. Hence, whether a society favours a sustainable pathway will depend largely on its sense of identity with posterity. Sustainability is unlikely to emerge as a major concern if brother and sisterhood is not extended to future generations. To some observers, the fact that future generations do not exist means they cannot be treated as moral subjects in the same manner as currently living people (Pasek, 1992). Smart and Williams have argued against such a premise by suggesting the following:

> Why should not future generations matter as much as
> present ones? To deny it is to be temporally parochial.
> If it is objected that future generations will only prob-
> ably exist, I reply: would not the objector take into
> account a probably existing present population on a
> strange island before using it for bomb tests? (Smart
> and Williams, 1973, page 63).

Barry (1977) implicitly supports Smart and Williams on the basis that the term "generation" is an abstraction from a continuous process of population replacement. Hence, according to Barry, conceptualising a generation is only possible if a generational cut-off point is arbitrarily instituted. Because this must leave some people living on either side of the cut-off point, one must ask how moral principles can apply and then not apply to the same person? The obvious answer is they cannot. Clearly, in the absence of any meaningful cut-off point, moral principles applying to everyone presently alive must hold intertemporally or not hold at all. Thus, if the principle of intragenerational equity carries sufficient moral weight to be universally applied to everyone currently alive, so must the principle of intergenerational equity carry equal moral weight and be applied to posterity. As such, the goal of sustainability should be viewed as a categorical imperative emerging from an appeal to the higher-order moral principle of intergenerational equity (Pearce, 1987).

That having been said, the equal moral worth of future people must be put into perspective. There are few who would doubt that the basic needs of the present should always be given priority over those of the future, even if it means fewer possible future generations. However, the extravagant desires of the present should not. Thus, it is only when the basic needs of present and future people are at issue, not extravagant desires, that future generations of people may be legitimately sacrificed. Of course, if meeting the basic needs of the present results in fewer possible future generations, it clearly indicates there are too many people alive in the present. Thus, on moral grounds alone, measures must be taken to limit human population numbers. How this can be fairly and efficiently achieved is discussed in Chapter 15.

4.1.2 *Sustainability as a development prerequisite*

Arguments in support of sustainability invariably rest on the future welfare effects of current actions. However, for two good reasons, the goal of sustainability can be considered a necessary condition to facilitate the development of those who already exist. In the first instance, it has been pointed out that the continuation of the evolutionary process and the subsequent extension of brother and sisterhood to posterity are required as part of humankind's higher-order or psychological needs (e.g., safety needs, and the need for belongingness and love). Furthermore, the following has also been identified by Boulding (1966, page 11):

> There is a great deal of historical evidence to suggest that a society which loses its identity with posterity, and which loses its positive image of the future, loses also its capacity to deal with present problems and soon falls apart.

Such evidence suggests that any society which fails to extend brother and sisterhood to future generations, particularly by way of invoking the principle of intergenerational equity, is likely to rapidly disintegrate. Because human well-being cannot possibly be sustained within the confines of a self-destructing society, the goal of sustainability emerges as a necessary condition for achieving any form of human development.

In the second instance, it is often believed that the costs of unsustainable human activity are borne largely by future people who, through improved knowledge, can effectively deal with them as they arise. This erroneously ignores the fact that unsustainable human activity has both an immediate and regressive impact upon the welfare of currently existing people. As any society operates unsustainably in the present, considerable time, effort, and energy needs to be expended to defend itself from the immediate and unavoidable welfare-diminishing consequences of both ecological and social decay (the latter being an important consideration given Boulding's evidence). There is, worldwide, an ever growing proportion of GDP being set aside for environmental as well as social defence and rehabilitation purposes. Ironically, these expenditures, which one is led to believe reflect an increase in human welfare, are an indication of the growing costs of economic activity. Indeed, so much so, emerging new evidence suggests the marginal cost of the current scale of human activity is exceeding the marginal benefits it yields and, in so doing, is rendering many nations poorer rather than richer (e.g., Daly and Cobb, 1989; Redefining Progress, 1995; and Figures 14.3 and 14.4 in Chapter 14).

Overall, it is clear that sustainability does matter — it is a desirable human goal. Because universalised moral principles must hold intertemporally or not hold at all, sustainability must be viewed as a categorical imperative emerging from an appeal to the higher-order moral principle of intergenerational equity. Second, sustainability is a development prerequisite because societies that lose their positive identity with posterity have a propensity to rapidly disintegrate. Moreover, unsustainable human activity imposes considerable and immediate welfare-diminishing costs that appear to be exceeding their welfare-increasing benefits. Operating sustainably cannot be considered a luxury or an afterthought. Nor can it be considered an unnecessary imposition on the development needs of present generations as some observers, such as Beckerman (1992), believe. It is, instead, an essential element of the human development process, thereby indicating the degree to which sustainability and development are inseparably related.

4.1.3   What needs to be sustained?

To appreciate what sustainability is, it must first be known what needs to be sustained. In the previous chapter, it was pointed out that development requires the growth of human-made capital to cease once a sufficient or optimal quantity is reached. Because the movement toward SD constitutes the primary concern of this book, it is clear that what needs to be sustained is the optimal stock of human-made capital. Of course, producing and sustaining human-made capital is only possible if macroeconomic systems are sufficiently well organised to physically transform low entropy matter–energy into human-made capital and allocate newly produced items efficiently. Thus, to know what is required to achieve sustainability, it is necessary to understand the physical nature of macroeconomic systems, the very systems that must exist to produce and maintain the optimal stock of human-made capital.*

There are essentially two contrasting views of macroeconomic systems. The first is the conventional *circular flow* representation of the macroeconomy. As Figure 4.1 illustrates, the circular flow model depicts the economic process as a pendulum movement between production and consumption within a completely closed and isolated system. The primary aim of this model is to show how exchange values (prices) travel back and forth between the macroeconomy's two main exchanging entities — firms and households. While the government is also a significant exchanging entity, it is excluded from the circular flow model for lucidity and because firms and households are responsible for most production and consumption-related decisions. Figure 4.1 shows the exchange value embodied in human-made commodities flowing from the firms that produce them to the households that purchase and consume them. At the national level, this is called the national product. Because of Say's Law, an equal amount of exchange value embodied in human labour and producer goods flows back to the firms from the very same households. This is called the national income.

Because the circular flow model depicts the macroeconomy as an isolated system, it assumes, even if implicitly, that there is no "throughput" of matter–energy to connect the circular flow of exchange values to the surrounding ecosphere. As Daly has argued (1996), so long as one is thinking only in terms of abstract exchange values, the circular flow representation of the

---

* Some commentators have pointed out that SD cannot be achieved without the sustainability of social and cultural systems (e.g., Redclift, 1992; and Norgaard, 1988). They are entirely correct. The sustainability of social and cultural systems is not only necessary to ensure a well-organised macroeconomy but, in view of Boulding's earlier mentioned observation, is necessary to ensure societies remain intact. This aside, it is our belief that biophysical factors constitute the sustainability "bottom line." Social and cultural systems may be critical to achieving sustainability but, in the final analysis, its achievement depends on how well societies or nations, indeed, the global community, recognise the limits imposed by biophysical constraints.

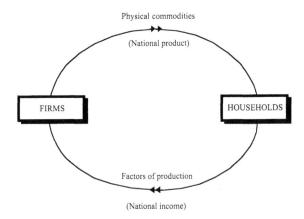

Figure 4.1 The circular flow model — the macroeconomy as an isolated system. (From Daly, 1996, page 47. With permission.)

macroeconomy is entirely reasonable, indeed useful, because it constitutes the basis of many important macroeconomic identities. If, however, one is thinking in terms of money, the physical token used to represent exchange values, the accuracy with which the circular flow model reflects concrete reality quickly diminishes. This is because money, or at least that part of the money stock that exists as notes and coins, wears out as it circulates. This means new money must be continuously minted and printed to maintain the stock of notes and coins intact. To do this requires the input of low entropy and leads, inevitably, to the generation of high entropy wastes. The circular flow representation of the macroeconomy now becomes technically deficient because a circular flow of money cannot continue if the macroeconomy remains an isolated system disconnected from the supporting ecosphere. Nevertheless, in view of the negligible rate of throughput required to maintain the stock of notes and coins, it is not unreasonable to ignore the throughput of matter–energy and think of the circular flow of exchange values as the dominant macroeconomic concern. Thus, in relation to money, the circular flow representation of the macroeconomy continues to be of great benefit.

If one goes a step further and views the economic process in terms of what is required to maintain intact the optimal stock of human-made capital, the focus of attention moves away from the circular flow of exchange values to the production and consumption of physical commodities. Because production and consumption are both physical transformation processes (the former involving the transformation of low entropy to human-made capital and the latter the transformation of human-made capital to high entropy waste), it is the linear throughput of matter–energy that becomes the dominant concern. Unfortunately, the overemphasis on the circular flow representation of the macroeconomy has led many observers to believe that physical commodities can, like abstract exchange values, circulate like a perpetual motion machine.* Consider, as Daly (1991b) has, the following error committed by Heilbroner and Thurow (1981, pages 127 and 135): "The flow of output is circular, self-renewing, and self-feeding" and, as such, "the outputs of the economic system are returned as fresh inputs." Note the very thing Heilbroner and Thurow focus on is output, that is, physical commodities and not abstract exchange values. One could be excused for believing the economic process can proceed without any exchanges of matter and energy with the supporting ecosphere. Like so many economists, and a great number of noneconomists, Heilbroner and Thurow incorrectly view the macroeconomy as an isolated, self-renewing system instead of a physical subsystem totally dependent on a linear throughput of matter–energy.

The deficiencies inherent in the circular flow representation of the physical aspects of the economic process have led ecological economists to devise the second of the two contrasting views of the macroeconomy, the *linear throughput model* (Figure 4.2). This alternative model depicts the macroeconomy as an open thermodynamic system existing within and dependent on the ecosphere. The linear throughput model does not seek to downplay the importance of the circular flow of exchange values. Hence, within the box representing the macroeconomy, one could imagine exchange values circulating between the firms and households responsible for the majority of production and consumption decisions. What the linear throughput model does do, however, is explicitly recognise that the economic process begins with an exaction from the ecosphere of valuable low entropy resources and ends with an insertion back into the ecosphere of valueless high entropy wastes. Hence, the linear throughput model is able to illuminate the constraints on production and consumption imposed by the limited dimensions of the supporting ecosphere and the first and second laws of thermodynamics.

4.2   The instrumental source and sink functions of natural capital

Because production and consumption involves the entropic dissipation of matter–energy, sustaining the optimal stock of human-made capital requires the following:

- A continued source of low entropy resources because, without it, new items of human-made capital cannot be produced to replace used-up and worn-out items,
- The continued availability of a high entropy waste-absorbing sink because, without it, pollution will degrade the ecosphere to the extent that it will eventually be incapable of providing the same flow of low entropy resources over time (e.g., an overpolluted river eventually provides fewer fish).

---

* Abstract exchange values are able to circulate perpetually because they have no physical dimension.

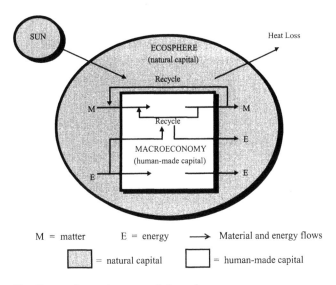

Figure 4.2 The linear throughput model — the macroeconomy as an open sub-system of the accommodating ecosphere. (Adapted from Daly, 1996, page 49. With permission.)

Because the ecosphere, or natural capital, is both the primary source of low entropy as well as the sole high entropy waste-assimilating sink, natural capital has two critical instrumental functions — its *source* and *sink* functions. Although both functions are renewable to some degree, neither can be considered limitless. The source function is limited by the available stock of nonrenewable low entropy and the regenerative capacity of renewable resources. The sink function is limited by the assimilative capacity of the waste-receiving ecosphere. Furthermore, and for reasons to be later explained, the capacity to augment both functions is itself limited. It is because of these constraints that an upper limit exists on

a. The sustainable rate of matter–energy crossing the macroeconomy-ecospheric boundary.
b. The maximum sustainable scale of macroeconomic systems.

Thus, in effect, the ecosphere has a limited carrying capacity.

4.2.1 *The complementarity of human-made and natural capital — how much natural capital is needed?*

Given that natural capital is necessary to sustain the optimal stock of human-made capital, exactly how much natural capital is needed? More particularly, is there a minimum quantity of natural capital required to achieve sustainability and, if so, what is it? The answers to these questions depend not only on the nature of natural capital itself, but also on the quantity of low entropy needed to maintain intact the stock of human-made capital. For example, if human-made capital can be sustained by an ever diminishing entropic rate of throughput, then, in principle, it may be possible to get along with a smaller stock of natural capital. Considerable debate surrounds this point, even amongst observers recognising the critical role played by natural capital, because it depends very much on the ability of the human-made capital, in particular, producer goods and human labour, to substitute for any erosion of the source and sink functions of natural capital. Clearly, to know how much natural capital is required to achieve sustainability, it is necessary to understand the extent to which human-made can serve as a natural capital substitute. To do this, however, the relevance of the first and second laws of thermodynamics must first be elucidated.

Consider the first law of thermodynamics, the *law of conservation of matter–energy*. It ensures that matter–energy embodied in the resources exacted from natural capital (which is then transformed into human-made capital) can never be less than the matter–energy embodied in the human-made capital produced. To suggest otherwise is tantamount to saying that energy and matter can be created through production which, of course, constitutes a breach of the first law. The second law, the *entropy law*, ensures that matter–energy is less available for future production following its transformation into human-made capital. In addition, the entropy law forbids 100-percent recycling of matter while categorically precluding any recycling of energy. When taken together, the first and second laws of thermodynamics decree the following in relation to the production of human-made capital. First, the ratio of matter–energy embodied in human-made capital to that embodied in the resources used to produce them, sometimes referred to as a measure of *technical efficiency* (E), must be less than a value of one.* Second, and as a consequence of the first implication, the production and maintenance of any given quantity of human-made capital requires the input of a minimum, irreducible quantity of low entropy resources. Exactly what the minimum low entropy requirement is at any point in time depends entirely on the human technology embodied in producer goods and human labour. For example, as technological gains are made, the amount of low entropy wasted in its transformation to human-made capital is reduced. Thus, as human technology improves, a given quantity of human-made capital can

---

* The most satisfactory measure of technical efficiency (E) is a second law of thermodynamics efficiency measure where:

$$E = \frac{\text{available work embodied in physical commodities produced}}{\text{available work embodied in resource inputs}}$$

The only difficulty with using this concept is the lack of any distinct conventional measure of *available work*. In the absence of a better alternative, Ayres (1978) suggests available work be measured in terms of an *electrical energy equivalent* (e.g., kWh) because electricity is one of only two forms of energy that is 100-percent available.

be produced from a smaller quantity of low entropy resource input. Ultimately, however, a point is reached where it is no longer thermodynamically possible to reduce production waste. This is where the technical efficiency ratio reaches its asymptotic limit of one. It is also where the production and maintenance of an optimal stock of human-made capital requires the continued input of a minimum, irreducible quantity of low entropy. This is a critical point in the substitutability/complementarity debate. It is also one that has been seriously overlooked by many economists who, in the past (e.g., Solow, 1974; Stiglitz, 1979), have employed inadequate neoclassical natural resource models and production functions to support their belief that any level of real GDP can be sustained regardless of how small the stock of natural capital becomes.

With the preceding in mind, it is my contention that, for the following reasons, natural and human-made capital are fundamentally *complementary* forms of capital and not substitutes. First, substitutability between natural and human-made capital does not strictly exist because any so-called empirical evidence of substitutability between the two forms of capital is a case of "implicit substitutability" (Lawn, 1999b). Implicit substitutability refers to an illusion created when the capacity to produce a given quantity of human-made capital from a lessened input of low entropy results, not from substitutability *per se*, but from a reduction in the amount of low entropy wasted in its conversion to newly produced human-made capital. This does not constitute an example of substitution because any genuine substitution requires human-made capital to "take the place of" the natural capital which, in effect, necessitates the low entropy needed to produce a given quantity of human-made capital to fall below the minimum irreducible requirement (i.e., to bring about a value of $E \geq 1$). Second, human-made capital and natural capital are fundamentally different elements of the production process. Natural capital, and the low entropy it provides, constitute the material cause of production. Human-made capital, as low entropy resource transforming agents, constitute the efficient cause of production. More of an efficient cause of production produces nothing without the input of the material cause of production. As Daly (1996, page 78) points out, a fish harvest is eventually limited by remaining fish populations, not by the number of fishing boats; timber production is limited by remaining forests and timber plantations, not by the number of logging trucks and sawmills; and agricultural production is limited by remaining topsoil and the availability of water, not by the number of tractors, harvesters, and even land area. Hence, only natural capital and the low entropy it provides can be considered the "true" input of the production process (Lawn, 1998). Third, human-made capital is itself the result of the production process requiring the input of low entropy resources exacted from natural capital. Consequently, to produce more of the perceived substitute requires more of the very thing being substituted, which is the defining condition of complementarity not substitutability (Daly, 1996). Fourth, if the substitutability of human-made capital for natural capital is as great as so many believe, why has humankind devoted so much time, effort, and energy to accumulate the former when its perceived substitute already existed in copious quantities? The simple answer is because they are complementary forms of capital and, to begin, production was limited by the lack of human-made capital — the efficient cause of production. Interestingly, in view of the current high rates of global deforestation and humankind's disproportionate reliance on nonrenewable resources, the greatest long-term threat to production is now the limits imposed by a lack of natural capital, the material cause of production. Finally, while a portion of the stock of human-made capital is used to recycle some of the high entropy waste matter generated through production and consumption, it can never fully replicate the waste assimilative and matter recycling role (sink function) played by natural capital.

With everything considered, it can be concluded that natural and human-made capital are fundamentally complements. Any observed substitutability between the two forms of capital is nothing other than implicit substitutability, the room for which diminishes as technical efficiency moves closer to the asymptotic limit of one. Clearly, in view of the complementarity between natural and human-made capital, achieving sustainability requires precepts to maintain intact the ecosphere's source and sink functions. This means having to establish and adhere to operational guidelines in relation to

1. The rate at which low entropy resources are exacted from the ecosphere that provides them.
2. The rate at which the quantity as well as the qualitative nature of high entropy waste is inserted back into the receiving ecosphere.

There is one other major implication of the complementarity between natural and human-made capital that requires mentioning. Because, as indicated in Chapter 2, the weak and very weak sustainability measures of Hicksian income assume the substitutability of human-made for natural capital, the most appropriate measure of Hicksian income is that of the strong sustainability variety. This is the only measure of Hicksian income that explicitly recognises the complementarity of natural and human-made capital and the need to maintain some degree of natural capital intactness (Lawn, 1998). Given the policy relevance of national income measures, this issue is revisited in Chapter 14 when SD indicators and a number of national accounting reform measures are considered.

4.2.2   *Sustainability requires the maintenance of the ecosphere's negentropic potential*

Before any sustainability precepts can be established, it is first necessary to understand the relationship between the rate at which humankind exacts and inserts low and high entropy and the ecosphere's regenerative and waste assimilative capacities. This is because the rates of the former *vis-à-vis* the latter ultimately affect the source and sink functions of natural capital. To gain a better understanding of this relationship, the physical nature of macroeconomic systems requires further examination.

As subsystems of the supporting ecosphere, macroeconomies are physically ordered systems that serve as media through which, initially, incoming matter–energy (low entropy) is transformed into human-made capital. This is the production phase of the economic process. As a consequence of using up or wearing out human-made capital, outgoing matter–energy (high entropy) is dissipated into the receiving ecosphere. This is the consumption phase of the economic process. Because 100-percent technical efficiency in production is impossible (i.e., E must be less than a value of one), high entropy dissipation also occurs during the production phase. Altogether, it is the gradual dissipation of low to high entropy as matter–energy passes through a macroeconomic system that makes it a *dissipative structure* (Norgaard, 1986). Dissipative structures are dynamic systems. In the process of a dissipative structure drawing in a portion of the low entropy energy made available by its "parent" system, it has the capacity to change its physical form, to grow, and, potentially at least, to develop. Thus, provided a dissipative structure is fulfilling its thermodynamic potential, it will tend toward a state of increasing order or increasing "negentropy." This is sometimes referred to as the augmentation of a system's *negentropic potential*. It occurs whenever a surplus of low entropy beyond that necessary to maintain a dissipative structure in a steady physical state is either intentionally or fortuitously "invested" toward a new interaction between itself and its parent system such that the carrying capacity of the latter is increased (Norgaard, 1984). However, for the augmentation of a system's negentropic potential to ensue, the investment must increase the structural organisation or negentropy of the dissipative structure without unduly increasing the disorder or entropy of the parent system. Should the opposite occur, the capacity of the parent system to maintain the dissipative structure in its present steady state will decline over time.

Until recently, there was a tendency to regard the increasing order of open systems as being outside the bounds of conventional equilibrium thermodynamics (Prigogine and Stengers, 1984). The entropy law does, after all, indicate a general tendency toward maximum disorder. Recent work by Prigogine and his colleagues has shown that increasing order is possible if only because dissipative structures are characterised by "far-from-thermodynamic-equilibrium" conditions (Prigogine, 1962; Nicolis and Prigogine, 1977; Prigogine and Stengers, 1984). Far-from-thermodynamic-equilibrium conditions occur when, as a consequence of a system being open with respect to an energy flux, the potential exists for the spontaneous generation of structural organisation (negentropy) within the system's own boundaries (Faber and Proops, 1990). The Earth is characterised by far-from-thermodynamic-equilibrium conditions because it is subject to a solar energy gradient that, through the agency of biogeochemical processes, permits the augmentation of its negentropic potential and, thus, its sustainable carrying capacity.

It would appear that the potential for further increases in the planet's structural organisation is far from exhausted (Norgaard, 1984). Further learning, mutation, cultural change, and technical innovation offer considerable scope to increase the ecosphere's carrying capacity.* This having been said, any such increase can only be generated slowly. Moreover, its augmentation requires increases in the ecosphere's source function to be matched by increases in its ability to assimilate waste. The capacity to exact more low entropy is of no long-run benefit if the subsequent generation of larger levels of high entropy waste overloads a static or declining sink capacity.

Macroeconomic systems, our main focus of attention, are also subject to far-from-thermodynamic-equilibrium conditions. As open thermodynamic systems, they have the capacity to ingest a portion of the low entropy made available by the accommodating ecosphere. By and large, macroeconomic systems have been extraordinarily successful dissipative structures. However, despite their recent success, it would be a mistake to consider the growth of macroeconomic systems as a concrete reflection of an increase in the ecosphere's sustainable carrying capacity. Not unlike the process of biological evolution, where the increase in order at the ecospheric level is possible because of the increasing disorder of the sun, so the maintenance or increasing order of macroeconomies can come at the expense of increasing disorder at the ecospheric level. But it need not do so. There is simply a need to ensure that, in attempting to increase or maintain the structural organisation of macroeconomic systems, the ecosphere's source and sink functions remain intact. Considerable doubt surrounds whether this is being achieved. Indeed, a number of observers, including many ecological economists, believe current levels of production and consumption are reducing the ecosphere's carrying capacity by reversing the negentropic potential established over eons by the process of biological evolution (e.g., Catton, 1980; Norgaard, 1984; and Ehrlich et al., 1980). Whether this is the case just yet is a moot point. Certainly, a tendency toward greater disorder must eventuate if production and consumption levels continue to grow. Above all else, however, one thing needs to be broadly understood. The mere existence on Earth of far-from-thermodynamic-equilibrium conditions beyond any anthropocentrically relevant time span, combined with the capacity of macroeconomic systems to increase in structural organisation, does not guarantee the intactness of the ecosphere's source and sink functions. Nor, therefore, does it guarantee the sustainability of the present physical scale of the global macroeconomy.

---

* The long-run potential for DNA mutations of the photosynthetic biomass to increase the planet's carrying capacity is enormous. At present, the existing ecological order is maintained by the photosynthesis of only 0.25 percent of the solar energy reaching the earth's surface. It has been estimated that up to 13 percent could be captured through photosynthesis (Norgaard, 1984, page 166). Exactly if and how human beings can facilitate the DNA mutation in a positive and ecologically nondestructive fashion is another matter.

4.2.3 *Establishing guidelines as precepts for operating sustainably*

As a result of the preceding discussion, two of the four sustainability precepts to be presented in this chapter can now be established. To assist in their establishment, reference will be made to Figure 4.3, a more detailed representation of the linear throughput model revealed earlier in the chapter. Figure 4.3 shows the ecosphere as a far-from-thermodynamic-equilibrium system fuelled by the importation of a solar flux. Within it exists the macroeconomy, a subsystem of the ecosphere. The sustainability of the macroeconomy and the human-made capital of which it is comprised is shown to be directly dependent on the input of low entropy resources and the output of high entropy wastes.

4.2.3.1 *The source function*

Consider, first, the source function of the ecosphere as represented in the left-hand side of Figure 4.3. Indicated here are two forms of natural capital. The first is renewable natural capital (RKn) which consists of fresh air, the soil, ground and surface water, the photosynthetic biomass (both land and water), and the herbivores and carnivores at the apex of the various food chains. Renewable natural capital has the ability to contribute to the maintenance or the increase in the negentropic potential of the ecosphere by generating a surplus of low entropy matter–energy (s_R) at a rate determined by its natural regenerative capacity.

If the rate at which low entropy is exacted or harvested (h_R) continuously exceeds the regenerative capacity of renewable natural capital (i.e., $h_R > s_R$), both the stock and the surplus low entropy it generates declines. Furthermore, the negentropic potential of the ecosphere is eroded. This demonstrates that even renewable resources are potentially exhaustible if exploited faster than their capacity to regenerate. If, on the other hand, the rate at which low entropy is exacted is less than the regenerative capacity of the stock of renewable natural capital (i.e., $h_R < s_R$), then, up to a point, the stock itself will increase. It is important to note that an increase in the stock of renewable natural capital does not always lead to an increase in the surplus low entropy it generates. Mature ecosystems, for instance, are generally less productive than immature systems and, as a consequence, generate much lower rates of surplus low entropy.* Hence, depending on the stock of renewable natural capital itself, harvesting low entropy at a rate less than its regenerative capacity does not always lead to a significant increase in the overall generation of low entropy surpluses nor, therefore, to any major increase in the ecosphere's negentropic potential.

The second of the two forms of natural capital is nonrenewable natural capital (N-RKn) which consists of such resources as minerals, oil, and coal. Nonrenewable natural capital cannot renew itself and thus has no capacity to generate a continuous surplus of low entropy matter–energy (i.e., $s_{NR} = 0$).

---

* The changing relationship between the size of the stock and its regeneration rate is usually depicted by an inverted U-shaped biological yield curve (Gowdy, 1994).

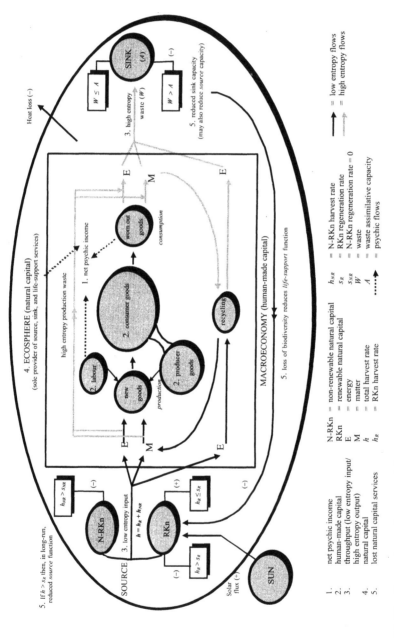

Figure 4.3. The linear throughput model

Consequently, the low entropy embodied in nonrenewable natural capital is reduced by an amount exactly equal to that exacted. Clearly, any continued harvesting and depletion of the nonrenewable natural capital stock must increase humankind's long-run reliance on renewable natural capital.

Altogether, the total low entropy harvested at any point in time (h) is equal to the sum of low entropy exacted from both varieties of natural capital (i.e., $h = h_R + h_{NR}$). Given the inability of nonrenewable natural capital to renew itself ($s_{NR} = 0$), the total low entropy surplus generated by the ecosphere is equal to the regenerative rate of the renewable natural capital stock alone (s_R). Thus, sustaining the total harvest rate (h) and, in the process, maintaining the rate of low entropy entering the macroeconomy is conditional upon offsetting the declining availability of low entropy brought about by a smaller stock of nonrenewable natural capital. This can be achieved by:

- Increasing the harvest rate of renewable natural capital (h_R).
- Discovering additional nonrenewable resource deposits.
- Expanding the stock of renewable natural capital through cultivation practices.
- Increasing the renewable natural capital/low entropy surplus ratio (RKn/s_R).

Considerable limits exist in relation to the first and second methods. The ability to increase the harvest rate of renewable natural capital is limited by the fact that the growth in the harvest rate must cease once it equals the regenerative rate of renewable natural capital (i.e., once $h_R = s_R$). The second method is limited by the fact that only so much nonrenewable natural capital exists. For these reasons, the third and fourth methods particularly are important. Unlike the first and second methods, both options expediently increase the surplus of low entropy generated by the stock of renewable natural capital (s_R) — one by expansion of renewable natural capital itself, the other through an increase in its productivity. Close examination of the third method reveals that, to be successful, nonrenewable low entropy must be exploited at a rate equal to the cultivation of renewable resource substitutes. As for the latter method, the renewable natural capital/low entropy surplus ratio can be increased by a variety of means. These include

- Improved forest and plantation management practices to increase their timber yield.
- Better management of pasture and cropland to increase crop yields and grazing stock densities.
- Augmentation of a river flow to increase available water supplies.
- Development of disease resistant and faster growing crops, timber varieties, etc.
- The release of new biological control agents to eradicate crop and forest pests.
- Improved resource exploitation and harvesting techniques.

With everything considered, the following can be concluded. First, where there is no deliberate cultivation of renewable resource substitutes to offset the depletion of the nonrenewable stock, maintaining the ecosphere's source function requires the rate at which low entropy is harvested (i.e., $h = h_R + h_{NR}$) to be no greater than the natural regenerative capacity of renewable natural capital (s_R). That is, there is a need to ensure $h \leq s_R$ so that as h_{NR} inevitably tends towards zero, h_R can be increased until it equals s_R. Any harvesting of renewable low entropy in excess of the renewable stock's natural regenerative capacity (i.e., $h_R > s_R$) is unambiguously unsustainable.

In the second instance, harvesting low entropy at a rate exceeding the natural regenerative capacity of renewable natural capital (i.e., $h = h_R + h_{NR} > s_R$) can be sustained, but *only* if the rate at which the nonrenewable natural capital is depleted (h_{NR}) is offset by a corresponding increase in the surplus low entropy generated by the stock of the renewable natural capital (s_R). This means some portion of the nonrenewable low entropy must be exploited at a rate equal to the cultivation of suitable renewable low entropy resource substitutes, and/or at a rate equal to the increase in the renewable natural capital/low entropy surplus ratio equal to the difference between the rate at which all natural capital is harvested ($h = h_R + h_{NR}$) and the regenerative capacity of renewable natural capital (s_R). There is, however, no necessity to exploit the entire stock of renewable natural capital at a rate equal to the cultivation of suitable renewable resource substitutes unless, of course, $h - s_R = h_{NR}$, or equivalently, if h_R already equals s_R. Why? Because until the natural regenerative capacity of renewable natural capital (s_R) is reached, sustaining the total low entropy harvest can be met by the stock of renewable natural capital itself.

Interestingly, the sustainable rate of low entropy harvestable from the combined natural capital stock is greater in the second instance than the first. This highlights an indispensable point. A lack of willingness to create renewable resource substitutes limits the future sustainable rate of harvestable low entropy, a point stressed some time ago by Georgescu-Roegen (1973) when warning of the danger of establishing a largely nonrenewable resource-reliant macroeconomy.

4.2.3.2   The sink function

Consider the sink function of the ecosphere represented in the right-hand side of Figure 4.3. The high entropy waste generated by the economic process is represented by (W). All waste inevitably finds its way back to the ultimate waste repository, the receiving ecosphere. The ability of the ecosphere to absorb high entropy waste is determined by its waste assimilative capacity (A) which, itself, is a function of the stock of renewable natural capital. The capacity of a given amount of renewable natural capital to assimilate waste is reflected by the renewable natural capital/waste assimilative ratio (RKn/A). The higher the ratio, the more high entropy waste a given quantity of renewable natural capital can safely assimilate. Again, and not unlike the source capacity of the ecosphere, it is possible through further learning, mutation, cultural change, and technical innovation to qualitatively trans-form the renewable natural capital/waste assimilative ratio. For instance, it is possible to

- Manipulate the stock renewable natural capital so as to increase its per-unit waste assimilative capacity.
- Cultivate additional high entropy waste-absorbing, renewable natural capital, particularly those elements best suited to the absorption and natural treatment of novel wastes.
- Develop more efficient high entropy waste-absorbing forms of renewable natural capital.
- Treat high entropy waste more thoroughly so it can be easily and readily absorbed by the ecosphere.
- Change the techniques of waste disposal.
- Change the qualitative nature of the high entropy wastes being generated by altering the methods of production and the nature of the human-made capital produced.

Two scenarios arising from the relationship between high entropy waste and the ecosphere's waste assimilative capacity will now be considered. To begin with, if the quantity and qualitative nature of high entropy waste being generated by the economic process is no greater than the assimilative capacity of the receiving ecosphere (i.e., if $W \leq A$), high entropy waste is benignly converted into ecologically nourishing products. These products are important because they assist in the maintenance of the ecosphere's negentropic potential. In the second instance, if the quantity and qualitative nature of the high entropy waste exceeds the ecosphere's assimilative capacity (i.e., $W > A$), the excess high entropy waste feeds back on the ecosphere and reduces the stock of renewable natural capital. In doing so, it reduces the future waste assimilative capacity of the ecosphere as well as its source function.

The latter observation reveals two very important facts regarding the relationship between the ecosphere's source and sink functions. First, the maximum sustainable quantity of low entropy entering macroeconomic systems is not only directly constrained by the limited source function of natural capital, but also indirectly by its limited sink function. Second, the source and sink functions of the ecosphere are interdependently related. Consequently, attempts to increase the renewable natural capital/low entropy surplus ratio independent of any high entropy waste implications are unlikely to augment the ecosphere's sustainable carrying capacity. Indeed, it is because of the potential impact of high entropy waste on the ecosphere's source and sink functions that some observers believe pollution limits are likely to impinge on the economic process sooner than resource limits (O'Connor, 1991a, 1991b; Daly, 1973). These fears may be well justified given the majority of environmental concerns are pollution- rather than depletion-related (e.g., urban pollution, lead contamination, acid rain, and global warming).

4.2.3.3   *Recycling*

Figure 4.3 indicates that, through recycling, human beings have the capacity to make some of the high entropy matter generated by the economic process available again for production purposes. The recycling of high entropy energy (the net increase in the quantity of low entropy energy resulting from energy reconcentration) is strictly forbidden (Georgescu-Roegen, 1971). Just how much high entropy matter can be technically reconcentrated depends on the recycling or reconversion pathways that exist for any specified substance. This, of course, is a function of the human technology available at any point in time (O'Connor, 1991c). Despite this, the feasibility of any reconversion pathway is always dependent upon its low entropy resource requirements because no amount of recycling is "resource free" and technology, like everything else, is subject to the entropy law. Therefore, while a profligate transformation of low entropy resources into high entropy waste may never technically preclude the reconcentration of high entropy matter, the higher the entropic dissipation rate, the larger the amount of low entropy energy required to power the recycling process.

   This final point is of particular importance. Because the maximum sustainable input of low entropy is constrained by the limited source and sink functions of the ecosphere, so, too, is humankind's capacity to reconcentrate high entropy matter. Clearly, while the recycling of matter provides an opportunity to increase the maximum sustainable scale of macroeconomic systems, it can never overcome the severe constraints imposed by the ecosphere's limited source and sink functions.

4.2.4   *Sustainability precepts to maintain the ecosphere's source and sink functions*

A position has been reached where the first two sustainability precepts can now be presented. They are

1. Maintain the ecosphere's source function intact and sustain the rate of low entropy entering the macroeconomy by:
   a. ensuring that if there is no deliberate cultivation of renewable resources to substitute for the depletion of nonrenewable natural capital, the rate of low entropy harvested from all natural capital (i.e., $h = h_R + h_{NR}$) is no greater than the natural regenerative capacity of renewable natural capital (s_R). This means ensuring $h \leq s_R$ so that as h_{NR} inevitably tends towards zero, h_R can be increased until it equals s_R.
   b. ensuring that if the harvest rate of low entropy from the total stock of natural capital is in any way greater than the natural regenerative capacity of renewable natural capital (i.e., if $h > s_R$), some portion of the harvestable nonrenewable low entropy is exploited at a rate equal to the cultivation of suitable renewable low entropy resource substitutes and/or at a rate equal to the increase in the renewable

natural capital/low entropy surplus ratio equivalent to the difference between the rate at which the total stock of natural capital is
harvested and the renewable stock's natural regenerative rate (i.e.,
$h - s_R$).

2. Maintain the sink function of the ecosphere by ensuring the quantity
   and qualitative nature of the high entropy waste generated by the
   economic process (W) does not exceed the waste assimilative capacity
   of the ecosphere (A). That is, by ensuring $W \leq A$.

Despite their restrictive flavour, it would be a mistake to believe that the
preceding precepts demand a static rate of low entropy exaction and high
entropy waste insertion over time. It is entirely conceivable for the rates of
exaction and insertion (h and W) to be progressively increased in line with
the gradual augmentation of the ecosphere's regenerative and waste assimilative capacities (s_R and A). However, given that increases in the ecosphere's
negentropic potential can only be generated slowly and, as such, s_R and A
are momentarily static, two things are guaranteed in the short-run. First,
barring any catastrophic or fortuitous event, the maximum sustainable rate
of matter–energy crossing the macroeconomy-ecospheric boundary is relatively constant. Second, the maintenance of the ecosphere's source and sink
functions demands natural capital intactness (Pearce and Turner, 1990). In
the long-run, the capacity to increase the ecosphere's negentropic potential
suggests natural capital need not remain intact over all time. Of course,
whether, under these circumstances, a diminution of natural capital puts the
sustainability goal at risk depends on whether natural capital has other
critical functions that, to be maintained, require natural capital intactness.
With this in mind, we now turn to the instrumental life-support function of
natural capital.

4.3   The instrumental life-support function of natural capital

The previous section considered the source and sink functions of the ecosphere and their capacity to sustain the economic process and the human-
made capital produced. However, as important as human-made capital is to
human well-being, of greater importance is the human habitability of the
planet. Hence, to confine the instrumental role of natural capital to its source
and sink functions is to adopt an inadequately narrow view of its full instrumental value. Clearly, natural capital has a further instrumental role — its
human *life-support* function.

The ability of natural capital to support existing life is owing only to the
fact that the ecosphere is the "fittest possible abode" for all presently living
organisms (Henderson, 1913). The ecosphere is currently fit for life because,
as a far-from-thermodynamic-equilibrium system characterised by a range of
biogeochemical clocks and essential feedback mechanisms, it has developed

a capacity to regulate the temperature and composition of the Earth's surface and atmosphere. It strangely possesses this ability despite lacking an inherent intelligence or purpose of its own (Wallace and Norton, 1992). Yet, it is the extraordinary self-regulating feature of the ecosphere that gives the impression that the Earth is in some way "alive." Indeed, so much so, it prompted Lovelock (1979) to develop his *Gaian hypothesis* — an hypothesis based on planet Earth, or Gaia, as an immense quasi-organism.*

There has, unfortunately, been a growing tendency for human beings to take the conditions for life for granted. Not only have modern creature comforts and intensifying urbanisation detached most people from the vagaries of the natural world, lurking in the background is a pervasive sense of technological optimism. It is a consequence of humankind's frivolous attitude toward the conditions for life that a number of falsely held beliefs have emerged. Two of them warrant mentioning. The first is a widely held belief that the ecosphere's current uniqueness for life was preordained. This is not so because, as Blum (1962) explains, had the Earth been a little smaller, or a little hotter, or had any one of an infinite number of past events occurred only marginally differently, the evolution of living organisms might never have eventuated. What is more, the path of evolution could easily have excluded human beings' participation in it. Second, it is widely believed that organic evolution is confined to living organisms responding to exogenously determined ecospheric conditions. However, it is now transparently clear that "fitness" is a reciprocal relationship between the ecosphere and its constituent species. Indeed, the ecosphere is as much uniquely suited to existing species as are the latter to the ambient characteristics of the ecosphere. Hence, according to Blum (1962, page 61), it is "impossible to treat the environment as a separable aspect of the problem of organic evolution; it becomes an integral part thereof." Clearly, just as current ecospheric conditions were not predestined, nor are the conditions of the future. They will always depend on the evolution of the ecosphere's constituent species and, in particular, the actions of a recalcitrant species.

An awareness of the aforementioned brings to bear a critical point. While human intervention can never ensure the ecosphere remains eternally fit for human habitability, humankind does have the capacity to bring about a premature change in its prevailing comfortable state. Indeed, global warming, ozone depletion, and acid rain may already be the first signs of a radical change in the planet's comfortable conditions. Nonetheless, there are some observers who argue that these events are merely symptoms of a "Gaian adjustment" brought on by the eccentricities of humankind. After all, and as Lovelock's Gaian hypothesis suggests, any maladies of Gaia do not last long in terms of its lifespan because anything that makes the ecosphere uncomfortable to live in tends to induce the evolution of a new and more comfortable environment. Does not the Gaian hypothesis offer humankind

---

* While the metaphor "Gaia" is used to represent the self-regulating nature of planet Earth, this should not be seen to imply that Earth is an organism in its own right, which it clearly is not (Boulding, 1985).

immunity to the consequences of its own actions and an open invitation to deplete and pollute at will? No, it does not. While Gaian feedback mechanisms ensure the quasi-immortality of the ecosphere, quasi-immortality prevails solely at the ecospheric level and only because of an informal association Gaia has with its constituent species. Quasi-immortality does not extend to any particular species. Indeed, historical evidence indicates the tendency for Gaia to correct ecological imbalances in ways that are invariably unpleasant for incumbent species. Hence, whilst Gaia has shown to be immune to wayward species like oxygen bearers in the past, individual species, including human beings, are in no way immune from the consequences of their own collective folly. As Lovelock so bluntly puts it (1962, page 212):

> Gaia is stern and tough, always keeping the world warm and comfortable for those who obey the rules but ruthless in her destruction of those who transgress. Her unconscious goal is a planet fit for life. If humans stand in the way of this, we shall be eliminated with as little pity as would be shown by the micro-brain of intercontinental ballistic nuclear missile in full flight to its target.

4.3.1   Biodiversity and the life-support function of natural capital

Natural capital may well have an important life-support function, but what is it about natural capital that bestows it with this quality? Is it the quantity of natural capital or is it some particular aspect of it? Lovelock (1988) leaves us in no doubt by emphasising that a minimum number and complexity of species are required to establish, develop, and maintain Gaia's biogeochemical clocks and essential feedback mechanisms. To wit,

> The presence of a sufficient array of living organisms on a planet is needed for the regulation of the environment. Where there is incomplete occupation, the ineluctable forces of physical or chemical evolution soon render it uninhabitable. (1962, page 63)

It is, therefore, a combination of species diversity, the convoluted interactions and interdependencies between species, and the complexity of ecological systems — in all, the *biodiversity* of natural capital — that constitutes its life-support linchpin. That is not to say that the size of natural capital is unimportant. It is, if only because the biodiversity needed to maintain the Earth's habitable status requires a full, not partial, occupation by living organisms. But size, itself, should never be equated with biodiversity.

In passing, We must make mention of the fact that biodiversity also contributes to the ecosphere's source and sink functions, particularly at the local or regional level. It does this in two main ways. First, greater biodiversity levels facilitate more productive biogeochemical relations among biological producers, consumers, and decomposers. Second, biodiversity increases the resilience of ecological systems to external shocks and stresses (Pearce, et al., 1990). As a consequence, greater biodiversity levels are able to increase the renewable natural capital/low entropy surplus and waste assimilative ratios (s_R and A) which, in turn, permit increases in the maximum sustainable rates of low entropy exaction and high entropy insertion (h and W). This is no better evidenced than by comparisons between the productivity of agricultural land in close proximity to areas of remnant vegetation and those that are not. Barring disparities in land management practices and natural fertility levels, in virtually all cases the former proves to be the most productive and the least likely to suffer from erosion or soil structure decline (Beattie, 1995; Biodiversity Unit, 1995; State of the Environment Advisory Council, 1996).

Returning to the life-support function of natural capital, if the sheer magnitude of natural capital is an inadequate indication of the effectiveness with which it can foreseeably support life, how is it possible to know the minimum level of biodiversity needed to maintain the ecosphere's life-support function? Unfortunately, the whereabouts of a biodiversity threshold is not known, although it is widely recognised that some semblance of a threshold must exist. What is known about biodiversity is that in the same way biodiversity begets greater biodiversity, so diminutions beget further diminutions.* In addition, the present rate of species extinction is exceeding the rate of speciation to the extent that biodiversity has become a nonrenewable resource on any relevant time scale (Daily and Ehrlich, 1992). Given that an increase in the global rate of extinction will unquestionably increase the vulnerability of human beings to its own extinction, should humankind continue to tolerate further losses of biodiversity on the basis that depletion has not yet rendered the ecosphere incapable of supporting life? In view of humankind's ignorance regarding the exact nature of the biodiversity threshold as well as the irreversibility of most of humankind's detrimental impacts on the ecosphere (to be fully explained in Chapters 10 and 11), a sensible risk-averse approach would be to draw and rigidly abide by a biodiversity "line in the sand." But where should this line be drawn? Ehrlich provides a hint by pointing out that despite what humankind does not know about biodiversity, it knows enough to operate on the principle that "all reductions in biodiversity must be avoided because of the potential threats to ecosystem functioning and its life-support role" (Ehrlich, 1993, page IX). Clearly, as a corollary of Ehrlich's dictum, humankind ought to draw a line in the sand at the currently existing level of biodiversity. Furthermore, it should confine natural capital exploitation to areas already strongly modified by previous human activities.

---

* It has been estimated that for every plant species lost, approximately 15 animal species will follow (Norton, 1986, page 117)

At the global level, Ehrlich's advice would indicate an urgent need to preserve and, if need be, restore and propagate large expanses of natural wilderness comprising much of the planet's remaining biodiversity. This view is similarly expressed by Gowdy and McDaniel (1995, page 189) in light of the fact that only 3.2 percent of the Earth's land mass is currently set aside for habitat preservation. Similarly, at the local or regional level, conscious efforts should be made to preserve remnant vegetation and important ecosystems even if the principle objective is to boost the productivity of local resources.

Of course, the mere preservation or "locking up" of large and small ecosystems will not by itself ensure biodiversity maintenance. Given the interdependent relationships between systems of all types, individual ecosystems are not entirely self-supporting (Lovelock, 1988). Their continued existence and the well-being of the biodiversity they contain is conditional upon the exchanges of both matter–energy with and between neighbouring and far-distant systems.* This applies to systems of all kinds, whether they be relatively pristine, moderately disturbed, or totally refined. In all, to maintain biodiversity, natural capital exploitation must be conducted on the basis of respecting the holistic integrity of geographical land and water resource units.

4.3.2   *Implications of the need to preserve biodiversity*

The need to adopt an holistic and risk-averse stance to natural capital exploitation has important ramifications for the two already established sustainability precepts. Because the maintenance of the ecosphere's life-support function demands biodiversity preservation, further restrictions are required on the exaction of low entropy ($h = h_R + h_{NR}$) and the insertion of high entropy waste (W). Moreover, there will almost certainly be restrictions on humankind's capacity to augment the ecosphere's regenerative and waste assimilative capacities (s_R and A). For example, an attempt to increase timber yields by manipulating previously unmanaged native forests or replacing low yielding native forests with high yielding exotic plantations will undoubtedly reduce biodiversity levels. So, too, will the damming of a river to provide additional water storage capacity impact on the ecology and biodiversity of the river and the ecosystems it supports. In addition, while genetic engineering promises increasingly productive forms of renewable natural capital and more useful biological control agents, it has the capacity — if sufficient care is not taken — to inflict significant damage on the stock of biodiversity.

In what way must the two previously established sustainability precepts be adjusted to make allowances for the maintenance of the ecosphere's life-support function? In the end, neither precept demands alteration. Instead,

---

* The exchange of matter–energy is particularly important in regard to the movement of essential trace elements and soil nutrients.

a further over arching precept is required. The third sustainability precept is the following

3. To the best of its ability, humankind should preserve and/or restore the ecosphere's rich biodiversity in order to limit any irreversible damage it might inflict on the ecosphere's instrumental functions. Because the exaction of low entropy resources and insertion of high entropy wastes must, on occasions, impact on ecosystems, there is a need to ensure both are limited to a rate that avoids humankind's need to exploit pristine ecosystems while permitting, where possible, previously impacted or sacrificed ecosystems to be readily restored.

4.4   *The intrinsic value of natural capital*

A failure to respond to any potential concern for the intrinsic value of the ecosphere and its constituent species does not, by itself, threaten the critical instrumental functions of the ecosphere. Nor, therefore, does it threaten the sustainability of the optimal stock of human-made capital. Nevertheless, if humankind values nonhuman species beyond their capacity to meet human needs and wants, recognising and acting on such values is very much a part of the SD process. To consider the potential implications of recognising the intrinsic value of natural capital, it is first necessary to know what is meant by intrinsic value. Second, it is necessary to know what intrinsic value means in terms of the "rights" of subhuman life and whether it restricts humankind's exploitation of natural capital for its own instrumental purposes.

Intrinsic value refers to the value pertaining to something because it exists or because there is something unique or peculiar associated with its existence. To acknowledge intrinsic value is to acknowledge that values can exist apart from humankind's knowledge of them (Cobb, 1973). For instance, one does not have to be consciously aware that something is experiencing pleasure or pain to recognise that the former is an intrinsic good and the latter an intrinsic evil. Hence, by virtue of some creatures being able to experience both pleasure and pain and the simplest enjoyment of life, sentient nonhuman beings clearly possess intrinsic value in their own right.

The issue now at hand is thus: Does the intrinsic value of sentient nonhuman beings mean humankind should grant limited "rights" to the subhuman world? Earlier in the chapter, it was argued that the principle of intergenerational equity should be invoked because present and future human beings accrue a number of universal rights and privileges. This call was put into perspective by indicating that although the needs of posterity should take priority over the extravagant desires of the present, they should always remain subordinate to the latter's basic needs. There is a good reason for this. People currently alive are able to experience the pain of being deprived of their needs but a person yet to exist cannot. The key issue here is sentience which, clearly, future people do not yet possess. Given that sentience is a feature of the subhuman world and serves as a means for determining what rights accrue to whom and when, it would seem entirely unjustifiable to overlook the moral concerns and rights of sentient nonhuman beings simply because they are incapable of expressing preferences in the same way as human beings (Pearce, 1987). Indeed, as Johnson (1991) suggests, the genuine interests of sentient nonhuman beings must carry at least some moral weight for human interests to carry any moral weight at all. Clearly, any recognition of the intrinsic value of sentient nonhuman beings warrants the limited rights of subhuman species to be included in the general domain of human rights — what would be, in a sense, an *extended principle of justice*. While the rights of sentient nonhuman beings would in no way equal the rights of humans, one would expect an extended principle of justice to include the dignified and, where possible, cruel-free treatment of sentient nonhuman beings.

4.4.1   An obligation to preserve the evolutionary process

Given the preceding, to what extent should the domain of rights be extended to the subhuman world? This depends on how far intrinsic value should be extended beyond human values which, according to Cobb (1973), is best determined with the aid of a satisfactory theory of intrinsic value. Because intrinsic value is largely a function of sentience, Cobb believes a theory of intrinsic value requires a theoretical measure of the "intensity of feeling" to appraise values and to serve as a yardstick for ranking human experiences in relation to subhuman forms of life. This still leaves the problem of having to gauge the intensity and inclusiveness of an experience and the feeling associated with it. Cobb suggests it can be overcome by applying the concept of a "biotic pyramid." A biotic pyramid describes the movement of life from microorganisms at its base through to human beings and other primates at the top. The pyramid's total value correlates roughly with the richness of its foundation, the number of different levels of species involved in the movement of life from base to top, the number and diversity of species at each level of the pyramid, and the complexity of the creatures existing at its apex. The biotic pyramid provides a useful perception of the hierarchy of feeling by ranking all creatures according to their contribution to the pyramid's total value. The most valuable biotic pyramids have human beings at the top. It is for this reason that human beings have the greatest intrinsic value and the most extensive rights. Below human beings, one would expect, for example, a chimpanzee to have greater rights than a rat which, in turn, would have greater rights than a microorganism. Of course, to some people, the idea of according rights to microorganisms is preposterous. In a roundabout sense they are right if only because the intrinsic value of microorganisms is negligible. However, the mere fact that microorganisms are necessary for a biotic pyramid to exist suggests microorganisms must have some intrinsic value, even if an individual microorganism has virtually none. As Daly (1980, page 124) has pointed out:

> Human beings may be at the apex of the biotic pyramid
> from which they have evolved, but an apex with no
> pyramid underneath is a dimensionless point.

Exactly how a society views the biotic pyramid depends on whether it views the species below it as having both instrumental and intrinsic value, or instrumental value alone. If it choses to adopt a purely egocentric view, the respective levels and diversity of organisms of the biotic pyramid are valued instrumentally according to their capacity to meet human needs and wants. In this case, only the source, sink, and human life-support functions of natural capital warrant concern. If a society adopts a biocentric view of the biotic pyramid and is genuine in its belief that all forms of life have varying degrees of intrinsic value, it cannot ignore the loss of intrinsic value entailed by its manipulation as it seeks to maximise its instrumental values.

Assume, for a moment, that a biocentric stance is adopted. How does humankind maximise the combined instrumental and intrinsic values of natural capital? Cobb (1973, page 317) believes the dilemma is best solved by asking another question, "Does humanity have any commitments or debts beyond the limits of human society?" Yes, according to Cobb, because, as a gift of the total evolutionary process, humankind is overwhelmingly obliged to service that to which it is clearly indebted. Naturally, how a society responds to this commitment will rest on how it perceives the total evolutionary process. It would seem, however, that basic to the evolutionary process is the "urge for life, for continued life, and for more and better life" (Cobb, 1973). Consequently, one would expect a society adopting a biocentric view to seek the continuation of the evolutionary process and, in doing so, aim to facilitate its inclusive work.

There are, however, two things left to be considered. First, what is required to ensure the continuation of the evolutionary process? Second, what is required to protect the intrinsic values of sentient nonhuman beings? As far as Pearce (1987) is concerned, natural capital intactness and biodiversity preservation and restoration are more than sufficient to continue the evolutionary process and protect the habitats of sentient nonhuman creatures. Hence, according to Pearce, there is no need to make allowances above what is already required to maintain the instrumental functions of natural capital (Pearce, 1987). To the extent that natural capital intactness fulfils the requirement to oversee the continuation of the total evolutionary process, Pearce is entirely correct. However, to the extent that it does not guarantee the humane treatment of sentient nonhuman beings, we take exception to Pearce. For instance, ensuring the survival of habitats does not prevent the unwarranted removal of sentient nonhumans from their habitats nor any ill-treatment that may arise out of their subsequent incarceration, transportation, or exploitation. Hence, the adoption of a biocentric stance demands more than natural capital intactness. As previously mentioned, it also requires the dignified and, where possible, cruel-free treatment of sentient nonhuman beings.

4.4.2   Implications of a biocentric stance

Should a biocentric view of the ecosphere be embraced, the application of an extended principle of justice must, in some way, restrict the rate at which low and high entropy can be exacted and inserted from and into the accommodating ecosphere. Restricted, also, must be the ability of humankind to augment the regenerative and waste assimilative capacities of the natural capital stock. Why? Consider low entropy exaction. While inhumane means of incarceration, transportation, and exploitation need not result in maximum sustainable yields of meat, dairy, and poultry products being exceeded, the confinement of livestock handling to dignified practices can greatly lessen maximum sustainable yields. And while various logging practices do not threaten sustainable timber yields, the banning of some because they result in unacceptable losses of wildlife and old growth forests can lead to sustainable timber yields being significantly reduced. As for the regenerative and waste assimilative capacities of natural capital, the transition to what many people believe to be grossly inhumane forms of animal husbandry have greatly boosted the productivity of natural capital. So, too, has the replacement of slow-growing native forests with rapidly growing exotic timber plantations. Clearly, any prohibition of these forms of natural capital exploitation on biocentric grounds will undoubtedly limit humankind's ability to augment the ecosphere's negentropic potential. Given these restrictions, a biocentric optimum — one where the optimal scale of the human presence in the ecosphere involves the maximisation of the combined instrumental and intrinsic values of natural capital — will be much smaller than an egocentric optimum because, in the latter instance, a society is unconstrained by the need to respond to potential concerns for the intrinsic value of natural capital and its constituent species. To what extent the size of a biocentric and egocentric optimum differ depends entirely on a society's valuation of the intrinsic worth of the subhuman world. The higher the valuation, the smaller is the biocentric optimum and the greater is the difference in the respective optima.

If it is assumed that, as a consequence of humankind's indebtedness to the evolutionary process, the intrinsic value of natural capital and the sentient nonhuman beings it comprises must be properly recognised, an additional sustainability precept is in order. The fourth precept is as follows.

4. The three sustainability precepts so far presented should be adhered to subject to the condition that an extended principle of justice recognising the intrinsic value of sentient nonhuman beings must not be violated. This means the human exploitation of natural capital must avoid the undignified and cruel treatment of sentient nonhuman beings.

In view of the four sustainability precepts presented in this chapter, achieving sustainability effectively requires human beings to adhere to what might best be described as a *principle of stewardship*. In a similar vein to the principle of limitation, stewardship can be regarded as the sustainability-based rule of right action. Stewardship of natural capital not only ensures the maintenance of its three critical instrumental functions, it enables human-kind to discharge its obligation to oversee and facilitate the continuation of the total evolutionary process.

4.5   What is sustainability? A summary

Sustainability is a point of conjecture if only because low entropy matter–energy is absolutely scarce and because the input of low entropy is required to sustain the optimal stock of human-made capital. Beyond biophysical realities, sustainability is a desirable human goal because, apart from being a development prerequisite, its achievement satisfies the moral principle of intergenerational equity.

As the sole provider and assimilator of low and high entropy mat-ter–energy, natural capital constitutes a prime factor in the sustainability equation. The importance of natural capital is reinforced by the fact it is also the sole repository of the rich biodiversity and self-organising mechanisms necessary to keep Earth uniquely comfortable for living organisms, including human beings. Hence, given the inability to humanly replicate the eco-sphere's full range of instrumental values, achieving sustainability is condi-tional upon the rate of low and high entropy exaction and insertion remain-ing within the regenerative and waste assimilative capacities of natural capital. Conceived in this manner, at least the first three of the four sustain-ability precepts outlined in this chapter must take precedence over individ-ual subjective valuations (Common and Perrings, 1992).

Whether natural capital must remain intact over all time depends on the extent to which humankind can augment the ecosphere's negentropic poten-tial. If it can, then, in principle, it can get by with a smaller stock of natural capital — although one should always be mindful of the potential impact a smaller stock can have on biodiversity levels. This having been said, the fact that the ecosphere's negentropic potential can only be generated slowly means there is a definite short-run need for natural capital intactness.

Apart from its instrumental source, sink, and life-support functions, natural capital has intrinsic value that arises from its uniqueness and the sentience possessed by much of the subhuman world. Because rights accrue to human beings partly because of their sentience, humankind cannot ignore the limited rights of sentient nonhuman creatures, even if they are signifi-cantly less than those accorded to human beings. While the failure to recog-nise the intrinsic value of natural capital does not, in itself, threaten the sustainability goal, its recognition does have implications in terms of the eventual size of the human presence in the ecosphere. Indeed, it is likely to be considerably smaller.

chapter five

What is sustainable development?

5.1 Reconciling sustainability and development

Chapter 5 is a short, synthesising chapter. Having considered what sustainability and development are and what is required to achieve them, the two respective conditions can now be reconciled to establish an ecological economic-based SD concept. We have so far argued that, in relation to development, a principle of limitation is the desirable rule of right action. With regard to sustainability, the desirable rule of right action is a principle of stewardship. In their own ways, both principles recognise the existence of absolute limits and the need for their adherence — one to existential limits, the other to biophysical constraints. This not only reinforces the interconnectedness between sustainability and development, it demonstrates the extent to which one begets the other. One cannot expect true human progress to be achieved if human activity is ecologically unsustainable, nor sustainable practices to be adopted if human well-being is in decline.

The interdependency of sustainability and development also points to the existence of an optimal physical scale of human activity or, more particularly, an *optimal macroeconomic scale*, something almost entirely overlooked in the SD debate. An optimal macroeconomic scale is achieved when the physical scale of the macroeconomy and the qualitative nature of the human-made capital of which it is comprised maximise a nation's sustainable economic welfare or, more particularly, the sustainable net benefits of economic activity. Logically, in view of the need to operate sustainably, an optimal macroeconomic scale must be a sustainable scale. However, the maximum sustainable scale need not be an optimal one, indicating sustainability is a necessary but insufficient condition for the attainment of the optimum.

It is unfortunate, therefore, that most observers advocating restrictions on the growth of macroeconomies do so largely because of ecological factors and rarely because of the potential for continued growth to become existentially undesirable. It is unfortunate for two reasons. First, it is increasingly obvious that the need to limit the physical expansion of macroeconomic systems, usually confined to sustainability-based (or means-based) arguments, also applies to development-based (or ends-based) arguments. Second, because many of the costs of operating unsustainably will be borne by future generations, appropriate institutional and policy reforms are more likely if the current generation can be convinced of the present undesirability of growth.

Specifically, two things would appear necessary to achieve SD. The first is poverty alleviation. Poverty alleviation is not only a moral imperative, it is also a prerequisite to human development given that higher-order needs cannot be met without first satisfying basic lower-order needs. The second is population growth control. Given that the development of individual citizens is conditional upon an equitable per capita ownership of human-made capital, a larger population implies the need for a larger aggregate stock. But therein lies the problem because an ever growing stock of human-made capital must eventually exceed the ecosphere's sustainable carrying capacity.

The time has finally arrived to define SD and to list and detail a set of SD-based parameters, guidelines, and minimum positions. By combining the summaries of Chapters 3 and 4, the following can be stated, "A society experiencing SD is one characterised by the qualitative improvement in the human condition where, to facilitate human betterment, brother and sister-hood is extended beyond presently living people to include future generations. It is a society that, in the process of maintaining the human-made capital required by it to develop, expends matter–energy at a rate that can be sustainably provided and assimilated by the source and sink functions of the accommodating ecosphere. In addition, it is one that seeks to ensure the survival of the biosphere and all its evolving processes while also recognising, to some extent, the intrinsic value of sentient nonhuman beings."

While the preceding is a general definition of SD, achieving it requires the satisfaction of the following conditions. These conditions constitute the objective value-based standards and principles or measuring rod against which the human evolutionary process subsequently can be judged and assessed. They include:

- A recognition of the need for and the subsequent establishment of an ordering or ethical guiding principle — an ultimate end — that best constitutes the end goal of life itself.
- The weeding out of erroneous images from human belief systems and the incorporation of truer ones as they emerge. This ensures a continual revision of the ultimate end. The ultimate end needs to be constantly updated — and the hierarchy of intermediate ends rearranged — to ensure that the satisfaction of human needs and wants best facilitates the human development process.
- The regeneration of moral capital, if and when it is depleted, to maintain the civilised standards of conduct necessary for a successful market-based economy.

- Operating within the carrying capacity of the accommodating ecosphere by limiting the exaction and insertion of low and high entropy matter–energy to that which can be sustainably provided and absorbed by the ecosphere's source and sink functions.
- Maintaining, at least in the short run, a constant stock of natural capital by exploiting nonrenewable resources at a rate equal to the creation of renewable resource substitutes.
- Preserving and/or restoring biodiversity to both maintain the ecosphere's life-support function and to increase its resilience to potentially catastrophic stresses and shocks.
- Confining human activities to areas already considerably modified by human beings.
- Conducting resource exploitation on the basis of regional and local geographical land and water resource units. This enables humankind to operate within the uniquely defined carrying capacities of the various ecosystems it exploits.
- Incorporating the limited rights of sentient nonhuman beings into an extended principle of justice by confining resource use practices to those that minimise their inhumane, undignified, and cruel treatment.

5.2 Justifying the concept of sustainable development

Before moving on to Part II of the book, there is a need to address the criticisms levelled at the concept of SD. Unless it can be demonstrated that SD is sufficiently meaningful and operational, it will prove inordinately difficult to suggest ways and means to achieve SD itself. In order to show that the SD concept has normative value, attention will be focused on the criticisms put forward by Beckerman because, in many ways, they represent the broad range of criticisms that exist. Beckerman insists that

- Because there is no blueprint to achieve SD, the SD concept is devoid of operational value.
- Because an adherence to policies prescribed by SD advocates involves unnecessary sacrifices on the part of presently living people, in particular the impoverished, the concept of SD is morally indefensible.

Let us address both criticisms separately. To the first, Beckerman bases his criticism on a particular interpretation of SD, namely, that SD involves a "requirement to leave future generations a stock of assets that provides them with some predetermined level of potential welfare, such as that existing today" (1992, page 492). In view of this interpretation, albeit a perfectly reasonable one, Beckerman believes achieving SD boils down to knowing what substitution possibilities permit the current level of welfare to be obtained from different combinations of assets. For example, if there are fewer insects in future, should there be more trees? If there are fewer fish, should there be more machines? Moreover, if the answer to both questions is yes, how many more trees and machines are required? It is because these questions have no precise answers that Beckerman believes the SD concept is devoid of any operational value.

As correct as Beckerman is regarding the lack of precise answers, his criticism falls apart on three counts. To begin with, Beckerman fails to understand that trees are not a very good substitute for insects, while machines are no substitute for fish at all. Second, while inexact knowledge of what can substitute for what constitutes a perennial problem, it is exact knowledge of what cannot substitute for what that makes SD both a meaningful and operational concept. It is, for example, the knowledge that human-made capital cannot substitute for natural capital that makes natural capital intactness a meaningful policy directive. Finally, why should an inability to know what is precisely required to achieve SD render the concept operationless? Does not the achievement of any condition suffer the same fate? Consider the conditions previously listed for achieving SD. They, themselves, do not specifically indicate how to achieve SD. Nor does continuous economic growth, the very thing that Beckerman believes is the key to achieving SD. For instance, how do we achieve the maximum rate of growth? With more bridges and fewer schools? Lower taxes and higher savings? In the end, because the criticism levelled by Beckerman at the SD concept can be just as easily extended to Beckerman, his criticism hardly seems warranted.

As for Beckerman's second criticism, that SD is a morally indefensible concept, Beckerman fails to recognise that the call for SD only questions growth as it relates to nations already possessing a sufficient stock of human-made capital. Except for population control, it does not argue for a halt to growth in impoverished nations. While it is true that poverty persists in nations with adequate quantities of human-made capital, to assert that a halt to growth is morally indefensible because it leaves the poor impoverished is to suggest that only growth can alleviate poverty. This simply is not so. Indeed, for the following reasons, it would appear that growth, not SD, is morally indefensible. First, growth appears to be rendering many nations poorer rather than richer (see Figures 14.3 and 14.4 in Chapter 14). Second, by seeing growth as the only genuine poverty elixir, growth advocates avoid the moral dilemma of wealth redistribution and population control. Finally, growth is reducing the number of people who can live over all time by permitting the extravagant desires of the present to take priority over the needs of future generations, a clear violation of the moral principle of intergenerational equity.

To sum up, Beckerman's criticisms of the SD concept can be categorically rejected. SD is an entirely justifiable, if not essential, intellectual concept. After all, what matters more than sustainable human progress and that which is required to achieve it?

II

Sustainable development, economic theory, and macro policy goals and instruments

chapter six

Sustainable development and an optimal macroeconomic scale

6.1   What is an optimal macroeconomic scale?

Having established an intellectually rigorous SD concept, we are well placed to consider some important theoretical aspects concerning the movement toward SD. These considerations will ultimately indicate the respective policy goals, instruments, and mechanisms relevant to achieving SD. From an ecological economic perspective, four economic-related considerations stand out. Three of them concern the microeconomics of SD, namely, economic efficiency, the allocative role of relative prices, and the role played by the market as a resource allocation mechanism. All three are respectively dealt with in Chapters 7, 8, and 9. For now, consideration is given to the macroeconomics of SD.

In the previous chapter, it was reiterated that an optimal macroeconomic scale is one that maximises a nation's sustainable net benefits. An optimal macroeconomic scale can be considered the primary macroeconomic objective in moving toward SD because sustainable net benefits (the difference between the benefits and costs of economic activity) is equivalent to a measure of sustainable economic welfare — a critical SD concern. Of course, this objective differs entirely with the standard macroeconomic objective of a low-inflation, full-unemployment, growth economy. It should therefore be stressed that the desire to attain an optimal macroeconomic scale in no way belittles the importance of price stability or, more particularly, full unemployment. To the contrary, it treats the problem of unemployment more seriously than usual given that, and unlike a measure of real GDP, the cost of unemployment is factored into a comprehensive calculation of sustainable net benefits. Indeed, it is only when the cost of unemployment and an unequal distribution of income are incorporated into a macroeconomic performance indicator that policymakers are forced to think earnestly about how to generate employment and reduce impoverishment in a potentially no-growth macroeconomic environment.

6.1.1   Standard economics' glittering anomaly

Why is there such a major disparity in the macroeconomic objectives of standard and ecological economics? While some of the underlying reasons have already been elaborated (e.g., the circular flow representation of the macroeconomy), the main reason is the extraordinary anomaly between micro- and macroeconomics. Glance through any standard undergraduate textbook and you would be excused for thinking that the micro- and macroeconomic sections were written by two people on entirely different wavelengths. Microeconomics, for example, is virtually an expanded variation on the theme of optimal scale. Whether it be a firm in production or an individual making decisions concerning his spending choices or the number of hours he should work, it is customary to identify, define, and often mathematically formalise a benefit-and-cost function. All along, the microeconomist is at pains to separate benefits and costs. The two basic economic laws of diminishing marginal benefits and increasing marginal costs are incorporated into the analysis to reflect the usual way that total benefits and costs vary in accordance with an increase in the scale of any given activity. In every instance, the recommended course of action is the same, that is, expand the scale of the activity until marginal benefits equal marginal costs. So long as the prescribed course of action is followed, net benefits are maximised and an optimal scale of activity is attained.*

Strangely, standard macroeconomics ignores the notion of optimal scale almost completely. Virtually no attempt is made to compare the benefits and costs of a growing macroeconomy nor, furthermore, to seek and attain an optimal macroeconomic scale. The continued expansion of the macroeconomy is deemed desirable simply because it involves an increase in per capita output. For instance, consider the following

> The growth of total output relative to population means a higher standard of living because an expanding real output means greater material abundance and implies a more satisfactory answer to the economising problem (Jackson et al., 1994, page 414).

---

* The owner of a firm, for instance, should increase production until the marginal revenue received from the sale of output equals the marginal cost of the last unit of output produced. At this point, the firm maximises its profit and should cease to increase the scale of production, even if it is less than the firm's maximum productive capacity. In a similar vein, an individual should increase the number of hours worked until the marginal utility gained from the income he earns equals the marginal disutility from having to work an extra hour. At this point, the individual maximises utility and should cease to increase the scale of employment.

Is it not unreasonable to ask why an increase in the scale of economic activity constitutes a more satisfactory answer to the economising problem when, from a microeconomic perspective, operating at maximum productive capacity is not necessarily the desirable course of action. One can only suggest that in moving from the microeconomic to macroeconomic sphere, most economists continue to celebrate the benefits of a growing macroeconomy while forgetting or ignoring the attendant costs. Yet, it is because of this oversight that it is presently impossible to ascertain whether any given macroeconomic system is nearing or has surpassed its likely optimum.

How, then, does one set about identifying an optimal macroeconomic scale? In particular, what are the relevant benefits and costs of economic activity that need to be identified and compared? To begin with, imagine the ecosphere as a far-from-thermodynamic-equilibrium system consisting only of natural capital — the original source of all economic activity. Human beings have the onerous task of transforming a portion of natural capital into human-made capital in order for needs and wants to be better satisfied than by low entropy resources left *in situ* (otherwise the transformation process would be a pointless exercise). This task is a never-ending one because, as physical commodities are either directly consumed or worn out, new commodities must be produced to keep the stock of human-made capital intact. Of course, the desire to increase the rate of transformation is in no way unconstrained. As Chapter 4 demonstrated, the source, sink, and life-support functions of natural capital are severely limited. Hence, there is only so much low entropy that can be sustainably transformed into human-made capital. In addition, the capacity of human-made capital to serve the ultimate end is likely to reach an asymptotic limit.

With these constraints in mind, the following question arises: To what extent should a nation continue to transform natural to human-made capital and expand the physical scale of its macroeconomy? Because achieving an optimal macroeconomic scale requires the maximisation of a nation's sustainable net benefits, there are two aspects to consider in answering this question. The first deals with the benefit of producing and maintaining a stock of human-made capital. The second deals with the cost of its production and maintenance in terms of instrumental natural capital services foregone.

6.1.2   The uncancelled benefits of economic activity

As has been mentioned a number of times previously, how well the stock of human-made capital is able to qualitatively improve the human condition depends on how effectively it serves an appropriately conceived hierarchy of intermediate ends — the ultimate end. The intensity with which physical commodities are able to serve the ultimate end depends on their individual *service-yielding* qualities. Unlike human-made capital itself, the service yielded by the stock is a "psychic flux" and, with no physical dimension of its own, cannot be accumulated (Daly, 1979). As a flux rather than a stock or flow, service closely corresponds to Fisher's (1906) notion of *psychic income*. Psychic income constitutes the true benefit of all economic activity and has three main sources. The first source of psychic income is that which arises from the consumption or wearing out of human-made capital. The second source emerges from being directly engaged in production activities (e.g., the enjoyment and the personal sense of contribution and self-worth obtained from work). The third source is the psychic income yielded directly from natural capital in terms of its existence values and its aesthetic and recreational qualities. Obviously, this final source of psychic income does not come from economic activity. Indeed, if anything, economic activity tends to destroy rather than enhance such values. It is, therefore, better that such values be taken as "givens" and their subsequent destruction be counted as an opportunity cost of economic activity.

This last point is a reminder that not all economic activity enhances the psychic enjoyment of life. The production and subsequent consumption of some portion of human-made capital can reduce the psychic enjoyment of life if consumers make bad choices or if needs and wants have been inappropriately ranked. In addition, while benefits can be enjoyed by individuals engaged in production activities, for most people, production activities are unpleasant. Unpleasant things that lower one's psychic enjoyment of life (which can also include noise pollution and commuting to work) represent the "psychic outgo" of economic activity. It is the cancelling out of psychic outgo from psychic income that enables one to obtain a measure of *net psychic income* — the final or "uncancelled benefit" of economic activity. Net psychic income is the uncancelled benefit of economic activity because, in tracing the economic process from its original source (natural capital) to its final, psychic conclusion, every intermediate transaction involves the cancelling out of a receipt and expenditure of the same magnitude. Only once a physical commodity is in the possession of the final consumer is there no further cancelling of transactions (unless, at some further stage, he or she opts to exchange it for something preferable).

Perceiving the final benefit of economic activity in this manner means one should never directly identify the quantity of human-made capital with benefit itself. The stock of human-made capital should always be construed as a benefit-yielding physical magnitude because there is no necessary correlation between the magnitude itself and the net psychic income it yields (Daly, 1979). It has been the direct identification of the quantity of commodities consumed with the intensity of net psychic income enjoyed that has led many observers to erroneously believe that consumption is a "good." Certainly, in the case of some items, service can only be derived through their direct consumption (e.g., food, drink, and petrol). Nevertheless, on the whole, service is not directly derived from the act of consumption because very few people derive much satisfaction from seeing the paint on their house crack, from viewing a deteriorating TV image as the picture tube nears the end of its useful life, or from wearing a shirt that has long begun to perish and fade. Clearly, while it is necessary to consume human-made capital to enjoy the net psychic income it yields, consumption should be viewed as a "necessary evil" and something to be minimised (Boulding, 1966). Only the service-yielding qualities of human-made capital, not the rate at which human-made capital is consumed, should be maximised.

6.1.3   The uncancelled costs of economic activity

In order to deal with the second aspect concerning the preceding question, it is useful to recall two things from Chapter 4. First, the production and maintenance of human-made capital requires a continuous entropic through-put of matter–energy. Second, no matter how benignly human beings conduct their economic activities, the subsequent disarrangement of matter–energy always has some deleterious impact on the ecosphere (Perrings, 1986). Consequently, human beings have no option but to acknowledge and accept some loss of the free source, sink, and life-support services provided by the ecosphere as they manipulate, refine, and directly transform natural capital and the low entropy it provides into human-made capital.

Because natural capital is the original source of all human activity, the loss of natural capital services is the final or "uncancelled cost" of economic activity just as net psychic income is the uncancelled benefit. That is, if one traces economic activity from its final conclusion (net psychic income) back to its original source, all intermediate transactions are cancelled except, finally, the natural capital services sacrificed in obtaining the low entropy required to produce and maintain human-made capital. Thus, in the same way human-made capital should not be directly identified with benefit, so the throughput of matter–energy should not be directly identified with cost. For, unlike the loss of natural capital services, the throughput constitutes nothing more than a cost-inducing physical flow that is necessary to keep the stock of human-made capital intact (Daly, 1979).

One final but very important point. Although, from an accounting point of view, lost natural capital services constitute the uncancelled costs of economic activity, of greater concern is whether the combined loss of natural capital services renders the current level of economic activity ecologically unsustainable. Should this be the case, it follows that the prevailing macroeconomic scale is, by definition, suboptimal. Clearly, it is necessary to limit the throughput of matter–energy resulting in the loss of natural capital services to a level that can be ecologically sustained.

6.1.4   Sustainable net benefits and an optimal macroeconomic scale

Given that net psychic income and lost natural capital services constitute the uncancelled benefits and costs of economic activity, sustainable net benefits can be represented by the following simple equation.

Sustainable net benefits = net psychic income − lost natural capital services   (6.1)

The extent to which a nation should increase the physical scale of the macroeconomy depends on whether its expansion leads to an increase in sustainable net benefits. In the event that it does, growth is desirable. In the event that it does not, the macroeconomy should cease to grow because its expansion becomes, in effect, "uneconomic." The notion of an optimal macroeconomic scale is diagrammatically represented by Figure 6.1.

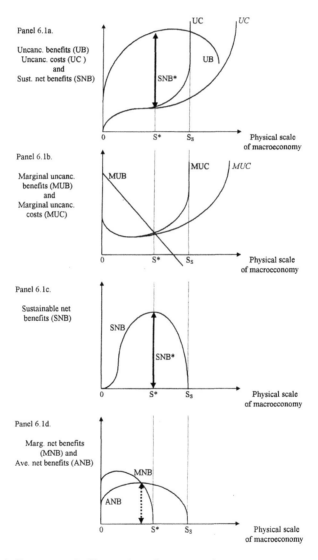

Figure 6.1 A diagrammatic illustration of an optimal macroeconomic scale. (Adapted partly from Daly, 1991b, page 28. With permission.)

The uncancelled benefits (UB) curve in Panel 6.1a represents the net psychic income yielded by a growing macroeconomy. The characteristic shape of the UB curve is attributable to the law of diminishing marginal utility which, barring technological improvements, is equally applicable to the total stock of human-made capital as it is to individual items. It is as a consequence of this law that the marginal uncancelled benefit (MUB) curve in Panel 6.1b is downward-sloping.

The cost of increasing the physical scale of the macroeconomy is represented in Panel 6.1a by an uncancelled cost (UC) curve. It represents the free instrumental services — the source, sink, and life-support functions — lost in the process of transforming natural capital and the low entropy it provides into human-made capital. The shape and nature of the UC curve is attributable to the law of increasing marginal costs, a reflection of the increasing costs arising from the macroeconomy growing relative to a finite natural environment. For instance, as the macroeconomy grows and nonrenewable resources are progressively depleted, human beings are forced to go to much greater lengths and expend more energy to discover and exact new and often poorer grade resources. Also, in a world where the scale of the macroeconomy relative to the ecosphere is small, humankind's presence has only a minimal impact on the source, sink, and life-support functions of natural capital. As the macroeconomy expands, and undesirable ecological feedbacks increase in their significance, the opportunity cost associated with every new disruption of natural capital is greater than the previous disruption. It is because of the law of increasing marginal costs that the marginal uncancelled cost (MUC) curve in Panel 6.1b is upward-sloping.

In Panels 6.1a and 6.1b there is a second uncancelled cost and marginal uncancelled cost curve (UC and MUC). Both are vertical at the macroeconomic scale of S_S. This is because S_S denotes the *maximum sustainable* macroeconomic scale — what is, for given levels of human know-how, the largest macroeconomic scale that a nation can continue to maintain at the maximum sustainable entropic rate of throughput.* The need for the additional curves arises because the UC and MUC curves merely represent the total and marginal uncancelled costs of maintaining a given macroeconomic scale. The UC and MUC curves do likewise and are, in most part, identical to the UC and MUC curves. However, they are vertical at S_S to take account of the fact that while, in the short-run, it is possible to sustain the uncancelled costs incurred from having a macroeconomic scale larger than S_S, they cannot be ecologically sustained in the long run. For this reason, the relevant curves in the determination of the optimal scale and the calculation of sustainable net benefits must be the UC and MUC curves.

For any given macroeconomic scale, sustainable net benefits (SNB) are measured by the vertical difference between the UB and UC curves. This difference is represented by the SNB curve in Panel 6.1c. An optimal

---

* By maximum sustainable rate of throughput, we mean the highest rate of matter–energy crossing the macroeconomy-ecospheric boundary consistent with the sustainability precepts established in Chapter 4.

macroeconomic scale is denoted by S.. It represents the physical scale of the macroeconomy that maximises the vertical difference between the UB and UC curves (SNB.). In Panel 6.1c, this is where the SNB curve is at its highest point. In Panel 6.1b, it is where the MUB and MUC curves intersect. Because the optimal macroeconomic scale of S. is to the left of S_S (the maximum sustainable scale), Figure 6.1 illustrates an important point made in the previous chapter. A sustainable scale is a necessary but insufficient condition for the achievement of the optimum. Operating at the maximum sustainable scale clearly does not, in this instance, maximise sustainable net benefits.

The marginal net benefit (MNB) and average net benefit (ANB) curves are to be found in Panel 6.1d. These curves have been included for two reasons. First, it is where the MNB curve cuts the horizontal axis (i.e., where MNB = 0) that sustainable net benefits are at a maximum and the macroeconomic scale is at the optimum. Second, because a measure of average net benefits is equivalent to the ratio of uncancelled benefits to uncancelled costs, the ANB curve represents the change in the efficiency with which natural capital is transformed into human-made capital.

Finally, Figure 6.1 demonstrates that growth is only a desirable macroeconomic objective in the early stages of a nation's developmental process. Continued physical expansion of the macroeconomic subsystem, which is equivalent to moving along the UB and UC curves, is uneconomic, in fact anti-SD, because it leads to the eventual decline in sustainable net benefits. Indeed, given that growth becomes uneconomic before the macroeconomy reaches its maximum sustainable scale, Figure 6.1 indicates that an economic limit to growth is likely to precede an inevitable biophysical limit to growth.

6.1.5   *Electing to operate at the maximum sustainable macroeconomic scale — when the maximum sustainable scale is the optimal scale*

In the preceding example, the optimal macroeconomic scale is reached prior to the macroeconomy becoming ecologically unsustainable. Hence, whether the UC or the preferred *UC* curve is used in the determination of the optimal macroeconomic scale is irrelevant because the result is exactly the same. This, however, is not always the case. Consider Figure 6.2.

Assume, for a moment, that the optimal macroeconomic scale is determined by maximising the difference between the UB and *UC* curves. In Panel 6.2a, the optimum exists at a macroeconomic scale of S* — what might be referred to as a "perceived" optimum in the absence of any sustainability considerations. In Panel 6.2b, the perceived optimum is where the MUB and *MUC* curves intersect. Because ecological sustainability is a necessary condition for the achievement of an optimal macroeconomic scale, the perceived optimum does not constitute the "true" optimum. Instead, the true optimum exists at the maximum sustainable macroeconomic scale of S_S where the UC and MUC curves are vertical. While a higher level of net benefits can be enjoyed at the perceived optimum (i.e., *SNB.* is greater than SNB.), they are

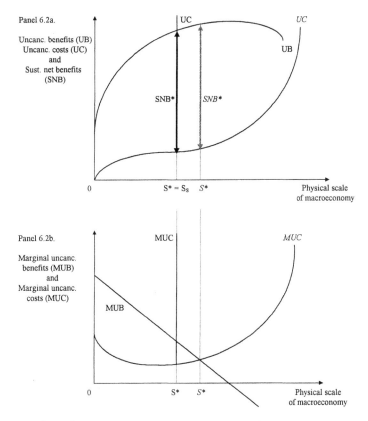

Figure 6.2 When the maximum sustainable scale is the optimal macroeconomic scale.

not sustainable net benefits in the sense that they cannot be ecologically sustained in the long run. Clearly, in the preceding circumstances, it is necessary to disregard the perceived optimum and operate at the maximum sustainable scale. For this reason alone, Figure 6.2 highlights an important point to be extensively dealt with in the following chapter. That is, what appears to be economically rational need not coincide with what is ecologically sustainable.

6.2 *Technological progress and an optimal macroeconomic scale*

One cannot ignore the role played by advances in technology. Technological progress can increase the net psychic income gained while also decreasing the natural capital services sacrificed in maintaining a given macroeconomic scale. This is because technological progress can beneficially shift the UB curve upwards and the UC curve downwards and to the right. In doing so, technological improvements can permit, over time, an increase in the overall size of the optimal macroeconomic scale. Furthermore, by also reducing the loss of natural capital services, technological progress can also increase the maximum sustainable scale.

6.2.1   Introducing the ecological economic efficiency (EEE) identity

To explain how the two curves can be positively shifted, the uncancelled benefits and costs of economic activity can be arranged to obtain a measure of *ecological economic efficiency*. Because efficiency is the ratio of benefits to costs, an overall measure of ecological economic efficiency (EEE) can be represented by the following ratio:

$$EEE = \frac{\text{net psychic income}}{\text{lost natural capital services}} \tag{6.2}$$

An increase in the EEE ratio indicates an improvement in the efficiency with which natural capital, and the throughput of matter–energy it provides, is transformed into benefit-yielding human-made capital. A multitude of factors can be shown to contribute to an increase in the EEE ratio. To demonstrate how, the EEE ratio is extended to reveal the following identity.*

$$EEE = \frac{\text{net psychic income}}{\text{lost natural capital services}}$$

$$= \underbrace{\frac{\text{net psychic income}}{\text{human - made capital}}}_{\text{Ratio 1}} \times \underbrace{\frac{\text{human - made capital}}{\text{throughput}}}_{\text{Ratio 2}} \times \underbrace{\frac{\text{throughput}}{\text{natural capital}}}_{\text{Ratio 3}} \times \underbrace{\frac{\text{natural capital}}{\text{lost natural capital services}}}_{\text{Ratio 4}}$$

$$\tag{6.3}$$

Starting from Ratio 1 and progressing through to Ratio 4, each component ratio cancels out the ensuing ratio. This leaves the basic EEE ratio on the left-hand side. The order in which the four component ratios are presented is in keeping with the nature of the economic process, namely, that net psychic income is the true benefit of economic activity; that human-made capital must be produced to enjoy the psychic benefits of economic activity; that an entropic throughput of matter–energy is required to produce and maintain human-made capital; that natural capital must be exploited to obtain the throughput of matter–energy; and that, as a consequence of natural capital exploitation, some of its source, sink, and life-support services are sacrificed. Each of the four ratios represents a different form of efficiency that can be increased via improved human knowledge in both technique

---

* The EEE ratio and its component ratios are adapted from an efficiency ratio posited by Daly (1996).

and purpose (the latter owing to the importance of ranking needs and wants appropriately).

6.2.2 Impact of a beneficial shift of the uncancelled benefits (UB) curve

Ratio 1 is a measure of the *service efficiency* of human-made capital. It increases whenever a given physical magnitude, though never the same population, of human-made capital yields a higher level of net psychic income. An increase in Ratio 1 causes the UB curve to shift upwards and can be achieved in the following ways: first, by improving the technical design of newly produced commodities; and second, by reducing the psychic outgo associated with the production and maintenance of human-made capital. For example, improvements in the way human beings organise themselves in the course of production can reduce such things as the disutility of work and the cost of commuting and unemployment. Of course, even if the service-yielding qualities of human-made capital improve over time, Ratio 1 can still fall and the UB curve can shift downward if, in the process of keeping the stock of human-made capital intact, psychic outgo costs rise sufficiently enough to more than offset the increased psychic benefits.

A third means of augmenting Ratio 1 is to increase the efficiency with which low entropy resources are allocated to the production of human-made capital. An increase in allocative efficiency leads to more of the incoming resource flow being allocated to the production of physical commodities embodying the highest service-yielding qualities. This can be achieved by improving the effective operation of markets (e.g., ameliorating existing instances of market failure and ensuring a more appropriate market domain). It can also be achieved through international trade. By permitting nations to specialise in the production of certain commodities (i.e., those they are most adept at producing), trade allows engaging nations to increase the service efficiency of the human-made capital they ultimately have in their possession.

Ratio 1 can also be augmented by increasing the distributive efficiency among a nation's citizens. Often overlooked, the redistribution of income from the low marginal service uses of the rich to the higher marginal service uses of the poor can increase the net psychic income enjoyed by society as a whole (Robinson, 1962). There is, however, a limit on how far the redistribution of income can increase Ratio 1. This is because an excessive rate of redistribution adversely dilutes a market-based incentive structure designed to reward efficiency-increasing efforts.

Finally, Ratio 1 can be increased by revising the ultimate end to ensure a more appropriate ranking of needs and wants. Because the demand for physical commodities on the basis of an erroneously conceived ultimate end leads to the production of illth under the guise of wealth, an appropriately revised ultimate end ensures the subsequent production of more wealth and less illth.

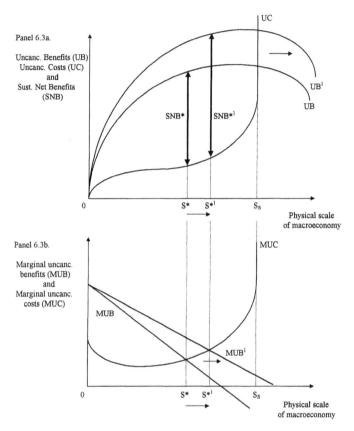

Figure 6.3 A change in the optimal macroeconomic scale brought about by an increase in the service efficiency of human-made capital (Ratio 1).

Figure 6.3 illustrates what happens to sustainable net benefits and the optimal macroeconomic scale when the UB curve shifts upwards. Because an increase in Ratio 1 augments the net psychic income yielded by a given physical magnitude of human-made capital, the UB and MUB curves in Panels 6.3a and 6.3b shift up to UB^1 and MUB^1. The UC and MUC curves do not move because the uncancelled cost of producing and maintaining a given quantity of human-made capital remains unchanged. Moreover, the maximum sustainable macroeconomic scale remains at S_S.

Prior to the increase in the service efficiency of human-made capital, sustainable net benefits are maximised by operating at a macroeconomic scale of S_* where sustainable net benefits equal SNB_*. Following the beneficial shift in the UB curve, sustainable net benefits are no longer maximised at the prevailing macroeconomic scale. They are instead maximised at a scale S_*^1 where sustainable net benefits equal SNB^1. It is now desirable for a nation to expand the physical scale of its macroeconomy to S_*^1.

Of course, because the UC curve remains stationary, it means the macroeconomy moves closer in size to the maximum sustainable scale. This is an important point. It indicates that while continued increases in the service efficiency of human-made capital can boost sustainable net benefits, unless there is a beneficial shift of the UC curve, an expanding optimum must eventually reach the maximum sustainable scale of S_s where further growth is suboptimal.

6.2.3   Impact of a beneficial shift of the uncancelled cost (UC) curve

Changes in Ratios 2, 3, and 4 cause the UC curve to shift. Ratio 2 is a measure of the *maintenance efficiency* of human-made capital. It increases whenever a given physical magnitude of human-made capital can be maintained by a lessened rate of throughput. This can be achieved by developing new technologies that reduce the requirement for resource input, whether it be through an increase in the rate of technical efficiency (i.e., a reduction in the low entropy wasted in its transformation to human-made capital), increased rates of product recycling, greater product durability, and/or improved operational efficiency (i.e., a reduction in the quantity of low entropy energy required to operate human-made capital). How does an increase in the maintenance efficiency of human-made capital shift the UC curve downwards and to the right? By lowering the required rate of throughput, an increase in maintenance efficiency reduces the need to exploit the same quantity of natural capital. This, in turn, means fewer sacrificed natural capital services and, therefore, a reduction in the uncancelled costs associated with maintaining a given macroeconomic scale.

Ratio 3 is the *growth efficiency* or productivity of natural capital. This form of efficiency is increased whenever a given amount of natural capital is able to sustainably yield and assimilate a greater quantity of matter–energy. An increase in Ratio 3 reduces the amount of natural capital requiring exploitation to obtain the rate of throughput necessary to keep the stock of human-made capital intact. This leads to fewer sacrificed natural capital services and to a downward and rightward shift of the UC curve. To increase Ratio 3, it is necessary to augment the ecosphere's negentropic potential by increasing the regeneration rate and the waste assimilative capacities of natural capital. The means to which the ecosphere's negentropic potential can be augmented were listed and discussed in Chapter 4.

Ratio 4 is a measure of the *exploitative efficiency* of natural capital. An increase in Ratio 4 means the exploitation of a given quantity of natural capital results in a reduced loss of natural capital services. Better techniques of natural capital exploitation (e.g., ecologically sensitive mining and lumbering practices) lead to increases in Ratio 4 which, as a consequence, shift the UC curve downwards and to the right.

Not unlike increases in the service efficiency of human-made capital, increases in Ratios 2, 3, and 4 can also be facilitated by the more effective operation of markets. By internalising a negative environmental externality, relative prices are able to more accurately reflect marginal social opportunity costs. This encourages operators to reduce the magnitude of the uncancelled costs associated with each economic activity which, in turn, leads to an increase in the EEE ratio and a downward shift of the UC curve. It must be said, however, that the more effective operation of markets does not ensure that the aggregate loss of natural capital services remains within an ecologically sustainable limit. More on this will be presented in the next chapter.

Once again, international trade can be of benefit. By enabling a nation to specialise in the exaction and insertion of the various forms of low and high entropy matter–energy that minimise the loss of natural capital services, trade can beneficially shift the UC curve. Consider a nation well endowed with hard and softwood timbers but, because of the nature of the ecosystems from which they are exacted, the harvesting of softwoods impacts on biodiversity levels significantly more than with hardwoods. Assume, also, that the nation's timber requirements are confined to its own domestic sources and that the combined loss of natural capital services from having to meet its annual timber demands is equal to X. Impact X can be significantly reduced if the nation in question could specialise in hardwood timber harvesting and exchange the surplus hardwood with softwoods from another nation.

Because the capacity to assimilate certain types of high entropy waste varies between different forms of natural capital, trade in wastes can also beneficially shift the UC curve. Just as some nations are better suited to growing cereal crops while others tropical fruits, so are nations likely to be better suited to assimilating certain kinds and quantities of high entropy waste. Hence, the usual restriction that a nation's natural capital places on its ability to assimilate high entropy wastes can be partially overcome by exporting difficult to assimilate wastes to a nation better able to absorb them. Some countries may even become waste assimilating "specialists." Naturally, because SD is a global issue, it would be important to ensure the waste levels being imported by the waste assimilating specialists were within their sustainable waste assimilative capacity.

Figure 6.4 illustrates what happens to sustainable net benefits when the UC curve shifts. Because increases in either maintenance, growth, or exploitative efficiency (Ratios 2, 3 and 4) reduce the uncancelled costs of producing and maintaining a given macroeconomic scale, the UC and MUC curves in Panels 6.4a and 6.4b shift down and out to UC^1 and MUC^1. However, the UB curve remains stationary because increases in either of these three efficiencies do not augment the net psychic income yielded by a given stock of human-made capital.

Unlike a shift in the UB curve, a shift in the UC curve results in an increase in the maximum sustainable macroeconomic scale (S_S to S_S^1). The logic behind this is quite simple. If there are now fewer natural capital services sacrificed in maintaining what was previously the maximum sustainable macroeconomic scale, it means a larger macroeconomic subsystem can be ecologically sustained from the same loss of natural capital services.

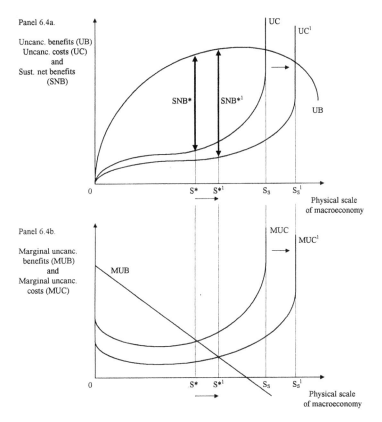

Figure 6.4 A change in the optimal macroeconomic scale brought about by increases in the maintenance efficiency of human-made capital (Ratio 2) and the growth and exploitative efficiences of natural capital (Ratios 3 and 4).

Prior to any increases in either maintenance, growth, or exploitative efficiencies, sustainable net benefits are maximised by operating at a macroeconomic scale of S. where sustainable net benefits equal SNB.. Following a beneficial shift in the UC curve, sustainable net benefits are now maximised at a scale of S¦ where they equal SNB¹. As in the case where the UB curve shifted upwards, it is desirable for a nation to expand the physical scale of its macroeconomy.

6.2.4 *Increasing efficiency — not always implying a larger optimal scale*

In the preceding examples, beneficial shifts in the UB and UC curves result in an increase in the optimal macroeconomic scale (Figures 6.3 and 6.4). For two main reasons, this will not always be the case. First, the new optimal scale depends not on the eventual positions of the new UB and UC curves but on their respective slopes. It is only at a scale where the slopes of the two curves are equal that the MUB and MUC curves intersect and a new optimal scale can be determined. This could just as easily be at a smaller macroeconomic scale. Second, an increase in service efficiency does not lead to a larger optimum if the maximum sustainable scale has already been reached. Consider Figure 6.5.

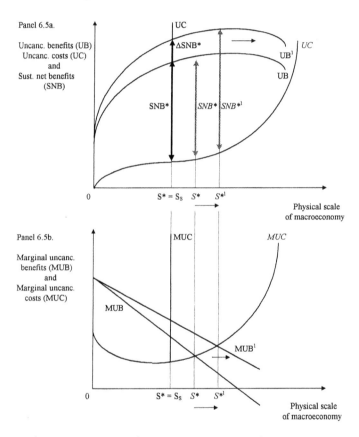

Figure 6.5 When an increase in the service efficiency of human-made capital does not result in a larger optimal scale.

In Panel 6.5a, sustainable net benefits are maximised at an optimal macroeconomic scale of S_S where sustainable net benefits equal SNB.. S. denotes a perceived optimal scale if one ignores the sustainability requirement. In Panel 6.5b, the perceived optimum is where the MUB and *MUC* curves intersect.

Following an increase in the service efficiency of human-made capital, the UB and MUB curves in Panels 6.5a and 6.5b shift upward to become UB^1 and MUB^1, respectively. The UC and MUC curves remain stationary. If one continues to ignore the sustainability requirement, the perceived level of sustainable net benefits increase from *SNB.* to *SNB.₁*. Because sustainability is a necessary condition for optimality, the shift in the UB and MUB curves does not lead to an expansion of the optimal macroeconomic scale. To maximise sustainable net benefits, the scale of the macroeconomy remains at S_S. The increase in service efficiency is, however, beneficial. The sustainable net benefits yielded by a quantitatively constant, but qualitatively superior, stock of human-made capital increases by an amount equal to ΔSNB_*.

6.2.5 *Efficiency-increasing and throughput-increasing technological progress*

Now is a good time to distinguish between efficiency-increasing and throughput-increasing forms of technological progress. *Efficiency-increasing* technological progress is that which increases any one of the four efficiency ratios described previously. Examples of this form of technological progress include better product design (increases Ratio 1); the development of recyclable materials (increases Ratio 2); improved land management practices and better methods of waste disposal (increases Ratio 3); and the development of ecosystem-sensitive methods of mining and timber harvesting (increases Ratio 4). Efficiency-increasing technological progress is always desirable because, by beneficially shifting either the UB or UC curve, it always results in an increase in sustainable net benefits.

Throughput-increasing technological progress, on the other hand, simply bestows a nation with the ability to exact from natural capital a greater quantity of low entropy. For this reason, its application brings to bear, at least in the short-run, a larger physical scale of a nation's macroeconomy. Examples of throughput-increasing technological progress include the development of a novel resource exploration method that leads to the discovery of a new oil deposit, a new resource exaction technique that allows a previously inaccessible mineral deposit to be exploited, and the development of a new use for a previously unwanted resource.

Unlike efficiency-increasing technological progress, the throughput-increasing variety is not always desirable. This is because the application of throughput-increasing technology merely brings about a movement along the two curves which, as Figure 6.1 shows, is only desirable in the early phase of a nation's developmental process. Eventually, its continued application leads to a decline in sustainable net benefits and a macroeconomic scale in excess of the optimum.

Of course, humankind develops both forms of technology and not the throughput-increasing variety alone. However, while efficiency-increasing technological progress provides breathing space for useful applications of throughput-increasing technology, its ability to do so is strictly limited. In the end, throughput-increasing progress must give way to efficiency-increasing technology because an increasing macroeconomic scale is biophysically limited by the carrying capacity of the ecosphere which no amount of throughput-increasing technology can continue to transcend.

6.2.6   *The ecological economic efficiency identity — breaking the ecological economic problem into manageable yet interdependent subproblems*

As useful as the EEE identity is in describing the efficiency with which natural capital is transformed into human-made capital, the transformation process and the purpose to which it serves are, as Daly (1996) has pointed out, far too complex to be fully reflected by the EEE ratio and its four component ratios. This aside, the ratios serve a couple of useful purposes. First, as evidenced previously, they allow the ecological economic problem to be categorised into well-defined and manageable subproblems. Second, they indicate the interdependencies associated with each of the various subproblems.

The last point is best understood by providing a couple of instructive examples. Consider the effect of replacing a slow growing native timber forest with a faster growing exotic timber plantation. While one would expect the growth efficiency of natural capital to increase (Ratio 3), one would also expect a loss of biodiversity and, therefore, a diminution of some of the ecosphere's life-support function. Hence, it would not be unreasonable to expect Ratio 4 to decrease. Should the decline in Ratio 4 exceed the increase in Ratio 3, the EEE ratio would, contrary to the desired intention, fall rather than rise. This would mean an undesirable shift upwards of the UC curve and a lowering of sustainable net benefits.

Next, consider the use of novel and highly toxic substances that increase the durability of physical commodities and, therefore, the maintenance efficiency of human-made capital (Ratio 2). Because these substances eventually re-enter the ecosphere as high entropy wastes, one would expect their dispersal to have a deleterious impact on the natural regeneration and waste assimilative capacities of natural capital as well as diminish the ecosphere's life-support function. If so, a reduction in the growth and exploitative efficiencies of natural capital (Ratios 3 and 4) would ensue. Not unlike the previous example, the net effect on the EEE ratio would depend on the extent of the impacts on each of the efficiency ratios affected. Should the EEE ratio decline, the original attempt to increase the maintenance efficiency of human-made capital would have an undesirable affect on both the UC curve and sustainable net benefits.

All told, one cannot deal with the efficiency ratios and subproblems individually if related subproblems go unrecognised both to the detriment of the remaining efficiency ratios and the EEE ratio generally. Clearly, policies designed to facilitate increases in any one of the four efficiency ratios without due recognition of the broader ecological economic problem are likely to have a perverse effect on the EEE ratio, the UB and UC curves, and the sustainable net benefits enjoyed by a nation's citizens.

6.3 Additional aspects of an optimal macroeconomic scale

Implicit in what we have so far demonstrated were the following assumptions. First, that natural capital has instrumental value only; second, that technological progress is unlimited in its capacity to shift the UB and UC curves; and finally, that it is safe to operate at the maximum sustainable scale should the perceived optimum exist somewhere beyond it. In what follows, consideration is given to the impact on the optimal macroeconomic scale as these assumptions are relaxed while others are brought into line with concrete reality.

6.3.1 Limits to the beneficial shift of the uncancelled benefits and uncancelled costs curves

Efficiency-increasing technological progress appears to support the arguments constantly presented by growth advocates, that is, increase the four efficiency ratios *ad infinitum* and there are no limits to the maximum sustainable scale of the macroeconomy, let alone limits to a desirable one. But appearances are deceiving. Severe limits exist in relation to the capacity for humankind to augment all four efficiency ratios and to continuously increase the sustainable net benefits generated from the transformation of natural to human-made capital.

Ratio 2, for instance, is limited by the first and second laws of thermodynamics [e.g., nothing is eternally durable, 100 percent recycling is impossible, and a measure of technical efficiency (E) cannot exceed a value of one]. Ratio 3 is limited by the inability to forever increase the ecosphere's negentropic potential (this includes the ecosphere's sink as well as source function), while Ratio 4 is limited by the fact that at least some of the ecosphere's instrumental functions are lost as a consequence of its exploitation (Perrings, 1986). Finally, increases in Ratio 1 are limited because, as mentioned a number of times previously, the capacity of human-made capital to serve the ultimate end is likely to reach an asymptotic limit.

Perhaps if there is any doubt about the limits to increasing the four component ratios it lies solely with Ratio 1 and only because service, as a psychic rather than physical magnitude, can seemingly grow forever. Yet there are at least two facts pointing to the probable limit on the service efficiency of human-made capital. First, as noted in Chapter 3, there are inevitable limits on the physiological capacity for human beings to experience service. As individuals, we can only feel so happy and contented. Second, regardless of human ingenuity, only so much use value can be added to a given quantity of matter–energy. Thus, even if we assume it to be possible to add enough use value to a small pot of soil for an entire family to be nourished from the food it yields (e.g., through the development of new plant strains), it is a bit much to expect it to eventually nourish the local neighbourhood, a small town, or a large city! Somewhere, a limit must be reached, just as it must in the case of many other services (e.g., shelter and clothing).

What, instead, if we assume that Ratio 1 can increase forever? Where does that leave growth advocates? Nowhere, other than having to concede that the need to focus on qualitative improvement rather than quantitative growth is more important than ever. Why? Because limits to Ratios 2, 3, and 4 imply an upper limit on the maximum sustainable scale of macroeconomic systems (a limit on growth), while a potentially unbounded Ratio 1 implies no upper limit on the ability of human-made capital to yield a higher level of net psychic income (no limit on development). Thus, as the limits to Ratios 2, 3, and 4 are approached, there is little point in striving for growth. Far better to focus time and effort on the qualitative improvement of human-made capital.

6.3.2   A sustainability buffer — allowing for uncontrollable fluctuations in the ecosphere's long-run carrying capacity

In the example given in Figure 6.2, the "perceived" optimal macroeconomic scale was larger than what the ecosphere could sustain in the long-run. As a consequence, the suggested course of action was to operate at the maximum sustainable macroeconomic scale (S_S) — what is effectively a sustainability threshold. However, this may not be the most appropriate course to follow. This is because a sustainability threshold is never as distinct as is represented in Figure 6.2. Moreover, for many uncontrollable reasons, it is subject to considerable fluctuations. To ensure sustainability, the maximum permissible scale is ideally limited to something less than the estimated maximum sustainable scale.

Clearly, a preferable option to operating at the maximum sustainable scale is to introduce a sustainability "buffer." A buffer is a self-imposed restriction on the rate of resource throughput set at less than the estimated maximum sustainable rate. Ideally, the sustainability buffer would be calculated as a percentage of the estimated maximum rate (i.e., 20 percent of S_S such that the maximum permissible rate would be $0.8 \times S_S$). To be of any genuine value, however, the percentage factor must be as large as is necessary to absorb the effect of an estimated worst-case decline in the ecosphere's carrying capacity.

Whether the imposition of a sustainability buffer affects the desirable scale of a nation's macroeconomy depends on where the prevailing optimum exists in relation to the newly imposed constraint. If the optimal scale is a perceived optimum and exists beyond the estimated maximum sustainable scale (S_S), the imposition of a sustainability buffer further reduces the desirable macroeconomic scale. If the optimal scale exists somewhere between the estimated maximum sustainable scale and maximum permissible scale ($0.8 \times S_S$), the imposition of a sustainability buffer reduces the physical scale of the optimum when, in fact, it would not have previously. Only if the optimum is equal to or something less than the maximum permissible scale is the optimum unaffected.

Notably, the maximum permissible scale need not remain static. In the same way that increases in Ratios 2, 3, and 4 shift the UC curve and increase the estimated maximum sustainable scale (e.g., S_S to S_S^1 in Figure 6.4), so they increase the maximum permissible scale. For example, if the sustainability buffer was set at 20 percent of S_S then, in Figure 6.4, the maximum permissible macroeconomic scale would increase from $0.8 \times S_S$ to $0.8 \times S_S^1$.

6.3.3   A biocentric versus egocentric optimum

Finally, the size of the optimal scale of the macroeconomy will also depend on the intrinsic value assigned to natural capital. As explained in Chapter 4, taking account of the intrinsic value of natural capital involves the dual consideration of the ecosphere's instrumental and intrinsic values — what we referred to as a *biocentric* perspective of natural capital. An *egocentric* view, on the other hand, focuses purely on the ecosphere's instrumental functions. Consequently, a biocentric optimum will, in most instances, be smaller than an egocentric optimum. Exactly how much smaller depends on whether intrinsic value is assigned to natural capital generally or more specifically to ecosystems and/or sentient nonhuman beings. For instance, if the assignment of intrinsic value is simply restricted to natural capital as a whole rather than individual sentient nonhuman beings, operating at or below the maximum sustainable entropic rate of throughput will have little or no effect on the optimal macroeconomic scale.

The same cannot be said if a society's biocentric perspective incorporates a belief that ecosystems and/or sentient creatures have intrinsic value in their own right. A more extensive assignment of intrinsic value will place an additional restriction on the maximum sustainable entropic rate of throughput. Furthermore, it may also limit a society's capacity to increase the service, maintenance, growth, and exploitative efficiency ratios, in particular, the latter two ratios where natural capital manipulation is involved. Hence, not only is a biocentric optimum likely to be much smaller than its egocentric counterpart but, in due course, future increases in the optimum are also likely to be restricted.

chapter seven

Economic efficiency and policy goals and instruments

7.1   Standard versus ecological economic efficiency

By demonstrating how increases in ecological economic efficiency can lead to the achievement of an optimal macroeconomic scale, the previous chapter highlighted the importance of economic efficiency as it relates to SD. But what is really meant by economic efficiency? More importantly, can we be certain that an increase in economic efficiency always advances us towards an optimal macroeconomic scale and toward SD? Indeed, could it be that ecological economic efficiency is simply a special case of efficiency?

By economic efficiency, economists generally refer to the effectiveness with which scarce means are being allocated to satisfy competing ends. If resources are allocated more efficiently, scarce means are able to satisfy a larger number of competing ends, or the same number more intensely, or both. Nevertheless, to increase the efficiency of resource allocation, the potential must exist for making someone better off without rendering another person worse off. Once this potential is exhausted, no improvement in economic efficiency is possible through allocation alone. At this point, a condition of *Pareto optimality* is reached. Only technological progress and a subsequent reallocation of scarce resources can improve a person's well-being at no one else's expense.

As important as economic efficiency is in achieving SD, there is, however, a need to determine the circumstances under which an efficient allocation of scarce resources contributes positively to the SD process. This need arises because although efficiency is, in many ways, independent of the choice of means and ends, the extent to which an efficient allocation contributes positively to the SD process is not. This may not be immediately obvious. If so, it is usually because it is not well known that the allocation process assists only in making the best of a given set of initial circumstances. Thus, while resource allocation can improve the overall set of circumstances, it cannot guarantee that the final outcome is commensurate with the SD-based parameters, guidelines, and minimum positions outlined in Chapter 5. In other words, it is possible for a condition of Pareto optimality to be achieved despite

1. The initial rate of low entropy resource exaction and the subsequent rate of high entropy waste insertion being ecologically unsustainable.
2. The intermediate ends served by human-made capital being inappropriately ranked in relation to the ultimate end.
3. The ultimate end being erroneously conceived to begin with.

In addition, while the allocation process can reduce the number of people in a state of impoverishment (i.e., by making some people better off without rendering others worse off), many people can remain impoverished if there are enough in such a state to begin with and if the potential to make someone better off at no-one else's expense is quickly exhausted. Hence, it is entirely conceivable for a Pareto optimal outcome to coexist with an inequitable distribution of income and wealth.

What does this reveal about economic efficiency as it relates to SD? It simply indicates that, if achieved, a condition of Pareto optimality is a case of having resolved the standard economic problem when, as we pointed out in Chapter 2, the successful resolution of the standard economic problem is a necessary but insufficient SD condition. Clearly, it is only insofar as a Pareto optimal outcome can be adequately qualified that an efficient allocation of scarce resources can be construed to have contributed to the movement toward SD. To wit:

> "If economic rationality is 'good' regardless of its content, and if economic rationality exhausts itself in the efficient pursuit of any goal regardless of its origin and content, there is no principle from which one can deduce the duty to examine the goal itself" (Weisskopf, 1971, page 91).

7.1.1   Recognising trade-offs that contribute to the movement toward sustainable development

The extent to which the economic efficiency concept is inadequately qualified is reflected by frequent references to equity and sustainability-efficiency trade-offs. Trade-offs are, of course, an integral feature of the allocation process. With the ultimate means absolutely scarce, there is a constant need to make choices and trade-offs to ensure low entropy resources are allocated to production processes generating the highest potential use values. However, because achieving SD requires a strict adherence to objective value-based standards and principles, it follows that the permissibility of trade-offs in the pursuit of increased efficiency is necessarily limited. Thus, from an ecological economic perspective, it is

important to recognise the distinction between a legitimate SD-enhancing trade-off on one hand and an increase in economic efficiency that occurs because of an erosion or violation of objective value-based standards and principles on the other.

A good example of the widespread failure to make this important distinction is an otherwise excellent book entitled: *Equality and Efficiency: The Big Trade-Off* (Okun, 1975). The book is largely devoted to the issue of efficiency trade-offs arising out of so-called infringements on the calculus of economic efficiency. Throughout the book, Okun refers to a number of infringements that represent genuine impediments to the achievement of economic efficiency (e.g., avoidable bureaucratic costs and inefficiencies, complicated distortionary tax laws, and unnecessary trading restrictions). However, a number of other so-called infringements involve rights, norms, and universally awarded privileges and entitlements that arise as a consequence of the institutionalisation of a society's objective value-based standards and principles. In what he more conveniently dubs the "domain of rights," Okun considers the institutionalisation of rights to be "inefficient" because it

> "...precludes prices that would otherwise promote economising, choices that would invoke comparative advantage, incentives that would augment socially productive effort, and trades that would potentially benefit buyer and seller alike" (Okun, 1975, page 10).

Hence, according to Okun, a trade-off exists because, in the process of any society providing its citizens with a vast array of free and universal entitlements and privileges, some degree of economic efficiency is foregone. Okun therefore believes an appropriate balance must be sought between the benefits gained through the provision of rights and those to be had through the efficient allocation of resources.

However, at close examination, it is clear that the majority of the trade-offs implied by Okun exist only because of his failure to distinguish between that which ought to be counted as a benefit or cost in the calculation of net benefits and that which transcends the efficiency calculus. Rights, customs, and norms, belong to the latter because, as a reflection of the institutionalisation of a society's objective value-based standards and principles, they constitute the civilised standards upon which a Pareto optimal outcome can be properly judged and assessed. Hence, rather than being infringements on the calculus of economic efficiency, rights, customs, and norms constitute the objective value-based framework — the bedrock — upon which the efficiency calculus is grounded. Consequently, when legislation is enacted to reduce ecological degradation or to protect legitimate individual rights, a society is not so much foregoing the net benefits it would otherwise gain from unfettered resource allocation. It is, in effect, setting in stone the objective value basis upon which it can qualitatively assess the effectiveness of the resource allocation process.

Of course, one must be careful not to interpret the adherence to SD-based parameters, guidelines, and minimum positions as an unwarranted limitation on the permissibility of all potential trade-offs. Despite the need to comply with objective value-based standards and principles, social inequalities and natural capital disruption should by no means be prohibited. Inequality, for instance, by promoting an incentive structure that rewards efficiency-increasing pursuits, is permissible to some extent because not all inequality is unjust or inequitable. Hence, up to an ethically tolerable limit, an efficiency-equality trade-off is perfectly legitimate. Similarly, ecological disruption, necessary for humankind to exact and insert low and high entropy matter–energy for its own instrumental purposes, is permissible up to an ecologically sustainable limit. Because not all natural capital exploitation is ecologically unsustainable, an efficiency-environment trade-off, at least until a sustainability threshold is reached, is entirely legitimate.

7.1.2   Ecological and standard economic efficiency

It should now be obvious that the qualitative nature of any Pareto optimal outcome is purely contextual and, as such, the extent to which any final outcome contributes to the movement toward SD depends not only on the effectiveness of the allocation process itself, but on the prior selection of ends and means. For this reason, it is necessary to know under what circumstances or in what efficiency "context" a Pareto optimal outcome contributes positively in the movement toward SD. Clearly, only the Pareto optimal outcomes that comply with the SD-based parameters, guidelines, and minimum positions listed in Chapter 5 contribute positively to the SD process. For the purposes of this book, these such Pareto optimal outcomes will be referred to as examples of *ecological economic efficiency* — a measure of the effectiveness with which human-made capital, produced and maintained from a sustainable entropic flow of matter–energy (the ultimate means), serves an appropriately conceived hierarchy of intermediate ends (the ultimate end).

There are three things that need to be said about an ecological economic efficiency context. First, it is an efficiency context that recognises the need to deal with the standard economic problem in light of its relationship with the larger ecological economic problem (i.e., it recognises the standard economic problem as just one of many manageable subproblems). Second, because an ecological economic efficiency context necessitates the strict adherence to SD-based parameters, guidelines, and minimum positions, any ensuing Pareto optimal outcome automatically incorporates the condition of ecological sustainability; intra- and intergenerational equity and justice; a deep respect for the continuation of the evolutionary process; and a limited recognition of the intrinsic value of sentient nonhuman beings. Finally, it is an economic efficiency context in which a Pareto improvement results in an increase in the EEE ratio and the movement toward an optimal macroeconomic scale.

For simplicity and convenience, instances of ecological economic efficiency will now be considered Pareto efficient outcomes existing within a "SD domain." The SD domain serves as a useful descriptive interpretation of a choice set bounded by a society's objective value-based standards and principles. Only within this domain is it possible to have a legitimate SD-facilitating trade-off. While trade-offs lying outside the SD domain can increase economic efficiency, they do so at the expense of violating or eroding a society's objective value-based standards and principles.

Because objective value-based standards and principles are, by nature, unique to each society, the parameters of the SD domain (assuming a society has in place an institutionalised set of SD-based parameters, guidelines, and minimum positions) are also fundamentally unique. Consequently, differing SD domains associated with each society or nation ensure the existence of no universal ecological economic efficiency context nor the likelihood of two identical contexts. This, as will be shown in Chapter 15, can have significant implications for international trade policy.

Contrasting sharply from an ecological economic efficiency context is an efficiency context pertaining to the standard economic problem. This form of economic efficiency will now be referred to as an example of *standard economic efficiency* — a measure of the effectiveness with which human-made capital, produced and maintained by any entropic flow of matter–energy (irrespective of whether such a flow is ecologically sustainable) serves any subjectively determined hierarchy of intermediate ends (irrespective of its objective value content). In a standard economic efficiency context, outcomes are not bound by the same choice set as an ecological economic efficiency context. This permits a Pareto optimal outcome to lie outside the SD domain. This in no way suggests that the standard efficiency context cannot give rise to efficient outcomes that serve positively in the movement toward SD. It simply means that the movement toward SD cannot be guaranteed because the allocation process is not subject to a set of SD-based parameters, guidelines, and minimum positions. Indeed, whereas in a standard efficiency context it would be economically rational to operate at a "perceived" optimal macroeconomic scale existing beyond the maximum sustainable scale (e.g., S_* in Figure 6.2 in the previous chapter), the same would not be the case in an ecological economic efficiency context.

With this in mind, it is interesting to compare the technological progress encouraged by the standard and ecological economic efficiency contexts. In a standard economic efficiency context, where there is no requirement to comply with a set of SD-based parameters, guidelines, and minimum positions, there is no particular preference for either efficiency or throughput-increasing technological progress. Conversely, in an ecological economic efficiency context, where all outcomes must lie within a SD domain, throughput-increasing technological innovation is only encouraged in the early phase of a nation's developmental process. Eventually, it is rendered unde-sirable and obsolete.

7.2   Macro policy goals and instruments

Given that an ecological economic efficiency context is required to genuinely foster and promote the achievement of SD, the EEE identity (Equation 6.3 from the previous chapter) has one final but highly consequential applica-tion. By breaking the ecological economic problem into manageable subprob-lems, it assists in identifying four key macro policy goals and the policy instruments pertaining to their resolution. As already explained, to achieve efficient outcomes of the ecological economic efficiency kind, they must comply with a set of SD-based parameters, guidelines, and minimum posi-tions (i.e., they must lie inside the SD domain). This means, in a sense, that to resolve the ecological economic problem and achieve an optimal macro-economic scale, four overarching conditions must be met. These conditions constitute four intermediate macro policy goals in moving toward SD. They are as follows.

1. An *ecologically sustainable rate of resource throughput* where throughput refers to the physical volume of matter–energy passing through the macroeconomy initially entering as low entropy resources and even-tually exiting as high entropy wastes. A sustainable rate of throughput is one where the exaction of low entropy resources and the insertion of high entropy wastes does not exceed the ecosphere's regenerative and waste assimilative capacities (i.e., where the prevailing rate of throughput satisfies at least the first three of the four sustainability precepts established in Chapter 4);

2. An *equitable distribution of income and wealth* where distribution refers to the relative division of the incoming resource flow as embodied in human-made capital among alternative people. An equitable distri-bution is not an equal distribution, but one where the difference between rich and poor is limited to what is considered fair and just (i.e., is limited to an ethically tolerable range of inequality).

3. An *efficient allocation of scarce resources* where allocation refers to the relative division of the incoming resource flow among alternative product uses (i.e., how much and what type of resources are allocated to the production of cars, tractors, shirts, television sets, and chocolate bars). An efficient allocation is one where the incoming resource flow is allocated to the production of physical commodities in conformity with individual consumer preferences, as weighted by the ability of each individual to pay for the means to their satisfaction.

4. *Moral growth* where moral refers, in the context of this book, to virtu-ous conduct according to civilised standards of right and wrong. Moral growth is required to ensure

a. That the stock of moral capital is kept at some necessary or mini-mum level.
b. That needs and wants are appropriately ranked in accordance with how well their satisfaction serves the ultimate end goal of life.
c. That the ultimate end is appropriately conceived and suitably revised.

While the identification of these four conditions is hardly earth shatter-ing, their relative independence is because it means each requires a distinct policy instrument — a policy absolutism discovered long ago by Tinbergen (1952).* Standard economists already recognise the considerable indepen-dence of both allocative and distributional considerations by admitting to the possible co-existence of a Pareto efficient allocation of resources and an inequitable distribution of income and wealth. Few, however, seem to be aware that a considerable degree of independence also exists between alloc-ative and sustainability considerations — that a Pareto optimal allocation of resources can similarly co-exist with an ecologically unsustainable rate of throughput. Hence, the majority of standard economists believe a sustainable use of resources can be achieved by "getting the prices right," that is, by ensuring markets reflect the full marginal costs of resource use and waste generation. Without denying the allocative importance of full cost pricing, ecological economists refute this claim (Howarth and Norgaard, 1990; Daly, 1992, 1996; Bishop, 1993; Gowdy and McDaniel, 1995; Costanza et al., 1997a). More will be presented on this shortly.

Unlike the problems of throughput, distribution, and allocation, the last of the four intermediate macro policy goals, moral growth, is not an eco-nomic-related goal because it does not directly involve benefits and costs. Indeed, it is the moral capital generated by a process of moral growth that determines the objective value-based standards and principles upon which benefits and costs are assessed. For this reason, the remainder of the chapter will focus predominantly on the first three policy goals along with their respective policy instruments.

Because the first three policy goals are more or less independent, they are best served by the following policy instruments.

- *Ecological sustainability* by the currently nonexistent policy instrument of direct quantitative controls on both the physical volume of matter–energy passing through the macroeconomy, and human population numbers.

---

* Strictly speaking, the previous four conditions are not entirely independent (Prakash and Gupta, 1994). In some way, shape, or form, the satisfaction of any one condition will affect one of more of the remaining conditions. For example, the rate of resource use and its initial distribution affects the qualitative nature of any Pareto efficient outcome. In addition, contestable markets, a mech-anism designed to increase allocative efficiency by reducing monopoly power, can bring about a more equitable distribution of income and wealth. Nevertheless, there is sufficient independence to ensure the need for a distinct, if not independent, policy instrument.

- *Distributional equity* by the policy instruments of transfers, that is, the redistribution of income and wealth by way of taxes, subsidies, profit sharing arrangements, and functional property rights; contestable markets, because by minimising the potential for the abuse of market power, they keep profits at the normal economic level and, in the process, act as an important if incomplete redistribution mechanism; and the free and universal distribution of rights, entitlements, and privileges.
- *Allocative efficiency* by the policy instrument of relative prices as determined by market demand and supply forces. To be fully effective, instances of market failure must be ameliorated to ensure relative prices reflect, as best as possible, the full marginal social benefits and costs of allocating a uniquely determined and distributed resource flow.

In order to achieve the preceding goals, the policy instruments pertaining to their resolution need to be embodied in appropriate institutional mechanisms. Depending on how they are designed, institutional mechanisms can, in some cases, embody more than one policy instrument (e.g., tradeable resource exaction permits and transferable birth licences). This simply indicates that the need for separate policy instruments does not imply the need for each to be embodied in a separate institutional mechanism. Nevertheless, to be effective, institutional mechanisms must forcibly separate the policy instruments embodied within them to ensure their relative independence. Since any institutional mechanism is limited in its capacity to embody and forcibly separate a multiplicity of policy instruments, it is necessary to ascertain their operational limits to ensure some goals are not subverted in the process of attaining others. While policy-resolving institutional mechanisms are covered in greater detail in Chapter 15, the limits to the effective domain of one important institutional mechanism — the market — will be closely analysed in Chapter 9.

7.2.1   *The erroneous belief that an efficient allocation solves the throughput problem*

It would come as a surprise to many standard economists that a sustainable rate of resource throughput would constitute a macro policy goal, indeed, for some, that it would constitute a predetermined goal at all. Certainly, there is little argument about an equitable distribution being such because, as noted above, the need to distinguish between distribution and allocation is widely accepted amongst economists. The widespread recognition of an equitable distribution as a distinct policy goal stems from the acknowledgement that a Pareto optimal outcome is achievable irrespective of the initial distribution of the incoming resource flow. However, as explained earlier, a condition of Pareto optimality is likewise independent of the entropic rate of resource throughput. Whether or not the incoming resource flow is ecologically sustainable, the market is able to churn out a Pareto optimal outcome so long as it is both devoid of externalities and sufficiently competitive or, as it is sometimes referred to, sufficiently *contestable*.* Yet standard economists seem completely unaware of this fact or simply choose to ignore it. Accept and acknowledge the co-existence of Pareto optimality and an inequitable distribution they will, but strangely, not the co-existence of Pareto optimality and an unsustainable rate of resource throughput.

Why, then, have economists failed to formally recognise a sustainable throughput rate as a distinct macro policy goal? The first reason stems from the standard preanalytic vision of the macroeconomy as an isolated circular-flow of exchange value. As explained in Chapter 4, this leads the greater majority of economists to believe that economic activity is, to a large degree, independent of the throughput of matter–energy. It also gives rise to the false belief that natural and human-made capital are perfect substitutes and, therefore, that any diminution of natural capital and the low entropy it provides can be offset by an increase in the stock of human-made capital.

The second reason can be attributed to the general insistence that a sustainable entropic rate of throughput is adequately met by dealing with the allocation problem alone. Achieve an efficient allocation, most economists say, and the resolution of one problem, the effective relative division of the incoming resource flow among alternative product uses, adequately resolves another, the sustainable volume of the total resource flow. The economists' insistence is based on a strong belief that should an increasing rate of resource throughput render scarce any previously abundant and cheap (even free) natural capital services, a shadow will be cast in the form of rising low entropy resource prices and/or rising high entropy pollution charges to reflect the concomitant rise in marginal opportunity costs. As a consequence, there is an ever increasing incentive to improve the efficiency

* The theory behind contestable markets was formally introduced in the early 1980s by Baumol, et al. (1982). The now well established theory contends that perfectly contestable markets can produce efficient outcomes equivalent to those predicted by the more traditional theory of perfect competition. The difference between a perfectly competitive and a perfectly contestable market is relatively straightforward. The former depends on the existence of a large number of actual competitors. The latter depends on the threat of potential competition. Whereas the traditional view is that a Pareto optimal outcome is only possible in the presence of many buyers and sellers, in a perfectly contestable market a Pareto optimal outcome is possible with two or more of either. The prerequisite for a perfectly contestable market is the nonexistence of sunk costs (i.e., no irrecoverable up-front costs). Sunk costs of zero maximise the contestability of markets by rendering the exit out of markets entirely free. This forces incumbent firms to keep their prices at the normal economic profit level because excessively high prices induce the hit-and-run entry of new firms seeking to capture a portion of the supernormal profits. Hence, by keeping prices down, incumbent firms are forced to behave as if they are operating in a perfectly competitive market. Not unlike a perfectly competitive market, the conditions necessary for a perfectly contestable market cannot exist — that is, some sunk costs, no matter how small, are always present. However, contestability theory is more relevant than competition theory insofar as the conditions necessary for a quasiperfectly contestable market are more likely to exist, or be created, than the conditions for a quasiperfectly competitive market. A casual observation on the policy front confirms this view (e.g., recent amendments to Australia's antitrust legislation).

of low entropy resource use and to reduce the harmful effects of high entropy waste by encouraging the development of cleaner, resource-saving technologies.

Of course, there is no denying the fact that whenever the marginal opportunity costs of natural capital exploitation increase, it is important for low entropy resources and high entropy wastes to obtain an appropriate price. After all, an efficient allocation is an intermediate macro policy goal in itself which, to be resolved, requires relative prices to adjust accordingly. However, it is a fundamental mistake to assume that "getting the prices right" will, alone, solve the throughput problem. Why? To begin with, the physical volume of the nonsubstitutable stuff to be allocated — low entropy matter–energy — cannot be determined by the very instrument that allocates it. It is one thing to allocate the physical volume of the stuff available for allocation, it is entirely something else to determine the physical volume of the stuff to be allocated. Second, while relative prices assist us in making the best of a given set of initial circumstances, the very set of initial circumstances — in this case, the rate of resource throughput — could be getting more unsustainable over time. If so, achieving successive Pareto optima over time is, as Daly (1996) has described it, like optimally allocating more and more cargo on a ship. Inevitably the ship sinks even if an optimal allocation of the ever increasing cargo slows the rate at which the ship descends. Clearly, it is better to be sailing afloat with a smaller macroeconomy than to be sinking under the weight of a macroeconomic scale that cannot be sustained. However, the qualitative difference in the two scenarios has nothing to do with efficiency in the Paretian sense because, as long as the relative prices in both instances are "right," both outcomes are efficient. The difference is simply a matter of throughput. For this reason, a sustainable rate of resource throughput must always be seen as a throughput problem, never an allocation problem.

However, the persistent standard economist has one last hurrah. Why is there any reason to believe a sustainability threshold should ever be traversed, or more particularly, that humankind should ever be forced to make the best of an ever worsening situation while the allocation of scarce resources remains efficient? Why, they ask, would an increasingly unsustainable rate of resource throughput ensue when macroeconomic outcomes are simply the aggregation of numerous but predominantly rational decisions made by individuals at the microeconomic level? After all, are individuals not only willing to forgo the loss of natural capital services if the marginal benefits derived from an increasing rate of throughput exceed the marginal opportunity costs incurred? To the last question — yes, individuals generally do. But as rational as individual behaviour may be, it still does not guarantee a sustainable rate of throughput for a number of reasons. First, neither the ecosphere or future generations bid for low entropy. Only presently living human beings do, whose decisions in relation to the rate of resource throughput are invariably shaped by individual subjective desires, not the objective desirability of operating sustainably in the long run.

Second, relative prices used to guide an individual's rational behaviour essentially reflect the *relative* scarcity of the different available forms of low entropy resources. Sustainability, however, has little to do with the amount of coal available for use *vis a vis* oil, or timber *vis à vis* iron ore, but the *absolute* scarcity of the total available quantity of low entropy.* Admittedly, the absolute scarcity of low entropy is, to some degree, always incorporated into resource prices if only because a limited quantity of low entropy flows through the market at any one time. Nevertheless, for resource prices to reflect the fact that a sustainability threshold has been traversed, they must boldly indicate that the physical volume of throughput has exceeded the maximum sustainable rate. But resource prices do not do this. That is, even if resource prices are rising, they do not simply change colour to indicate that the absolute quantity of low entropy passing through the market for allocation cannot be sustained in the future. In the end, the only way to recognise when a sustainability threshold has been breached is to be technically informed of the fact, in which case there is no need to rely on relative price information. And, if you happen to be someone who is not suitably informed? Just ask someone who is (Norgaard, 1990). Ironically, the only way to guarantee that resource prices are genuine "sustainability" prices is to impose direct quantitative controls on the rate of resource throughput — the very policy instrument suggested previously that standard economists believe is unnecessary.

Third, many individual micro decisions involving future costs and benefits are based on present value maximisation and relative price information made available by futures markets. By reflecting the future costs of present low entropy resource use, futures markets are able to facilitate an intertemporally efficient allocation of resources. However, because present value maximisation and futures market prices are also shaped by individual subjective desires then, not unlike a temporally efficient allocation of resources, an intertemporally efficient outcome can also be unsustainable and inequitable (Howarth and Norgaard, 1990; Bishop, 1993).

Finally, there is still the considerable problem of individual decisions, even if rational and objectively motivated, being distorted by the pervasive nature of externalities — a problem that always remains while individuals are inaccurately informed of the full opportunity costs of natural capital depletion (either because no markets exist for many natural capital services or because it is impossible to assign them a correct price). Indeed, because relative prices do not generally reflect full social costs and benefits, what often appears rational at the individual micro level can, when aggregated, result in "macro irrationality" (Daly, 1996).

---

* This does not mean that the ability to substitute one type of resource for another is not of great value, particularly if it is a renewable resource that can substitute for a non-renewable variety. However, substitutability between different resources does not increase the overall quantity of low entropy available for humankind's instrumental purposes. This second point is important and seems to have been overlooked by some observers who have only chosen to emphasise the importance of the first (see, for example, Solow, 1997).

In all, just as there is nothing inherent in the price system to identify an equitable distribution of the incoming resource flow, so there is nothing inherent in the price system to identify a sustainable rate of resource throughput. Clearly, relative prices must be seen as nothing more than a wonderfully effective device for obtaining the optimal allocation of whatever the rate or ownership distribution of the incoming resource flow happens to be. Contrary to the standard position, relative prices should never be considered a policy instrument to determine the sustainable rate of resource throughput.

7.2.2   *Price-determined outcomes and price-influencing decisions — addressing the intermediate macro policy goals in the correct order*

Because relative prices are unable to determine a sustainable rate of throughput or an equitable distribution, throughput and distributional decisions must be guided by ecological and ethical criteria (Daly, 1991b). Unlike the allocation process, which relies entirely on decisions made at the micro or decentralised level, determining an appropriate rate of resource throughput and an ethically tolerable range of inequality requires social or collective decisions. Furthermore, these decisions must remain antecedent to the market. This suggests the order in which each of the intermediate macro policy goals are addressed is particularly important (Costanza, et al., 1997b). For instance, do we resolve the allocation problem first and then make the necessary adjustments to ensure the incoming resource flow is ecologically sustainable and distributionally just? The answer is no. Because allocation involves the relative division of the incoming resource flow among alternative product uses, it is too late to adjust the physical volume of the resource flow should it be unsustainable. Moreover, because an individual's command over the allocation of the incoming resource flow depends on the ability to pay for the means to want and need satisfaction, it is too late to adjust the distribution of the incoming resource flow among alternative people should it be inequitable. Of course, a system of taxes and transfers is (and should) be imposed to ensure that, upon allocation, the distribution of income and wealth is within an ethically desirable range of inequality. However, income and wealth redistribution following resource allocation is much less efficient and considerably more incentive diluting than distributional measures taken prior to it.

Clearly, it is necessary to address and resolve the intermediate macro policy goals in the following order. First, establish a minimum stock of moral capital to ensure the unpalatable measures required to achieve SD are widely accepted. Second, establish a sustainable rate of resource throughput by imposing quantitative throughput controls as well as limits on human reproductive rights. Next, establish a fair and just distribution of the incoming resource flow by putting in place an appropriate combination of transfers, functional property rights, and universal entitlements and privileges. Finally, with a minimum necessary stock of moral capital in place and the throughput and distribution problems resolved, let market-based mechanisms perform their rightful role of efficiently allocating the incoming resource flow. Thus, by addressing the macro policy goals in this order, markets can be utilised to maximise sustainable net benefits and facilitate an adjustment toward the optimal macroeconomic scale.

This says something very important about collective decisions. Because they must be instituted prior to the allocation process, collective decisions bring to bear a society's objective value-based standards and principles on relative prices and, ultimately, on allocative decisions as well. That is, by modifying such things as the initial rate of resource throughput and the initial distribution of the incoming resource flow, collective decisions expediently alter the prevailing market supply and demand conditions and bring about an entirely different set of relative price signals. Provided the collective decisions are of an appropriate nature (which depends very much on a society's general belief system), more desirable forms of behaviour can be induced (e.g., they can encourage efficiency rather than throughput-increasing technological innovation, a movement away from nonrenewable to renewable resource reliance, and a transitional shift from quantitative growth to qualitative improvement). For this reason, collective decisions are fundamentally *price influencing*, not price determined, and should always remain so. They not only play an essential role in the determination of the prevailing economic efficiency context but also in the qualitative nature of any ensuing Pareto optimal outcome. More than this, because, to appropriately accommodate collective decisions, one must scrupulously separate the price-influencing problems of throughput and distribution from the price-determined problem of resource allocation, collective decisions can give explicit recognition to the objective values that transcend the efficiency calculus while also respecting the efficiency-enhancing role of individualistic subjective desires by way of decentralised market processes.

chapter eight

Sustainable development and the role of relative prices

8.1 What are the main determinants of relative prices?

Because the primary concern of the previous chapter was the role of economic efficiency as it relates to SD, it is necessary to consider, more rigorously, the role of relative prices in facilitating the movement toward an optimal macroeconomic scale. To do this, there is a need to gain a much clearer understanding of the following.

- The main determinants of relative prices.
- How relative prices are likely to differ between a standard and ecological economic efficiency context.
- How relative price variations between a standard and ecological economic efficiency context are likely to induce different forms of human behaviour.

To achieve these objectives, it is important to recognise that relative prices are, as Brown (1984, page 231) has pointed out, "a species of the genus *assigned value*, which belongs to the family of *value*." Relative prices are indications of assigned value because, in a market context, they reflect the expressed relative importance or worth that an individual assigns to different objects given his or her *held values* (objective values). The role that held values play in determining the appropriateness of relative price signals is critical. Held values belong to what Brown refers to as the "conceptual realm" of value and are very much a reflection of the objective values pertaining to a society's belief system. By shaping social and collective decisions as well as individual preferences, held values are price-influencing and play a significant role in encouraging suitable forms of human behaviour.

Exactly what, then, are the factors that determine relative prices? It would be a pointless exercise to mention them all if only because there are so many in number. Nevertheless, it is possible to isolate the main factors by recognising that an object's price or exchange value is a reflection of how well, at the margin, it serves a particular end or set of ends. Hence, if the likes of macroeconomic influences such as the nominal money supply and government taxes and subsidies are ignored, the two main price-determining factors of any physical object are its concrete *use value*, that is, its service-yielding qualities, and the *scarcity* of the object, both in a relative and an absolute sense. These two major price-determining factors are implicitly reflected in the following Pareto optimality identity:

$$\frac{P_X}{P_Y} = \left(\frac{MU_X}{MU_Y}\right)_A = \left(\frac{MU_X}{MU_Y}\right)_B = \left(\frac{MP_{Kh}}{MP_{Kn}}\right)_X$$

$$= \left(\frac{MP_{Kh}}{MP_{Kn}}\right)_Y = \frac{MC_X}{MC_Y} \tag{8.1}$$

where:

- P_X/P_Y equals the price ratio of commodities X and Y.
- $(MU_X/MU_Y)_A$ equals the ratio of the marginal utility (service) enjoyed by individual A from the use or consumption of commodities X and Y.
- $(MU_X/MU_Y)_B$ equals the ratio of the marginal utility (service) enjoyed by individual B from the use or consumption of commodities X and Y;
- $(MP_{Kh}/MP_{Kn})_X$ equals the ratio of the marginal products of the producer goods (Kh) and natural capital (Kn) used in the production of commodity X.
- $(MP_{Kh}/MP_{Kn})_Y$ equals the ratio of the marginal products of the producer goods (Kh) and natural capital (Kn) used in the production of commodity Y.
- (MC_X/MC_Y) equals the ratio of the marginal cost of producing commodities X and Y.

The preceding identity indicates that the price of an individual object is a reflection of both the marginal utility it yields and the marginal cost of producing it. The first aspect is, of course, very much dependent upon the object's use value. The second is very much a function of its scarcity of which the Entropy Law is its fundamental "taproot" (Georgescue-Roegen, 1971).

To simplify matters, use value will hereon include, not only the ability of an object to serve human needs and wants through its direct or indirect use, but also its *existence* value (intrinsic value) and *option* value (the value of the object as a potential future benefit as opposed to the actual benefit derived from its present use).

While an increase in the use value or scarcity of any type of object inflates its relative price — because, at the margin, it becomes more useful — one must not confuse the market price of an object with its service-yielding qualities. Price is strictly an abstract marginalistic concept. Too often, observers are of the false belief that relative prices reflect total values and equate the price of an object directly with the amount of service it yields. The diamonds-water paradox is a clear example of how price has little to do with an object's service-yielding qualities. Of course, an increase in wealth always seems apparent to the owners of increasingly scarcer forms of capital (whether they be human-made or natural capital) if only because their inflated price leads to an increase in the possessor's private riches. Nonetheless, it often comes at the public expense — a private riches–public wealth paradox first recognised long ago by Lauderdale (1819). This having been said, the marginalistic basis of relative prices does not constitute grounds for ruling out their usefulness as resource allocation signals. To the contrary, it is only through the consideration of marginal values, such as those reflected by relative prices, that total values can be maximised (Daly, 1968).

8.2 Relative prices — comparing the standard and ecological economic efficiency contexts

We will now consider the comparative role of relative prices in both the standard and ecological economic efficiency contexts. While this investigation will not indicate exactly what relative prices will be like in either context, it is nonetheless possible to gain an appreciation of their respective differences, as well as the different modes of behaviour they are likely to induce. To conduct this investigation, the relative prices of low entropy resources and human-made capital are examined.

8.2.1 Low entropy resource prices

The price of a low entropy resource is imputed, at the margin, according to how well it serves in the production and maintenance of human-made capital as determined by its own relative scarcity and the absolute scarcity of the total incoming flow of low entropy. In a sense, the price of a unit of any given low entropy resource reflects the marginal opportunity cost of having to exploit natural capital in order to maintain intact the stock of human-made capital. Because low entropy resources come in various forms (oil, coal, iron ore, timber, etc.), as well as various grades, each unit of low entropy serves differently in its transformation into human-made capital. These various forms and grades of low entropy also exist in varying degrees of abundance. Furthermore, because there is only so much low entropy available at any point in time, low entropy resources are absolutely scarce. All of these factors contribute to the final price of each unit of low entropy insofar as they directly or indirectly affect the respective marginal opportunity costs of their initial exaction and their subsequent transformation into human-made capital.*

As low entropy is used in the production of human-made capital, there are two classes of opportunity costs to consider in the determination of low entropy resource prices. The first is the opportunity cost of producing physical commodities in terms of alternative commodities forgone. As a unit of low entropy X is used to produce commodity A then, because the same resource cannot be reused, one forgoes the production of commodities B, C, D, etc. The second class of opportunity costs arises because, in exploiting natural capital to produce commodity A, the capacity of the ecosphere to perform its future source, sink, and life-support functions is diminished. This, in turn, reduces the ability to produce commodities A, B, C, D, etc., in the future. In the end, the relative prices of the various forms and grades of each unit of low entropy used should, or at least ought to, reflect the larger of the two classes of opportunity costs — the "true" opportunity cost. I say ought because although the first class tends to be thoroughly valued by the market, the second class does not. Hence, unless there is a genuine attempt to "internalise" the opportunity cost of lost natural capital services, by imputing and internalising an appropriate shadow price, the second class, should it be the highest, will not be fully reflected in the price of a low entropy resource. If so, a divergence between the true opportunity cost of an increment of low entropy used in the production of human-made capital and the opportunity cost reflected in its price constitutes a negative externality. Of course, even if the externality is internalised, there still remains the previously mentioned problem of relative prices, as an allocative instrument, being unable to indicate whether the total resource flow is ecologically sustainable. Thus, while the internalisation of the externality can facilitate an allocatively efficient outcome, it cannot guarantee an ecologically sustainable one.

8.2.1.1   Comparing the standard and ecological economic efficiency contexts

Given the above, let us now compare the likely difference in low entropy resource prices in a standard and ecological economic efficiency context. While, in both contexts, the presence of relative and absolute scarcities will similarly condition the relative prices of the various available forms and grades of low entropy, there is the potential for the absolute scarcity factor in the two contexts to differ considerably. This is because the physical volume of the incoming resource flow is restricted to the maximum sustainable rate in an ecological economic efficiency context (or the maximum permissible

---

* By the serviceable qualities of a low entropy resource, we mean two things. The first is the ease with which it can be transformed into human-made capital. The second is the physical qualities it gives human-made capital upon its transformation (i.e., its strength, durability, pliability or rigidity, and eventually, its reusability or recyclability).

rate if a sustainability buffer is imposed). There is, however, no such explicit restriction in a standard efficiency context. Any price-conditioning absolute scarcity factor is simply an arbitrary one, whether it be below, at, or above the maximum sustainable rate. Consequently, low entropy resource prices will only be the same in both contexts if the volume of the incoming resource flow is identical, which, of course, means the flow must be no greater than the maximum sustainable rate.

Consider an ecological economic efficiency context where the incoming resource flow is restricted to the maximum sustainable rate. What does a self-imposed sustainability constraint imply in terms of the two classes of opportunity costs? It implies that, for any rate of throughput beyond the maximum sustainable rate, the opportunity cost of lost natural capital services is, by necessity, the larger of the two classes. Indeed, beyond the maximum sustainable rate, the opportunity cost of lost natural capital services is effectively deemed to be infinite. This is why, in Chapter 6, the uncancelled cost curve was vertical at the maximum sustainable scale and why any optimal macroeconomic scale beyond the sustainable limit can only be classed as a perceived optimum.

Assume, for a moment, that the incoming resource flow would have exceeded the maximum sustainable rate if not for the imposition of a sustainability constraint. What does the sustainability constraint imply in terms of low entropy resource prices? Clearly, any restriction on the incoming resource flow means a lessened rate of throughput which, in turn, means the incoming resource flow is made scarcer relative to the stock of human-made capital that needs to be maintained. Consequently, a sustainability constraint will exert upward pressure on the relative prices of the different forms and grades of low entropy. However, any increase in resource prices would not, as is normally assumed, reflect a decline in economic efficiency. It would simply reflect the fact that a prior collective decision has been made to ensure the incoming resource flow to be subsequently allocated is ecologically sustainable.

Consider, on this occasion, the difference between an unsustainable rate of throughput in a standard economic efficiency context and a lesser yet sustainable rate in an ecological economic efficiency context. What does a lack of any self-imposed sustainability constraint mean in terms of the two classes of opportunity costs? Comparatively speaking, the opportunity cost in terms of physical commodities foregone is less in the former instance (i.e., even if a unit of resource X is used to produce commodity A, there is more low entropy immediately available to produce additional units of commodities A, B, C, etc.), but the opportunity cost in terms of lost natural capital services is greater (i.e., making more low entropy available at any point in time requires a more intensive exploitation of natural capital and a diminished capacity to produce future commodities). While this increases the possibility of the true opportunity cost being the latter of the two classes, should it be so, it is unlikely to be fully reflected in low entropy resource prices. This is because the opportunity cost in terms of lost natural capital services tends to be greatly undervalued for three reasons already outlined but worth repeating.

First, it invariably escapes market valuation. Second, the opportunity cost of lost natural capital services does not usually manifest itself until some time in the future when, of course, decisions regarding the current rate of resource use have already been made. Third, should the previously nonvalued opportunity costs be internalised by way of an imputed shadow price, they tend to be heavily discounted by the present value maximisation considerations of the current generation making current low entropy resource use decisions (i.e., future people cannot bid for low entropy and thus play no active part in current low entropy resource use decisions). As a result, low entropy resource prices in a standard economic efficiency context will generally be lower than resource prices in an ecological economic efficiency context. Moreover, the divergence between the two will normally increase as the macroeconomy gets progressively larger and the rate of resource throughput needed to maintain the macroeconomy at its larger physical scale exceeds that which can be ecologically sustained in the long run.

Of course, this does not mean that an ecological economic efficiency context is immune from externalities or that the true opportunity cost of resource use will always be accurately reflected in low entropy resource prices. The undervaluing of lost natural capital services can occur just as easily as it can in the standard efficiency context. But there is one crucial distinction between the two. In an ecological economic efficiency context, the difference between low entropy resource prices fully or partially reflecting the true opportunity cost of resource use means the difference between achieving an efficient or inefficient outcome. It can never mean the difference between a sustainable or unsustainable outcome, as is possible in a standard economic efficiency context, because the predetermined volume of the incoming resource flow never exceeds the maximum sustainable rate.

Given that low entropy resource prices are likely to be higher in an ecological economic efficiency context, what forms of behaviour are they likely to induce? If we assume that markets are sufficiently competitive to ensure firms in any given industry minimise the cost of production, higher low entropy resource prices are likely to induce more rapid increases in the service and maintenance efficiencies of human-made capital (Ratios 1 and 2). Improved maintenance efficiency (Ratio 2) is encouraged because a reduction in the low entropy used in the production of new commodities offsets the now higher cost of maintaining intact the stock of human-made capital. In addition, higher resource prices, by raising the cost of replacing human-made capital, provide an additional incentive to increase the durability of all newly produced commodities. Greater service efficiency (Ratio 1) is encouraged because the higher cost of producing new commodities translates to a higher price paid by the customer for the service it yields. Because buyers will demand at least the same value for their money spent, firms in a competitive environment will be forced to increase the service-yielding qualities of all newly produced commodities or run the risk of losing out to their competitors.

In a standard efficiency context, where the quantity of low entropy made available can exceed the maximum sustainable rate, and where, in the short-run at least, there is less upward pressure exerted on low entropy resource prices, there is less incentive to increase Ratios 1 and 2. Moreover, because increasing the rate of throughput can keep resource prices down, there is often an incentive to develop and apply throughput-increasing technologies even when it is no longer desirable (i.e., even when it leads to the growth of the macroeconomy beyond the optimal scale). The incentive to develop and apply throughput-increasing technologies also applies in an ecological economic efficiency context but only in the initial developmental phase of a nation, and only because, as a consequence of the macroeconomy being small relative to the ecosphere, the opportunity cost of lost natural capital services is the lower of the two classes of opportunity costs. Any concern about the continued application of throughput-increasing technological progress is allayed in an ecological economic efficiency context because the incoming resource flow is always kept to an ecologically sustainable rate.

It is important to recognise that increases in low entropy resource prices do not, by themselves, induce increases in the growth and exploitative efficiencies of natural capital (Ratios 3 and 4). This is because an increase in low entropy resource prices only serves to increase the cost of maintaining intact the stock of human-made capital. It does not increase the cost to resource owners and suppliers of making available a given incoming resource flow, which is necessary to induce increases in Ratios 3 and 4. This is largely why "getting the resource prices right" through the internalisation of spillover costs does not guarantee a sustainable rate of resource throughput. Internal-isation, which increases the market prices of low entropy resources, only encourages a lessened rate of throughput per production activity. While this is certain to reduce the demand for low entropy per unit of human-made capital produced, the total quantity of low entropy used at any point in time can still rise if there is a sufficiently large enough increase in the total demand for low entropy by producers to cater to an increase in demand for physical commodities by consumers. For example, a 5 percent fall in the low entropy allocated to produce each commodity does not result in a decline in the total quantity of low entropy used if there is at least a 5 percent increase in the number of commodities produced. In fact, a rise of more than 5 percent in the number of physical commodities produced leads to an increase in the total incoming resource flow. Naturally, this possibility never arises in an ecological economic efficiency context where the incoming resource flow is restricted to the maximum sustainable rate.

Interestingly, it is the explicit restriction of the incoming resource flow in an ecological economic efficiency context that induces more rapid increases in the growth and exploitative efficiencies of natural capital (Ratios 3 and 4). It does so by providing a profitable incentive for resource suppliers to find ways to increase the rate of resource throughput for the same loss of natural capital services. Remember, the maximum sustainable or permissible rate of resource throughput is set in reference to the first three sustainability precepts outlined in Chapter 4 which are based on ensuring the loss of natural capital services is kept to a sustainable level. Adhering to these precepts still permits the rate of resource throughput to be increased, however, once the maximum sustainable rate has been reached, it can only be done so by increasing the productivity of natural capital (increases Ratio 3) and/or by reducing the loss of natural capital services associated with its exploitation (increases Ratio 4). Because there is no explicit throughput restriction in a standard efficiency context, one also has the option of exploiting natural capital more intensely or applying a newly developed throughput-increasing invention. These two alternative options are also available in an ecological economic efficiency context but only in the early stages of a nation's developmental process when the rate of resource throughput is less than the maximum sustainable rate.

8.2.1.2   *Renewable versus nonrenewable resource reliance*

Apart from the differences in general low entropy resource prices between a standard and ecological economic efficiency context, differences can also be expected between the two contexts in relation to the prices of renewable and nonrenewable resources. These price variations are important because they ultimately determine a society's perceived reliance on the two categories of low entropy and the rate at which it is likely to make the transition away from nonrenewable to renewable natural capital.

Because outcomes in an ecological economic efficiency context subsume the sustainability condition, it follows that whenever the total low entropy harvested from all natural capital exceeds the regenerative capacity of renewable natural capital (i.e., when $h = h_R + h_{NR} > s_R$), nonrenewable natural capital must be exploited at a rate equal to

a. The cultivation of suitable renewable resource substitutes.
b. To the increase in the renewable natural capital stock/low entropy surplus ratio.

Because this requires some of the proceeds from nonrenewable resource depletion to be invested in renewable natural capital, adhering to this condition increases the opportunity cost of nonrenewable resource use and, therefore, the price of nonrenewable low entropy. The same does not necessarily occur in a standard economic efficiency context because, with no sustainability constraint imposed, there is no requirement for conditions (a) and (b) to be fulfilled, Thus there is likely to be a greater disparity in the price of nonrenewable *vis à vis* renewable low entropy in an ecological economic efficiency context compared with standard efficiency context.

Clearly, as the opportunity costs and relative prices of nonrenewable low entropy increase *vis à vis* renewable low entropy in an ecological economic efficiency context, one is likely to see a reduced rate of nonrenewable resource depletion matched at all times by the cultivation and/or increased productivity of renewable natural capital. Hence, there is likely to be a much earlier and more orderly transition from the current reliance on nonrenewable natural capital to that of renewable natural capital. A smooth transition from nonrenewables to renewable resources is unlikely to occur in a standard economic efficiency context for two related reasons. First, with nonrenewable resource prices kept artificially low in the short run, there will be an excessive demand for nonrenewable low entropy and a more rapid depletion of the nonrenewable resource stock. Second, because of a lack of investment in renewable low entropy (owing to the entire proceeds of nonrenewable resource depletion being used to finance present consumption), the stock of renewable natural capital will be insufficiently augmented to sustain the required incoming resource flow.

8.2.2   The relative prices of human-made capital

As indicated in Chapter 2, the physical commodities that make up the stock of human-made capital (producer and consumer goods) constitute ends with respect to the entropic flow of matter–energy but means with respect to the hierarchy of intermediate ends — the ultimate end. For this reason, the price of a physical commodity reflects how well, at the margin, it serves human needs and wants as determined by its relative and absolute scarcity.

There are two aspects to consider in relation to the relative prices of human-made capital. The first concerns *demand-side* factors which are largely a function of the service-yielding qualities of human-made capital. Whenever a physical commodity better serves the hierarchy of intermediate ends and thus becomes more useful at the margin, the demand for it increases. *Ceteris paribus*, an increase in its price should follow. But the increase in price does not constitute an increase in opportunity costs. Despite the fact that a higher price must be paid for service-yielding commodities, it is not for an individual commodity that one chooses to pay a higher price — because a commodity itself does not constitute a benefit — but for the higher psychic income it yields. Therefore, given that opportunity costs are correctly viewed in terms of a sacrificed psychic benefit and not a sacrificed physical magnitude, more is paid for the commodity itself but less is paid to enjoy each unit of psychic income it yields. This, of course, reflects a decline and not an increase in the opportunity cost of each unit of psychic income enjoyed. What is more, it represents an improvement in economic efficiency.

The second area concerns *supply-side* factors, which are a function of the cost of producing and maintaining the stock of human-made capital. There are two factors contributing to the cost of producing physical commodities. The first is the cost of the low entropy resources used in their production. The second is the cost of the producer goods and labour power used in the physical transformation of low entropy resources to service-yielding commodities — i.e., the wage payments to labour and the rent payments to the owners of producer goods. A rise in both cost factors increases the marginal opportunity cost of commodity production which, by inducing a leftward shift in the supply curve for a given type of physical commodity, leads to an increase in its price.

8.2.2.1   *Comparing the standard and ecological economic efficiency contexts.*

Because low entropy is required for human-made capital production and maintenance, the differences previously outlined in relation to the relative prices of low entropy resources flow on to the relative prices of human-made capital. Beyond these variations there are still a number of other differences to consider.

First, consider demand-side factors. Because, in a standard economic efficiency context, there is no attempt to rank human needs and wants in accordance with an appropriately conceived and ultimate end, the effectiveness with which each commodity supposedly serves in the human development process reflects the subjective desires and preferences of individuals. Should a lack of psychological need satisfaction lead to an overemphasis on the accumulation of human-made capital at the expense of higher-order needs or should consumers be cajoled through advertising into desiring goods that do little to increase their well-being, it is possible for illth to be excessively demanded. Illth can, therefore, obtain a price exceeding the true marginal service that it yields. In a similar vein, items of wealth, despite serving an appropriately conceived ultimate end, can be potentially undervalued simply because they might otherwise be ranked lower in relation to a subjective, but ill-conceived, hierarchy of intermediate ends. Because higher prices for illth increase the opportunity for the producers of illth to earn economic profits, it is possible for low entropy resources to be undesirably allocated toward the production of illth at the expense of wealth.

In an ecological economic efficiency context, intermediate ends are ranked hierarchically in accordance with the ultimate end. Provided the ultimate end is conceived appropriately, items of wealth, because of their high use value, obtain higher relative prices than in a standard economic efficiency context. Items of illth, having little or no concrete use value, obtain a lower price *vis à vis* a standard efficiency context. For this reason, one can expect limited resources to be allocated toward the production and maintenance of wealth, as opposed to illth, because the higher relative prices they command increase the profitable opportunities available to wealth creators — even more so if producers are able to augment the service-yielding qualities of human-made capital. Few, if any, profitable opportunities lie in illth creation. Hence, the ultimate end, by setting objective value-based guidelines as to how human-made capital does or does not serve the human development process, can have an important price-influencing effect. In doing so, it can, if appropriately revised and updated, facilitate increases in the service efficiency of human-made capital (Ratio 1).

From a supply-side perspective, it has already been made clear that low entropy resource prices are likely to be lower in a standard economic efficiency context because the incoming resource flow is not restricted to the maximum sustainable rate. Therefore, in terms of the expenditure of low entropy, the marginal cost of producing and maintaining human-made capital will be, *ceteris paribus*, lower in a standard efficiency context. So, too, will be their relative prices. This gives the impression that the production and maintenance of human-made capital will be more efficient in a standard economic efficiency context. This would be a mistake, for the supposed improvement in economic efficiency occurs only at the expense of a sustainable incoming resource flow, meaning that any ensuing outcome will automatically be incommensurate with the SD objective.

Any false impression of improved economic efficiency procured in a standard efficiency context need not be confined to undesirable sustainability-efficiency trade-offs. A reduction in the cost of producing human-made capital could also be achieved by violating or eroding occupational health and safety standards, by permitting incomes to fall below an acceptable lower limit or by cutting taxes as a trade-off for reducing the size of the public infrastructure and the welfare safety net. Indeed, failing to institutionalise appropriate objective value-based standards and principles in the first place could also reduce costs. This would not be permissible in an ecological economic efficiency context or at least not regarded as efficiency-increasing because any cost-reducing impact of these measures would be gained only by having eroded or violated the objective value-based framework upon which the efficiency calculus must be grounded.

8.2.3 *Relative price variations — often reflecting disparate efficiency contexts*

The previous examination of relative prices highlights an indispensable point. Differences in relative prices do not always emerge because of, nor necessarily reflect, differences in the efficiency with which resources are allocated. Put simply, lower prices do not automatically indicate a qualitative improvement in the allocation process. Frequently, relative price variations reflect nothing but a difference in the set of circumstances that exist prior to the allocation process taking place, that is, because of a difference in the market's *price-determining parameters*.* Clearly, one cannot rely on relative prices as appropriate resource allocation signals without first considering the qualitative nature of the initial set of circumstances from which the price signals have emerged (i.e., the initial rate of resource throughput; the initial

---

* Price-determining parameters are the binding market constraints that, in the absence of externalities, manifest themselves as either explicit or implicit prices. Binding market constraints constitute price-determining parameters because everything obtaining an explicit market price has a well-defined connection with price-determining factors such as natural capital services, human technology, cultural norms, and beliefs, as well as individual tastes and preferences (d'Arge, 1994). For example, as environmental factors affect the supply of leather, they also affect the price of leather-based products. In this way, environmental factors affecting the supply of leather obtain an implicit price reflected through the explicit price of leather goods.

distribution of income and wealth; and the initial ranking of needs and wants by individual market participants). For example, it is perfectly conceivable for relative prices to reflect nothing more than the effectiveness with which limited means, free from any recognition of a sustainability threshold, serve competing ends free from any recognition of an ultimate end. While, in these circumstances (a standard efficiency context), a fall in the relative prices of physical commodities would indicate an apparent improvement in the efficiency of resource allocation, it would not indicate anything meaningful about how well it contributes to the sustainable qualitative improvement in the human condition.

In a similar vein, comparatively higher prices in an ecological economic efficiency context do not necessarily reflect an inferior allocation of limited resources. Higher prices can simply reflect that a society, by making a number of price-influencing decisions in relation to such things as the sustainable use of natural resources and the distribution of income and wealth, has chosen to incorporate objective values into the price-determining parameters of its markets. As previously explained, this does not constitute an efficiency concession. Rather, it recognises that objective values constitute the bedrock upon which an efficiency calculus is rendered meaningless without.

Given that the market is the primary price generating and resource allocation mechanism, the following questions need to be answered. What are the implications for the market from having to forcibly separate the price-influencing problems of throughput and distribution from the price-determined problem of resource allocation? In other words, what is the effective role and appropriate domain of the market at and to what extent should decisions be price-influencing and outcomes price-determined? We will deal with these questions in the next chapter. Suffice it to say now, a number of things require careful consideration. They include

- What are some of the characteristics and virtues of the market place that make it such an important resource allocation mechanism?
- What are some of the limitations of the market?
- How do these limitations relate to the fact that relative prices cannot adequately deal with all SD-related problems (e.g., ecological sustainability and distributional equity)?
- What is required to deal with these limitations — that is, what is required to maximise the virtues of the market?
- Does external intervention reduce the free and effective operation of the market place or can it, under certain circumstances, increase it?
- To what extent is the institutional setting of the market important in terms of its effective operation?

chapter nine

Sustainable development and the role of the market

9.1 What is a market and what is required for it to operate freely and effectively?

As was mentioned at the conclusion of the preceding chapter, the need to forcibly separate the price-influencing problems of throughput and distribution from the price-determined problem of resource allocation raises a number of important questions concerning the role of the market in moving toward SD. In order to gain a more transparent picture of what constitutes an appropriate role and domain of the market, it is first worth considering the following. What is a market? If one were to rely on the conventional view of a market, they would be led to believe, as Hodgson (1988, page 178) has argued, that markets are nothing but a mere "aggregation of bilateral exchanges between individuals." For instance, consider the following definitions of a market.

- A market is "a collection of buyers and sellers that interact, resulting in the possibility of exchange" (Pindyck and Rubinfield, 1995, page 10).
- A market is "a set of arrangements by which buyers and sellers of a good or service are in contact to trade that good or service" (Fischer et al., 1988, page 11).
- A market is "an arrangement whereby buyers and sellers interact to determine the prices and quantities of a commodity" (Samuelson, et al., 1992, page 709).

Surely a market cannot be as simple as that described above? For although simple definitions of a market sufficiently describe something akin to an Oriental bazaar (Lowi, 1985), they do not suffice as descriptions of a highly complex market as envisaged by the modern economist. Indeed, even the preceding definitions allude to the necessity of at least some kind of "arrangement" between buyer and seller. The shortcomings of the preceding definitions arise because neither makes an explicit reference as to how and in what way arrangements between buyers and sellers facilitate mutually advantageous exchange, nor how such arrangements facilitate the effective operation of markets. Such explicit references are of utmost importance when describing or defining a market because it is the very qualitative nature of the arrangement between buyer and seller that determines the potential for a market to emerge as well as the extent of the benefits generated through exchange.

9.1.1   Why do markets exist?

Perhaps, in attempting to determine what a market is, it is best to begin by asking the following question. Why do markets exist? According to Hodgson (1988), in the same way Coase was able to explain the existence of organisational structures, such as firms, one can also explain the presence of markets. Coase (1960) argued that firms emerge because co-ordinated activities and the close proximity of relevant agents and information within a given organisation overcome the *transaction costs** associated with exchange. But, of course, co-ordinated activities are not costless. They come at a logistical cost that increases in line with the growing dimensions and dynamic complexity of organisations themselves. Hence, at a particular scale of operation, determined largely by the economies of scale associated with the particular activity in question, the logistical costs of co-ordinated activities eventually exceed the transaction costs they seek to avoid. Thus, ultimately, decentralised mechanisms offer a better and more efficient alternative to organisational structures. It is at this point that markets come into their own.

The ability of markets to confer benefits above that which can be had from co-ordinated activities arises for a number of reasons. First, as a place where people are able to convey information that only they are aware or knowledgable of, the market is able to collect, assimilate, and communicate masses of scattered information that would otherwise be useless (Daly and Cobb, 1989).** This permits the independent and decentralised decisions of individuals to give rise, not to chaos, but to a spontaneous order — an order that makes the market the most efficient of all known resource allocation mechanisms (Hayek, 1945).

---

* Transaction costs include the value of the resources used in the process of exchange and/or in the process of obtaining information about the external parties one is about to deal with and the characteristics and prices of the various goods that one is considering purchasing. Transaction costs also include the opportunity cost of the time required to engage in the process of exchange.

Second, there is much to be said for the manner in which decisions made within the market context are largely decentralised and participatory in nature. This stamps the market as a very democratic institutional mechanism. Third, markets, whenever sufficiently contestable, serve as a useful if not incomplete income redistribution mechanism. By keeping economic profits at the normal level, contestable markets are able to reduce the economic power vested in certain individuals, groups, and organisations. Moreover, because contestable markets improve the distribution of income and wealth, they reduce a government's reliance on the redistributive mechanisms that suffer from the "leaky bucket" phenomenon (Okun, 1975). Fourth, the market is infinitely more flexible than its bureaucratic counterparts. Because the market is able to make the most sense of a complex array of information, it is able to recognise and communicate external and exogenous shocks much faster than bureaucratic mechanisms. As a consequence, the market is often in a better position to respond to changing circumstances than are remote bureaucrats (Daly and Cobb, 1989). Finally, the market has a uniform resource allocation signal — relative prices — which greatly simplifies the resource allocation process. In contrast to markets, centralised planning relies on a vast array of different and, in many, instances incongruent signals.

9.1.2   *The virtues of the market — a function of the market's institutional setting*

While, at some point, markets are more efficient than co-ordinated activities, it is important to recognise that a market's virtues depend almost entirely upon its institutional setting. As O'Connor (1989a, page 4) reminds us:

> "A functioning market is not something that falls from the sky, nor does it float freefall in a void. It has a history and a context, material and cultural in both respects. The performance of a market system is intimately conditioned by institutional factors that lie outside the determination of the market itself."

** It is important to recognise that the information conveyed by individual market participants is limited in its comprehensiveness. Buyers, for instance, can provide information regarding their individual tastes and preferences but not what constitutes a fair and equitable distribution of income and wealth. Resource suppliers can provide information regarding resource extraction costs and relative scarcities but not the absolute scarcity of low entropy matter–energy. Finally, the sellers of physical commodities can provide information about the cost of production and their service-yielding qualities but not always what differentiates wealth from illth. It is for this reason that markets cannot ensure distributional equity and ecological sustainability and that information related to these two separate problems is best conveyed by way of collective decisions made prior to the allocation process.

The importance of institutions lie not so much in their physical nature and form, but in the formal and informal rules they embody that attribute consciousness and purposefulness to all parties engaged in a market transaction. Institutions do this in two ways. First, they serve as a cognitive framework for interpreting reality and understanding sense data upon which choices and exchanges are made (Hodgson, 1988). Second, institutions play an important role as information guidelines without which there would be no meaningful or purposeful action in a complex economic environment (McLeod and Chaffee, 1972). Hence, it is only through a culturally defined institutional framework that market-based arrangements between buyers and sellers can be of a qualitative nature sufficient to facilitate mutually advantageous exchange. What is more, it is the intimate relationship between institutional arrangements and the market itself which enables the information and nonprice rules that institutions embody to be reflected in the market by way of relative prices. Consequently, institutions have an important price-influencing role in the SD process because it is they which are largely responsible for the kinds of human behaviour that market signals encourage.

In all, markets exist and function, not in spite of institutional structures, but because of them. Not only do institutions set the boundaries or limits to the market domain, institutions constitute the social and cultural substance of all economic and market activity. Unquestionably, any definition of a market must take account of the fact that exchange between two transacting parties occurs within an institutional context. As such, markets can be defined as a set of social and cultural institutions within which a large number of commodity exchanges between buyer and seller take place and which are, to some extent, facilitated and structured by those institutions. That is, markets are culturally organised and institutionalised exchange mechanisms.

9.2   The conventional view of the market

The preceding definition of a market stands in stark contrast to the conventional view of a market. The conventional "free market" stance is based on a belief that institutions are constraints or an impediment to the free and effective operation of the market. It is hardly surprising, therefore, that free market-based policy prescriptions usually involve the removal of supposed institutional barriers. Intervention is not altogether ruled out but is considered permissible only insofar as a market system requires the establishment of legally enforceable property rights. However, the type of property rights legislation advocated by free marketeers almost always stops short of conditioning such rights on the basis of the performance of duties and obligations. Thus, by privileging the sovereignty of individuals above objective value-based standards and principles, the policies of free market protagonists have greatly eroded the institutional setting of markets, thereby diminishing the normative values they have sought to harness.

9.2.1 The conventional free market view — falsely positing a conflict between freedom and authority

The erroneous evaluation of institutions by free market protagonists is, according to Lowi (1985), the result of falsely positing a conflict between freedom and authority. It is too often believed that, because coercive authority restricts the freedom with which economic agents are able to operate, barter, and exchange, institutions impede the free and effective operation of the market. The false antinomy between freedom and authority is, in our view, owing to an incorrect perception of what constitutes freedom. This, of course, begs the following question. What is meant by freedom, particularly in a market context, and how can it be maximised?

Long ago, Hegel pointed out that freedom is not simply a matter of being free from all formally imposed constraints. Because freedom also depends on how the actions of others affects one's own ability to act, Hegel believed that freedom required a "recognition of necessity" or, as Hardin more recently put it, "mutual coercion, mutually agreed upon" (Hardin, 1968, page 1247). The need to accept some semblance of self-restraint arises because one's freedom depends on two conflicting aspects of the total freedom equation (Berlin, 1967). These are "negative" and "positive" freedom. Because they have far-reaching implications for what constitutes a free market, both will be briefly explained.

The first notion of freedom is *negative freedom* or "freedom from." To maximise "freedom from" is to allow someone the freedom to engage in any activity or pursuit of his or her liking. It is what most people mean when they use the term freedom. From the perspective of a market, maximising "freedom from" is equivalent to allowing economic agents the freedom to pursue any form of economic activity or market exchange of their own choosing. In practical economic terms, maximising "freedom from" requires an unfettered market place free of institutional structures and other formal rules, laws, and regulations. Maximising "freedom from" is what most free-marketeers have in mind when they refer to the proper and effective operation of the market place.

The second notion of freedom is *positive freedom* or "freedom to." This form of freedom is the all too neglected aspect of the freedom equation. To be free in this regard is to exist within an environment that allows one to fulfil the opportunities made possible by an absence of external obstacles and regulations. It is here where the two notions of freedom conflict. While a dearth of regulations, rules, and other restrictions increases one's freedom to engage in whatever activity one desires, so too does it increase one's freedom to infringe upon the freedom of others. When magnified, the freedom to infringe upon others often leaves everyone in a state of relative paralysis.

Overall, to increase one's total freedom it is necessary to forego some "freedom from" in order to gain some "freedom to." Only an appropriate balance between the two categories of freedom can maximise one's total level of freedom. The same can be said of markets. For a market to be truly free and effective, there is a need for an appropriate balance between coercive macro control and individual micro freedom which, as Hodgson (1988, page 253) points out, requires the "continuous juridical, political, and institutional meddling by central authorities." It is, therefore, patently clear that, come what may, markets are never free in the sense of being independent of institutions. Equally, markets can never be free of price-influencing decisions. Nor, as explained, should markets ever be so because the market is only truly free after having been firmly grounded within the limits set by a sustainable rate of resource throughput and an equitable distribution of income and wealth.

9.2.2   *The conventional free market view — falsely positing a natural economic order driven by individual self-interest*

The negative evaluation of institutions by free market protagonists has strong historical and ideological roots that have since given rise to two false assumptions underlying the classical liberalist tradition. The first is a deeply held belief in a natural economic order whereby each and every individual can find moral as well as physical absolutes by dedicating life solely to economic pursuits (Bunke, 1964). The second is a view that individuals should be at liberty to pursue their own destiny free from any concern over the social consequences of their actions. It was essentially from these two assumptions that the so-called "invisible hand" of the free market became firmly entrenched — a perception that whatever fostered the good of the individual must, in principle, foster the good of all. Whereas an unfettered market was considered part of the natural economic order, coercive controls were considered to be, as Bunke (1979, page 12) describes

> ....wicked to the extent that they impede the forces of nature which, if permitted to operate freely, would automatically carry man and his society ever higher and ever closer to the state of human sublimity.

It is, therefore, instructive to examine the preceding assumptions in light of more recent evidence and a more accurate interpretation of the work of Adam Smith, the father of modern economics, and the one to whom free marketeers draw inspiration and support in their belief of a natural economic order. First, however, consider the following. Should it be possible to show that a natural economic order does not exist, it logically follows that the market has no inherent moral direction or purpose except for what human-kind is able to create for it itself. This, according to Bunke, is exactly what Keynes and the Great Depression were able to demonstrate. Both highlighted that the market had no objective purpose of its own — that, in spite of the market being a wonderful mechanism for generating efficient outcomes, it had no room for objectivity or human teleological purpose. Indeed, Keynes was able to show the market was (Bunke, 1979, page 13):

> [...] nothing more nor less than any other unruly ma-
> chine and, rather than being a divine instrument de-
> signed to automatically implement some grand
> teleological plan, was without purpose or goal except
> as human intellect gave it direction and meaning in
> helping man to fulfil his potential grandeur.

Because the market is unable to provide answers to even the simplest of moral questions, it is clear that humankind should never look to the market for direction and purpose in life. Oddly enough, the classical liberals need not have waited for Keynes and the Great Depression to discover the nonexistence of an inherently purposeful economic order. Recourse to the writings of Adam Smith shows he, at no stage, attributed moral and physical absolutes to the market itself. Instead, Smith considered the market's invisible hand to be entirely dependent upon built-in restraints derived from "shared morals, religion, custom, and education" (Daly, 1987, page 333).

It is also somewhat curious to discover that Adam Smith never assigned a critical role to the pursuit of self-interest in any of his writings. A more accurate interpretation of Adam Smith does not lend support to the believers in, and advocates of, a narrow interpretation of self-interested behaviour (Sen, 1987). When referring to self-motivated behaviour, Smith had in mind the idea of "prudence" — what Smith considered to be the union of "reason and understanding" on one hand and "self-command" on the other (Smith, 1790, page 189). Because it is unambiguously clear that Smith's notion of prudent behaviour was based on an ethics-related view of human motivation that reflected the Stoic roots of Smith's understanding of moral sentiments, it is plainly wrong to identify these two qualities with pure self-interest. Indeed, as Smith (1790, page 140) once put it:

> ...man ought to regard himself, not as something sep-
> arated and detached, but as a citizen of the world, a
> member of the vast commonwealth of nature, [...] and
> to the interests of this great community, he ought at all
> times to be willing that his own little interest should
> be sacrificed.

It is quite apparent, therefore, that in eschewing the influence of moral considerations on human behaviour, Smith's call for prudent and ethically motivated behaviour has tended to be lost in the writings of so many championing the so-called Smithian position on self-interest and its achievements (Sen, 1987). Contrary to the classical free market view, there is little doubt that the principle of self-motivated behaviour remains utterly incomplete as a social organising device. It can only operate with any degree of effectiveness if it operates in tandem with supporting social and moral capital (Hirsch, 1976). Clearly, the market must always remain a servant to humankind where, in order to harness its normative values, its direction and general course must be appropriately steered by humankind's price-influencing institutions — its collective nonprice rules.

9.3   The propensity for the market to erode its own requirements

The failure of the market to operate freely and effectively cannot simply be attributed to a lack of appropriate institutional structures and arrangements. For instance, assume the market's functional prerequisites have been put in place in recognition of the importance of a market's institutional setting. Should market intervention no longer be necessary? The answer is a resounding no. There still remains the problem of dealing with the market's propensity to undermine the many foundations upon which its successful operation depends. These include (Daly and Cobb, 1989):

- The tendency for competition to be self-eliminating.
- The tendency for an individualistic ethos to erode the moral capital presupposed by the market.
- The tendency for an unconstrained market to deplete the natural capital that the economic process depends on for sustenance.
- The tendency for markets to "fail".

9.3.1   The tendency for competition to be self-eliminating

Markets work best when no one or no select group of buyers and sellers are able to manipulate the prices and quantities of goods bought and sold for personal gain. This requires markets to be competitive or contestable. A contestable market is one where any attempt by buyers to procure large consumer surpluses or sellers to earn excessively large profits is thwarted by the ever presence of actual or potential rivals. As mentioned a number of times previously, contestability is what limits economic profits to the normal level and keeps market prices somewhere near the marginal cost of production. It also forces firms to constantly innovate in order to increase the efficiency of production (lower production costs) and to ensure the service-yielding qualities of the goods they sell do not fall below that of their rivals.

While, at a particular point in time, it is possible to have a macroeconomic system dominated by contestable markets (as was generally the case for most Western nations at the turn of the twentieth century), competition inevitably leads to winners (the most efficient and profitable operators) and losers (those who are not). Because winning and losing tends to be a cumulative process, market winners gain strength while losers soon vanish. Unless there is a concerted attempt to arrest the cumulative process, markets eventually lose the contestable status they require to keep economic profits at the normal level and to ensure resources are efficiently allocated.

Thus, in order to guarantee the continuation of contestable markets, it is necessary for governments to intervene by prohibiting contestability-eroding mergers. It is also necessary to ensure that *sunk costs*, the principal determinant of market contestability, are kept to a minimum wherever it is both practicable and feasible.* Economists, on the whole, are largely in agreement in terms of the need to maintain competitive or contestable markets. There is, however, growing concern that some economists have overstated the perceived benefits of economies of scale and, in so doing, have neglected their trust-busting origins. The issue of market contestability, sunk costs, and mergers are dealt with in more detail in Chapter 15.

9.3.2   The tendency for the market to deplete moral capital

It has been pointed out many times that a successful market economy requires various social virtues to serve as necessary agents in the efficient production, maintenance, and equitable ownership of human-made capital. Fortunately, when markets first emerged as a prominent institutional mechanism, the moral capital presupposed by a market economy was already in place — a legacy of a pre-capitalist past when morality played a critical role in the establishment of built-in restraints on individual self-motivated behaviour. However, the individualistic ethos that has since become an integral part of the modern capitalist environment has slowly undermined its moral capital foundations. It is for this reason that markets do not accumulate moral capital, they instead have a tendency to deplete it. As a consequence, the continued success of any market economy depends on society itself to regenerate moral capital, just as it relies on the biosphere to regenerate natural capital (Daly and Cobb, 1989). How and in what way deliberate steps can be made to keep moral capital intact is discussed in Chapters 15 and 16.

9.3.3   The tendency for the market to deplete its natural capital foundations

In a similar vein to the way markets undermine their moral capital foundations, markets have a tendency to encourage the depletion of natural capital. As was made very clear in Chapter 4, the complementarity between natural and human-made capital means ecological sustainability requires natural capital to be kept intact. While the market has the wonderful ability to facilitate the efficient allocation of the incoming resource flow, it is, as pointed out in Chapter 7, hopelessly inadequate at ensuring the incoming resource flow is kept to the maximum sustainable rate (necessary to ensure natural capital intactness). Thus, without population growth control and restrictions on the rate of resource throughput, the market is unable to discourage the

---

* A sunk cost is a cost that, once incurred, cannot be recovered. When the sunk costs associated with operating in a particular market are large, it can deter market entry and reduce its contestability status.

depletion of the natural capital base upon which the economic process depends for sustenance.

9.3.4   The tendency for the market to fail

Finally, there are always instances where, for technical reasons, markets are unable to efficiently allocate scarce resources. These are instances where the market "fails" and where government intervention can assist markets to allocate resources more efficiently. Market failure arises because of

1. The "public good" characteristics of certain types of human-made and natural capital.*
2. Natural monopolies and other imperfectly competitive market structures.
3. Positive and negative externalities.

Each of these examples of market failure requires a different policy response. Exactly what response is best in either instance is hotly debated by economists who, on the whole, have expended considerable energy in this area. Some of these responses and the debate surrounding them are left to Chapter 15.

All in all, the establishment of an appropriate institutional setting to facilitate the free and effective operation of the market is, by itself, insufficient. One must forever contend with the fact that the successful operation of an individualistic-based market economy has the propensity to erode its functional prerequisites which, as a consequence, leads to the undesirable annexing of what legitimately belongs to the price-influencing domain to the price-determined domain of the market. Accordingly, any society must be on guard to constantly keep the market "in its place" and to implement

---

* Public goods are tangible physical objects characterised by the nonexcludability and nonrivalry of their consumption. Nonexcludability occurs when one cannot deprive another from enjoying the benefits of a good should they choose not to pay for them. It is nonexcludability that gives rise to the problem of "free riding." Nonrivalry exists when the enjoyment of the benefits of something by someone does not affect the ability of someone else to do likewise. What happens with or to public goods when markets are left to deal with them depends on whether they occur naturally or are human-made. Consider human-made public goods, such as lighthouses, roads, and bridges. If their provision is left entirely to the private sector, they are generally underprovided. This is because the consumers of public goods can avoid paying for the benefits they receive which means a private firm will not be adequately renumerated for their production and maintenance. Human-made public goods are usually provided by governments because governments can recoup the costs of public goods by raising taxes or issuing licences. In the case of naturally-occurring public goods, there is no problem of underprovision. The good already exists. The problem with naturally occurring public goods is that, when exposed to market forces, they tend to be overexploited. This is because the users of naturally occurring public goods do not have to pay the full cost of their impact on them. To avoid the abuse of naturally occurring public goods, government ownership is usually required although with adequate rules and penalties governing their common ownership, it is possible for government ownership to be avoided (Ostrom, 1990).

the necessary policy measures when the market fails to allocate resources efficiently.

9.4 What is an appropriate market domain?

It is now the task to determine what constitutes a market domain that best facilitates the movement toward SD. To do this, the following will be considered. How big should the market's domain or ambit be? Or equivalently: to what extent should nonprice decisions and other institutional arrangements remain price influencing while outcomes themselves remain price determined?

9.4.1 Rejecting the economic imperialist and minimalist market positions

There are two extreme positions regarding the role and domain of the market that need to be outlined. They are the economic imperialist and minimalist market positions. As explained in Chapter 2, economic imperialism involves the reduction of everything to the level of intermediates and relativities in order for everything to obtain a relative price. This amounts to a belief that the market domain should be as large as possible and that the number of price-determined outcomes relative to price-influencing decisions should be maximised. The economic imperialist position is, in effect, equivalent to the conventional free market position as explained earlier.

At the other end of the spectrum are those who harbour a deep mistrust of the market and, as such, believe that the market's role and domain should be kept to an absolute minimum. These such observers often lend support to their claim by arguing, falsely in my view, that allocative efficiency is of little importance in achieving SD. Of course, given that issues of sustainability and distribution should never be determined by market forces, there is every reason to be apprehensive about the role and domain of the market. The need to limit the domain of the market is, after all, what this and the previous two chapters have sought to emphasise. Unfortunately, what the minimalists fail to recognise is that their argument is not so much one against the market per se (despite their direct attack on the market), it is ostensibly an argument against the inappropriateness of its institutional setting. Unbeknown to them, their belief that the market should be abandoned as the principal resource allocation mechanism is tantamount to suggesting that humankind is incapable of establishing a more appropriate institutional framework within which the market can operate. Yet, if humankind is in any way capable of making the necessary steps to move toward SD, it is surely capable of making the maximum use of the market's normative values. Indeed, it is difficult to envisage the social and institutional changes required to move humankind toward SD being anything less than that which is required to give the market an appropriate institutional setting.

Clearly, when it comes to ascertaining an appropriate market domain, both the conventional free market and minimalist market positions must be rejected. Because the former overlooks important moral, distributional, and sustainability considerations, and the latter the potential problems caused by allocative inefficiencies, neither can be relied on to bring about an optimal macroeconomic scale, let alone move a nation toward one. Only an appreciation of the important role played by both the market and society's institutions, as well as the need to find a correct balance between the two will lead to a market domain best able to facilitate the movement toward SD.

9.4.2   *The evolution of a more appropriate market domain*

Assume, for a moment, that an appropriate balance between price-influencing decisions and price-determined outcomes currently exists. It follows that anything either side of an appropriate balance constitutes a market domain that is either too large or too small. Consider then, a market domain which is too large, that is, one where there are too many price-determined outcomes and an insufficient number of price-influencing decisions. Not only would the market be wrongly called upon to resolve the intermediate goals of ecological sustainability and distributional equity, there is a strong likelihood that the market, left to its own devices, would rapidly erode its functional requirements. Potentially, the goal of allocative efficiency would still be achievable but, because it could not be guaranteed that all market outcomes would be commensurate with the SD-based parameters, guidelines, and minimum positions outlined in Chapter 5 (i.e., be of the ecological economic efficiency kind), one could not guarantee that such an outcome would serve positively in the movement toward SD.

Consider now a market domain that is too small. In this instance, there are too many price-influencing decisions and an insufficient number of price-determined outcomes. The bureaucratic and institutional network required for collective decision making would overwhelm the allocation process and impede the market's ability to efficiently allocate the incoming resource flow. While there is every chance that the chosen rate of resource throughput would be ecologically sustainable and the distribution of income and wealth would be equitable, it is possible, depending on the extent to which the market is undesirably constrained, for resource allocation to be inefficient enough to greatly reduce a nation's sustainable net benefits, indeed, enough to leave many citizens in an impoverished state despite an ethically tolerable disparity between rich and poor. This highlights an important point. While it is critical that equity and sustainability considerations be kept separate from the market domain, one must never lose sight of the potential danger associated with the establishment of institutional structures and arrangements over and above what is necessary to ensure that both considerations are met.

One must also be careful not to believe that all at stake is the size of the market's domain. There is no use in having a suitable balance between nonprice rules and price-determined outcomes if price-influencing decisions are inconsistent with a society's objective value-based standards and principles. This leads to a set of inappropriate and, in many instances, misleading price signals.

Given the importance of the qualitative nature of the collective decisions that constitute many of the market's price-determining parameters, an appropriate market domain is very much a function of a society's shared sense of and dogmatic belief in objective value — its ultimate end — which is a function of a society's general belief system. Because belief systems vary between different societies and cultures, what is considered an appropriate market domain will differ from one society to another. Some will rely very heavily on markets while others will opt for a greater reliance on institutional and other non-price arrangements. Of course, no society will have the ideal market domain even if some have a more appropriate domain than others. As indicated in Chapter 3, belief systems always contain some erroneous beliefs, images, and objective values. Thus, to ensure a more appropriate market domain, a society will need to take particular note of the messages and signals that feed back through its institutional structures and arrangements, including the market itself. Provided they are interpreted correctly and acted upon properly, a society should be well placed to suitably modify the balance between price-influencing decisions and price-determined outcomes and to endow the market with a more appropriate set of price-determining parameters. In doing so, it should ensure the gradual evolution of a more appropriate market domain.

III

---

*Sustainable development and the
co-evolutionary paradigm*

chapter ten

The neoclassical economic framework as a product of the Newtonian world view

10.1 The need to assess the neoclassical paradigm's atomistic-mechanistic foundations

This book has so far established a SD benchmark and outlined the macro- and microeconomics of SD. While this would seem sufficient to start pre- scribing specific policy measures, there is still the need to ask whether conventional economic frameworks — in particular, the neoclassical eco- nomic framework — are sufficiently robust to deal with policy-related issues. In other words, do they provide the necessary means for prescribing policy measures to achieve SD? The need to consider these questions arises because a necessary preamble for effective policy setting is, as many ecological econ- omists point out, the application of a conceptual framework commensurate with reality (O'Connor, 1989a).

Not all observers share the ecological economists' concern. Friedman, for instance, has argued that the realism of a model's assumptions are largely irrelevant because, in the event that a particular model or framework is widely accepted and used, its assumptions become "sufficiently good approximations for the purpose at hand" (Friedman, 1953, page 15). There is a fair degree of truth in what Friedman says. Models are simplifications of the complex reality they seek to describe and, to some degree, replicate. Hence, the assumptions underlying any model or conceptual framework cannot possibly deal with the real world in totality and cease to be of any practical value the moment an unmanageable array of underlying assump- tions are incorporated to cover all potential circumstances. This aside, it would seem a rather naive exercise to apply any model to any problem when, in most cases, a model's underlying assumptions predetermine the range or nature of the findings resulting from its application. Clearly, before applying the neoclassical economic framework to the problem of SD, one must at least compare its basic assumptions with the most fundamental aspects of the real world (Norgaard, 1984).

What if, following a comparison between neoclassical assumptions and real world fundamentals, the neoclassical framework is found wanting? There are basically three alternatives to choose from.

- Ignore the inadequacies and apply the neoclassical framework in its present form — alternative 1.
- Make cosmetic changes to the neoclassical framework and apply it in a modified form — alternative 2.
- Apply an alternative conceptual framework that incorporates the more applicable assumptions of the neoclassical paradigm — alternative 3.

A conceivable fourth alternative, the application of a conceptual framework utterly divorced from the neoclassical paradigm, will not considered for the simple reason that an ecological economic approach always remains firmly wedded to many of its relevant principles.

It has already been shown in Part II that the assumptions of the standard neoclassical paradigm — many of which are anything but sufficient approximations of reality — can lead to erroneous and harmful conclusions. Given, therefore, the inordinate number of shortcomings already associated with the neoclassical paradigm, we believe many fundamental changes need to be made. As such, it is unwise to apply neoclassical economic models in their present form. Hence, at this early stage, alternative 1 can be rejected.

What about alternative 2? Is it adequate enough to make cosmetic changes to the neoclassical paradigm and apply it in an altered form? If all that was required before confidently applying the neoclassical framework was a reformulation of its more basic assumptions, the answer would be a resounding yes. However, there is one important aspect of the neoclassical paradigm that that has so far been deliberately concealed. It is one that now warrants serious consideration. It concerns the fact that a number of the basic assumptions of the neoclassical paradigm are methodologically wedded to the principles underlying a Newtonian, atomistic-mechanistic (A-M) world view (Hinterberger, 1994). Because this aspect of neoclassical economics has deeply troubled many ecological economists, it requires considerable attention before any clear-cut choice between alternatives 2 and 3 can be made.

In determining whether it is necessary to develop and apply an alternative conceptual framework, there are a couple of factors worth considering. First, if SD is eventually achieved, it will essentially be the end-product of an ongoing evolutionary process. Second, any SD-based policy conclusions must involve the consideration of how the world, its components, and the relations between them, endlessly evolve over time. Both of these factors, it can be argued, immediately call into question the adequacy of an A-M-based methodological paradigm as a means of describing what is clearly an evolutionary and not mechanistic world. With this in mind, the relative merits

of the neoclassical paradigm can only be considered once the basic characteristics of the Newtonian, A-M world view have been outlined. It is this to
which we now turn our attention.

10.2 Characteristic features of the Newtonian world view and the neoclassical economic paradigm

The Newtonian world view emerged and later developed after the presix-
teenth century "organic" view of nature was radically transformed by the
work of two highly influential figures, namely, Descartes and Newton. Following on from the findings of Copernicus, Galileo, and Bacon, Descartes
formulated analytical constructs around his emergent conception of nature
as a machine. Descartes believed nature operated according to mechanical
laws and that everything could be explained in terms of the juxtaposition
and mechanistic movement of its component parts. It was because of its
tractable methodological appeal that Descartes' mechanical interpretation of
reality quickly established itself as the dominant epistemological paradigm.
In effect, Descartes gave scientific thought its "general conceptual framework
— a view of nature as a perfect machine governed by exact mathematical
laws" (Capra, 1982, page 46).

Despite creating a conceptual framework that was later to become the
springboard for the Scientific Revolution, Descartes could do little more than
sketch the basic outlines of his theory. It was not until Isaac Newton com-
pleted the mathematical formulation of the mechanistic view of reality that
Descartes' dream was eventually realised. In the process of completing the
Scientific Revolution initiated by Copernicus and Bacon, Newton developed
the methodology upon which the sciences, including the life and most social
sciences, effectively have been based ever since.

Going beyond Descartes, Newton not only viewed nature as a perfect
machine, he also conceived the universe as "absolute space" within which
minute, solid, and indestructible material particles were attracted by the
immutable forces of gravity. Moreover, Newton believed the world consisted
of a multitude of atomistic components, each of which could be understood
and known independently of all others. It was for this reason that Newton's
view of the world has since been labelled an atomistic-mechanistic (A-M)
world view (Norgaard, 1992).

10.2.1 Characteristic features of the Newtonian world view

As a product of an A-M view of the universe, the Newtonian world view is
characterised by a number of trademark features. To begin with, it considers
the entire universe to be rigorously *deterministic* insofar as all events are the
product of an identifiable cause that give rise to an equally identifiable effect.
So long as the state of the universe or any part of it is known *a priori*, it is
considered possible for humankind to predict, in principle, the future state

of the universe with absolute certainty. For this reason, the Newtonian world view assumes the potential for human omniscience and omnipotence — the latter implying the potential for full human control over future states of the universe as well as an ability to make appropriate corrective responses (Faber et al., 1992).

Second, the Newtonian world view is based on quantification rather than qualification. Qualitative terms are not so much excluded, but are recognised only if they can be measured by way of quantitative indicators. Even today, the epistemological view is that the test of any concept, model, or theory lies in its ability to accurately predict the future and to provide the observer with a quantitative-based solution (Norgaard, 1985). Newtonian models continue to be admired and applied for their simplicity, their tractability, and the quantitative definitiveness of their predictions.

Third, because it is assumed that ecological, social, and economic systems consist of independent and self-autonomous parts and components, it is assumed that their characteristics do not alter. Similarly, the very relationships between such parts remain unchanged. Any changes initiated within the universe are confined to variations in the relative numbers of its different components or to modifications in the relative strengths of the relations between them (Norgaard, 1992). Hence, actions of any type, including those initiated by humankind, do not change the underlying parameters of the universe because these remain immutable across space and time. It is for this reason that Newtonian models correspondingly assume the possibility of *reversibility* and, thus, an ability for humankind to restore a previously existing equilibrium by re-establishing the relative numbers of the universe's components and the relative strengths of the different relations between them.

The overall Newtonian world view is represented diagrammatically in Figure 10.1. It shows that humankind's actions typically begin with the observation of ecological, social, and economic systems and the derivation of theories to explain and further understand the perceived universal characteristics of system components and their relationships. This achieved, human beings typically test their theories against reality in order to reinforce their view of the world. They then proceed to design new technologies and institutions on the basis of their reinforced theories; they later select the most suitable technologies and institutional structures and arrangements; and finally, they modify the world as is deemed necessary. But, in modifying the relative proportions and importance of the components that make up ecological, social, and economic systems, it is assumed that the universal nature of the parts and components remain unchanged. This is reflected by the assumed barrier (dotted line) at the top of Figure 10.1. This barrier separates the perceived reality from which human beings draw theories and design technologies and institutions from the reality they incessantly impact upon through their application (Norgaard, 1992). As such, the assumed barrier is a key element of the Newtonian epistemological stance. It not only reflects its deterministic stance, it reinforces the Newtonian presumption that knowledge is universal across both space and time.

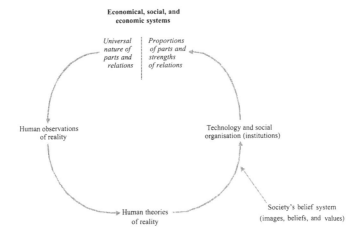

Figure 10.1 A diagrammatic representation of the Newtonian world view. (Adapted from Norgaard, 1988, page 613. With permission.)

Note also from Figure 10.1 that society's belief system — its interacting set of images, beliefs, and values — is separated out from everything inside the system. Because of the Newtonian belief in universal knowledge, the dynamics of the system and the manner in which humankind conducts itself are regarded as more or less independent of human belief systems. What is more, human belief systems are only called upon to explain certain phenomena at distinct and convenient times of need.

10.2.2 The neoclassical economic paradigm as a product of the Newtonian world view

Of all the disciplines that systematically have embraced the Newtonian world view, perhaps none have done so with as much warmth as the discipline of economics, at least in relation to its standard neoclassical form. There is a very good historical explanation for this. Almost from the beginning, economists attempted to gain respectability and notoriety by adopting the Newtonian paradigm to develop theories, concepts, and models around the popular principles of classical mechanics. They set out to achieve this in the belief that it would make economics "scientific" in much the same way as it had done for the natural sciences. Early and highly influential economists such as Cournot, Jevons, Pareto, Walras, and Marshall — the principal founders of the neoclassical economic paradigm — all formalised economic models in terms of the principles underlying the Newtonian, A-M, world view (Norgaard, 1992). A glance at any reputable economics journal highlights the degree to which economists continue to be pre-occupied by the neatness and elegance of the Newtonian paradigm.

Given that contemporary neoclassical economic models are supposedly Newtonian in nature, one is left to consider the following: In what way do neoclassical models resemble the A-M principles upon which they presumably are based? First and foremost, the neoclassical paradigm envisages macroeconomic systems as being independent of the ecological and social systems in which they are embedded. Hence, in accordance with the Newtonian world view, conventional neoclassical economists ignore the indisputable fact that macroeconomies and the markets they embody are merely one aspect of a larger ecological and social ensemble (Capra, 1982). The circular flow representation of the macroeconomy (Figure 4.1 in Chapter 4) is testimony to this Newtonian-based perception.

Second, neoclassical models are *atomistic*. All important economic components are treated like individual atoms that can be easily combined but never fundamentally altered during the transformation of natural to human-made capital. In addition, these very components are related to each other only in terms of their relative prices as determined through exchange (Norgaard, 1992). As a consequence, neoclassical models assume that all components are perfectly divisible and amenable to private ownership. This includes the source, sink, and life-support functions of natural capital which are assumed to be in no way deleteriously affected by their divisibility or supposed amenability to private ownership.

Third, neoclassical models are *mechanistic*. The macroeconomy is seen not only as a system independent of the ecological and social ensemble in which it is embedded, but the characteristics of the individual parts and components that comprise the macroeconomy, as well as the relations between them, are assumed to remain constant. Thus, neoclassical economic models assume that all systemic changes are confined to alterations in the relative numbers of the different components (prices and quantities of labour, producer goods, low entropy resources, and service-yielding commodities), and the relative strengths of the different relationships which tie the components together (market demand and supply factors as well as their respective demand and supply elasticities). Individual consumer tastes and preferences, technological possibilities, and a range of institutional, cultural, and ecological factors, all of which impinge on the macroeconomy and the dynamics of markets, are regarded as exogenous and universally given (Hodgson, 1988).

Fourth, because the only thing that changes is the relative numbers of the different components and the relative strengths of the relationships between them, the macroeconomy is assumed to be a system which

1. Has the capacity to move freely and frictionlessly back and forth between different positions
2. Is characterised by a range of stable equilibria.
3. Has the ability to operate in equilibrium at any position along a continuum.

Thus, as is normally assumed by the Newtonian paradigm, neoclassical economic models are characterised by *reversibility* (Norgaard, 1992). A

previously existing equilibrium position supposedly can be restored by altering the relative prices and quantities of human-made and natural capital as
well as the intensity of the market demand and supply forces between them.

Fifth, the neoclassical economist is considered to have an intimate knowledge of the mechanistic dynamics of the macroeconomy. Through economic
policy formulation and implementation, it is argued that the economist has
the ability to do any one of the following.

- Change the relative numbers of the macroeconomy's parts (the prices
  and quantities of human-made and natural capital) and the strengths
  of the relations between them (the respective market demand and
  supply conditions) without altering the characteristics of the parts
  and relations themselves.
- Identify and isolate existing causalities between the macroeconomy's
  components.
- Where the current state of the macroeconomy is known *a priori*, predict
  the macroeconomy's future state with quantitative precision, including all relative prices, market quantities, demand and supply forces,
  and the respective demand and supply elasticities (Faber and Proops,
  1990).

Furthermore, should the neoclassical economist suffer from ignorance
or uncertainty, both are easily reducible through study, learning, and the
application of the Newtonian-based scientific method (Faber et al., 1992).
Almost everywhere, it seems, one is able to observe past and present evidence of the assumed omniscience of economists. For instance, between
World War II and the end of the 1970s, there was an unprecedented attempt
to fine-tune economies through the deployment of Keynesian macroeconomic policies. While it is true that Keynesian policies were a radical departure from the neoclassical theories of the time, they still were, in effect,
extensions of the neoclassical paradigm. Certainly, there is little doubt that
the theories and policies proposed by Keynes and his disciples were strictly
Newtonian in nature and clearly assumed an omnipotent ability to direct
macroeconomic outcomes toward desired targets. More recently of course,
and despite the fact that Keynesian policies have lost favour with most
economists, one still regularly observes the adjustment of macroeconomic
policy instruments (e.g., interest rate modification through open market
operations, exchange rate manipulation through the buying and selling of
foreign currencies, alterations in tax rates, and variations in government
expenditure) on the premise that governments are able to desirably "manage" the behaviour of national economies.

Finally, standard economic models are exclusively concerned with quantitative outcomes. Only quantitative-based values and measures are utilised
to indicate the state of the macroeconomy and the success or failure of
economic policies. Indeed, the true test of any economic theory or concept
continues to depend on the ease with which it can be modelled formally and

mathematically. The extraordinary extent to which economists have incorporated celestial-based mathematics and econometrics into their analyses is a testimony to this quantitative obsession (Galbraith, 1970). This is not to say that qualitative factors have been overlooked or ignored altogether by neoclassical economists. Qualitative terms such as development and human welfare are key elements of the economist's system of thought. Nonetheless, they tend to be measured by quantitative indicators (e.g., GDP) that have little or no logical link with the qualitative factors they supposedly reflect.

10.3 The shortcomings of both the Newtonian world view and the neoclassical economic paradigm

To recapitulate for a moment, it was pointed out that there is need to assess the neoclassical economic paradigm before a choice could be made between alternatives 2 and 3 at the beginning of the chapter. This, it was argued, required the basic characteristics of the Newtonian world view to be outlined. That achieved, it was then deemed necessary to indicate the extent to which the principles underlying the neoclassical economic paradigm are a product of the Newtonian world view. As demonstrated previously, they are considerably so. If we now proceed with an examination of the neoclassical economic paradigm by focussing specifically on its Newtonian foundations, a number of gross inadequacies are immediately revealed.

To begin with, it is now widely accepted that twentieth century advances in thermodynamics and evolutionary-based theories have greatly exposed the limitations inherent in the Newtonian world view. There has been, in the scientific community at least, a widespread dethroning of Newtonian mechanics as the fundamental theory of all natural and human phenomena (Hinterberger, 1994).

It was pointed out previously that a Newtonian world view assumes that the nature of the parts of any system and the relations between remain unchanged and, as a consequence, within system activity does not alter its underlying parameters, nor those of other systems. However, in what is clearly a nonmechanistic world, within system activity does change, in a fundamental way, the characteristics of the components and the relations between them in a typically *irreversible* manner (Norgaard, 1988). What is more, the irreversible alteration of the underlying parameters of any particular system forever changes the characteristics of its existing components. In doing so, it also changes the way in which all newly created and emerging components relate to each other. Indeed, it is becoming clear that, in spite of the universality of many physical laws, even the rules governing interdependent relationships evolve over time (Norgaard, 1985). Hence, the assumed barrier (dotted line) in Figure 10.1, a key element of the Newtonian epistemological stance, does not exist.

The evolution of the rules governing the relationships between systems and their components brings to light another very important aspect of reality totally ignored by the Newtonian world view. This particular aspect involves the *positive feedback* of system dynamics over time. Positive feedback is a form of dynamic disequilibrium and occurs when the creation and introduction of new components and/or the modification of existing components alters the underlying parameters of a system which, in turn, alters its future dynamics. Newtonian models rule out positive feedback by assuming the existence, over time, of homeostasis or negative feedback (dynamic equilibrium). However, as much as it is possible for a system to maintain a homeostatic state in the short run, it cannot do so in the medium and long-run. It is for this reason that Newtonian models are overwhelmingly deficient in their ability to explain and describe the dynamism of systems.

The incessant presence of positive feedback has a number of other ramifications for the Newtonian paradigm. Because the change in the underlying parameters of a system is equivalent to a form of genotypic mutation, both the changes themselves and the future dynamics of the system cannot be forecasted, that is, genotypic change is strictly unpredictable (Faber and Proops, 1990). Hence, the future outcomes of a given system are, in many instances, entirely novel and unforseen. Invariably, they prove to be hostile. Thus, contrary to the Newtonian view, the novelty surrounding system dynamics ensures any observer of a given system remains, to a significant degree, ignorant. Furthermore, the supposed ability to reduce or even eliminate all forms of ignorance through research and the application of the Newtonian-based scientific method is a Faustian illusion. Quite simply, many forms of ignorance remain irreducible. This means the omnipotent "full control" over the future states of any system does not exist.

Last, but not least, the dynamics of the macroeconomy and the social and ecological fabric within which the macroeconomy is embedded are inextricably linked to the human belief systems that underlie all their dynamic manifestations (Capra, 1982). Contrary to the Newtonian perspective, human belief systems cannot be separated from everything else and ignored.

10.3.1 The inadequacies of the neoclassical economic paradigm

Having revealed the limitations and weaknesses of the Newtonian world view, the inadequacies of the neoclassical economic paradigm become recognisable all too easily. Quite clearly, the macroeconomy is a continuously evolving system. In addition, because the dynamics of macroeconomic systems are, in part, a function of the changing ecological and social systems in which they are embedded, macroeconomies cannot be treated as if they are machine-like with their atomistic components altering only in terms of their relative numbers and the relative strengths of the relations between them. Nor can they be treated as if they are independent of, and disembedded from, the ecological and social ensemble that accommodates them (e.g., in terms of an isolated circular flow of exchange value). Indeed, it would seem that the neoclassical economic approach has been "successful" only because ecosystems and social relations and institutions can be subdivided into parts that conveniently fit the neoclassical model. Yet, while natural capital, for instance, can be subdivided and privately owned, it is clear from Part II that markets are unable to deal sustainably with the instrumental services that natural capital provide. Why? Because ecological systems, unlike the low entropy resources that emanate from them, are indivisible; because ecological systems rarely, if ever, reach equilibrium positions (because, they too, are evolving systems); and because changes to ecological systems are frequently irreversible.

It was pointed out previously that the creation of new components within any system irreversibly alters its underlying parameters and its future dynamics. In what is clearly another inadequacy of neoclassical economic models, the production of new commodities and the generation of new forms of high entropy waste means that macroeconomic systems do not have the capacity to move freely and frictionlessly back and forth between different positions, nor are they characterised by a range of stable equilibria. Accordingly, a pre-existing equilibrium position as envisaged by the neoclassical economist cannot be restored by altering the relative numbers of the different components (prices and quantities of labour, producer and consumer goods, low entropy resource inputs, etc.) and the strengths of the relationships between them (the intensity of market demand and supply conditions).

In view of the preceding, the assumed omnipotent ability of the neoclassical economist to predict exact economic outcomes with precision; to control the qualitative and quantitative nature of economic outcomes by simply fine-tuning economic variables; and the ability to make simple adjustments to correct unforeseen but presumably exogenous circumstances, is an illusion. Because the future state of the macroeconomy is forever unpredictable, the economist can, at best, make tentative and imprecise predictions regarding a range of possible future macroeconomic states — never a specific future state.

Finally, neoclassical economic models fail to acknowledge the powerful relationship that exists between human belief systems and the functional operation of economic systems and the markets of which they are comprised. Given, also, that institutional, cultural, technological, and ecological factors are assumed to be exogenously determined and universally given, neoclassical models inexorably exclude an array of highly influential factors from the domain of economic activity. Yet, it is because macroeconomies are complex evolving systems embedded within a larger ecological and social ensemble that meaningful policy conclusions are unlikely to emerge while the economic process continues to be described as a process separated from history, culture, social structure, and the greater ecosphere in which it is embedded.

10.3.2 Bridging the gap between theory and reality — the need for a paradigm shift

From an ecological economic perspective, the influence of the Newtonian world view on the development of the neoclassical economic paradigm has resulted in a tremendous gap between economic theory and concrete reality. Unless economists are willing to participate in a paradigmatic shift toward an holistic and evolutionary view of the world, the existing gap between theory and reality is unlikely to be adequately bridged. Because this demands much more than cosmetic changes to the neoclassical paradigm, alternative 2 put forward at the beginning of the chapter must be rejected.

However, in embracing alternative 3, the search for a suitable alternative approach must begin with a recognition of the need for an epistemological framework that explicitly acknowledges the evolutionary properties of all systems and the interdependent links between them. In light of these requirements, some observers have called for the development and application of a *systems-based* approach (e.g., Ekins, 1994). Others have argued in favour of an *evolutionary* approach (e.g., Dosi and Metcalfe, 1991). Because both approaches contain elements that importantly transcend the limited nature of the neoclassical paradigm, there is much to recommend their use. Nevertheless, while a systems-based approach is capable of dealing with system interdependencies, it need not incorporate evolutionary properties. And, while an evolutionary approach accounts for the evolutionary properties of systems, it can overlook the importance of system interdependencies. Indeed, recognising the evolutionary nature of the economic process is only a marginal improvement on the Newtonian paradigm if, in the end, the macroeconomy is viewed as a system evolving independently of social and ecological systems. What is ultimately required to draw the necessary policy prescriptions to achieve SD is the application of a dynamic, evolutionary, and systems-based conceptual framework — one that has been conveniently referred to by Norgaard (1985, 1988, 1992, 1994) as a *co-evolutionary-based* conceptual framework. As a consequence, it is the development of a co-evolutionary paradigm that constitutes the task upon which the following chapter is based.

chapter eleven

The development of a co-evolutionary-based conceptual framework

11.1 The basic principles underlying a co-evolutionary-based conceptual framework

The conceptual framework to be revealed in this chapter is based on the principles underlying the biological concept of co-evolution. Co-evolution involves the interdependent relationship and reciprocal feedback responses of two or more closely interacting species. It is used to explain, for example, the relationship between bees and the distribution of flowering plants. The concept has been extended beyond the biological domain to describe the evolving relationships and feedback responses typically associated with any two or more interdependent systems.

Because systems of any kind can only be maintained through the existence of numerous feedback mechanisms, co-evolution takes place in the broader sense when at least one feedback loop is altered by within system activity which, in turn, initiates an ongoing and reciprocal process of change (Norgaard, 1985). By coalescing the theories variously held to be the best scientific explanations of all natural phenomena (e.g., thermodynamics, evolution, systems theory, etc.), co-evolution is fast becoming a central phenomenological theme. A co-evolutionary world view not only provides a more realistic understanding of the dynamics of the universe, it allows one to recognise and describe the many critical relationships that bind together the various systems that make up the global system (Earth).

11.1.1 What are the basic features of a co-evolutionary world view?

In contrast to the Newtonian paradigm, the co-evolutionary world view begins from the premise that the Earth is a system comprised of closely interacting and interdependent subsystems. Second, it recognises the Earth and its constituent systems as dissipative structures — the Earth as a dissipative structure open with respect to energy (a solar gradient), and its constituent subsystems as dissipative structures open with respect to energy, matter, and information.* In view of the openness of the Earth and its constituent subsystems, the co-evolutionary world view perceives the relationships between all systems as inherently dynamic. Each system is, in some sense, connected to and dependent on all others, and everything is evolving together over time. Furthermore, the rules which govern the evolution of relationships between systems are also constantly evolving.

This last point reflects a number of important aspects pertaining to the co-evolutionary world view. First, it precludes universal principles. Second, it never reduces the "whole" to equilibrium relationships amongst its constituent components. It always perceives the whole as being fundamentally different to the mere sum of its parts. Third, the co-evolutionary world view regards disequilibrium and change as the rule rather than the exception. As such, the notion of unfettered evolutionary processes being able to bring forth an increasingly stable and utopian macrostate (as is often the belief) is considered a myth. For many people accustomed to the Newtonian paradigm, this sounds at best unsettling, and at worst debilitating. But it need not so. As Norgaard (1985) has pointed out, disequilibrium and change should be seen as an ongoing process offering a plethora of opportunities for humankind to engage in *positive co-evolution*, which, for the purposes of this book, can be construed as a co-evolutionary process commensurate with the SD objective.

Fourth, from a SD perspective, the co-evolutionary world view considers the applicable unit of evolution to be the global system, or what Berry (1981) has referred to as the Great Economy — the "total web of life." SD should never be seen as a function of the state of the macroeconomy or any other system in isolation. For despite the fact that national well-being depends primarily on the immediate state of the macroeconomy, a nation's sustainable economic welfare depends as much on the macroeconomy's institutional setting and the natural and moral capital that supports it.

Finally, the co-evolutionary world view is based on a principle of system embeddedness, sometimes referred to as the *logos* of nature. Metaphorically, logos is a term used as a principal concept embracing the natural order of the universe. By acknowledging the logos of the global system, the co-evolutionary world view recognises firstly, that the world is characterised by self-organisation. Second, it recognises that systems exist at varying levels of complexity and, as such, are characteristically stratified and multilevelled (Laszlo, 1972). The logos of the global system and the embedded relationship between the three major spheres of influence — the macroeconomy, the sociosphere, and the ecosphere — are illustrated by way of Figure 11.1.

---

* In the natural world, information exists as genetic information coded in the DNA molecule. In the anthropocentric world, information exists as knowledge encoded in various institutions and organisations.

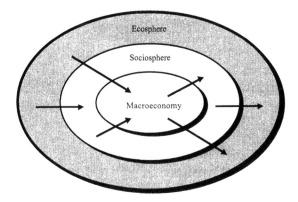

Figure 11.1 1 A co-evolutionary interpretation of the interdependent relationship between the macroeconomy, sociosphere, and ecosphere.

In Figure 11.1, the three major spheres of influence represent different systems at varying degrees of complexity. Each can be considered a *holon* insofar as they manifest the independent and autonomous properties of wholes and the dependent properties of parts.* That is, each sphere consists of smaller parts while simultaneously acting as the part of a larger whole (i.e., the macroeconomy acts as a component of the sociosphere which, in turn, acts as a component of the ecosphere). In a sense, this makes the sociosphere something akin to the interface between the macroeconomy and the greater ecosphere (Berkes and Folke, 1994).

The overall co-evolutionary world view is represented diagrammatically in Figure 11.2. Although the co-evolution of constituent parts and relations can be portrayed at any stratified level of complexity, Figure 11.2 describes the co-evolutionary world view by focusing on the following subsystems — the ecosphere (natural capital), human know-how (knowledge), social organisations (institutions), technology, and human belief systems. It shows that co-evolution is an intertwining of these chosen elements and is very much symmetrical. Neither dominates another as do neither provide a more obvious starting point for understanding the whole. Clearly, because each subsystem can only be properly understood in the context of all others, it is meaningless to analyse the nature and the dynamics of each subsystem in isolation. The same can be said of human belief systems. While they have been, in some sense, separated out in Figure 11.2, this does not imply that belief systems exist independently of the co-evolutionary process. Nor is it an attempt to portray belief systems as the dominant influence or, furthermore, the most useful starting point for understanding the whole. The relative separation of belief systems serves to emphasise the degree to which human belief systems play a key role in all human teleological and

---

* A holon is a term made popular by Arthur Koestler (Capra, 1982, page 303).

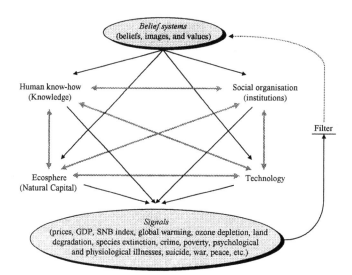

Figure 11.2 2 A diagrammatic representation of the co-evolutionary world view. (Adapted from Norgaard, 1992, p. 80. With permission.).

choice-oriented behaviour — a critical element of the co-evolutionary process (Boulding, 1956).

Figure 11.2 reflects what is now generally accepted in the natural sciences. That is, the global system has evolved over approximately three and one-half billion years with human beings having played an active part over the past three million years. By putting selective pressure on the ecosphere through its initial hunting and gathering and its more recent agricultural and industrial practices, humankind has always contributed to the extinction of some biological species and the enhancement of others. In turn, the ecosphere has placed selective pressure on human beings and its social, cultural, and economic systems. Because, in the end, all sub-systems place selective pressure on each other and influence the way they evolve, all systems come to reflect one another (Norgaard, 1988). Thus, when any within system activity affects the characteristics of its parts and the relationships between them, the survival of the newly created parts depends not only on their "fitness" with respect to the subsystem's existing components and relations, but also with respect to each of the subsystems that make up the global system.

This so-called reflection of one another highlights three important aspects of all co-evolutionary processes. First, co-evolutionary processes are characterised by the principle of co-evolutionary harmony or *co-evolutionary balance*. Second, each subsystem is fundamentally contextual. For instance, knowledge does not simply reflect a dominant world view. It also reflects the influences of cultural beliefs, images, and values; the technologies human beings use to perceive, transmit, and retain information; the forms of social organisation (institutions) peculiar to a particular society or culture; and immediate ecological and cultivated resource systems. Technology, on the other hand, changes the way human beings relate to each other and the surrounding ecosphere. Through reinforcement and rejection, the application of certain technologies underlies the importance of some beliefs and images. Others are simply banished altogether. In the case of human forms of social organisation, the co-evolutionary world view explains why individual human beings have little identity apart from the values reinforced and impressed on them by the organisational, institutional, and cultural systems of the society in which they live.

Finally, the co-evolutionary world view reveals why the beliefs, images, and values pertaining to human belief systems are embedded in everything that human beings operate from and within. As Capra (1982, page 196) points out, the standards and principles embodied in human belief systems, once expressed and codified, "constitute the framework of a society's perceptions, insights, and choices for innovation and social adaptation." Consequently, a society's beliefs, images, and values determine, to a very large degree, its choice and development of institutions, norms, routinised habits, and technologies.

Given also, that humankind has always played an active role in global co-evolutionary processes — a role which is increasing over time in line with its growing technical prowess — it is clear that humankind has, and will continue to have, a considerable influence over the eventual characteristics of all future social, economic, and ecological systems, as well as the relations between them. As such, the co-evolutionary paradigm recognises that economic, ecological, and social systems eventually reflect the assumptions of the predominant world view. Indeed, so much so, the global system ultimately rewards those who adopt and embrace the assumed values.

Of course, as much as human belief systems are influencing the co-evolutionary process, so is the co-evolutionary process, by emitting conflicting and reinforcing signals, feeding back and reshaping human belief systems (right-hand side of Figure 11.2). To what extent human belief systems are revised depends on how much human societies "filter" the endless stream of signals (Troub, 1978). The more conservatively resistant are human belief systems to change, the lower is the tendency for conflicting signals to lead to their revision. Moreover, the greater is the tendency for the global system to reflect the beliefs, images, and values pertaining to human belief systems.

11.1.2 The global system as one large evolutionary process

The relative symmetry of Figure 11.2 has one final important application. It allows one to portray the global system as one large evolutionary process comprised of, but never equal to, the sum of the co-evolutionary processes occurring within it. Imagine, for a moment, the evolution of the global system being equivalent to moving along a particular pathway toward point *A* in Figure 11.3. As it reaches this point, the underlying parameters of the global system change such that the system itself, depending on how the parameters

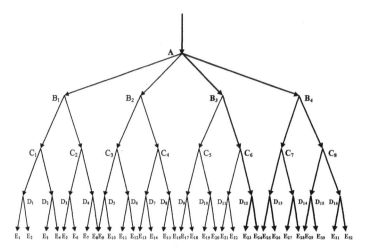

Figure 11.3 A diagrammatic representation of the evolutionary pathway of the global system. (Adapted from Mokyr, 1992, p. 9. With permision.)

evolve, moves along a new pathway to one of either points B_1, B_2, B_3, or B_4. As the global system reaches one of these four points, it then moves along a new pathway to any one of two possible points (i.e., from B_1 to either C_1 or C_2; from B_2 to either C_3 or C_4; and so on). Eventually, the global system has the potential to reach any of 32 possible future states (E_1 to E_{32}), although, in reality, the number of possible future states of the global system is potentially unlimited.

The pathways leading to points E_{23} to E_{32} are highlighted to simulate a range of evolutionary pathways commensurate with the SD objective. Moving along these pathways would be equivalent to a nation experiencing an increase in sustainable net benefits. Because only one of these pathways maximises a nation's sustainable net benefits, moving along it would be akin to achieving an optimal macroeconomic scale. Hence, while the remaining SD pathways would be ecologically sustainable and distributionally equitable, they would be allocatively less efficient than the optimal pathway. This does not mean that outcomes associated with any one of the non-SD pathways (E_1 to E_{22}) cannot be allocatively efficient. It simply means they are ecologically unsustainable or distributionally inequitable or both, and would, if they were efficient, be equivalent to making the best of a worsening set of circumstances.

What if a society happens to be moving along one of the non-SD pathways? Can it get onto a SD pathway by simply fine-tuning a couple of policy instruments? It would seem not. According to Mokyr (1992), extricating itself onto a desirable pathway requires a society to "jump" from its present pathway which, in a world where everything reflects each other, and where the system rewards those who adopt the assumed values, is much easier said than done. More on this as the chapter unfolds.

11.2 Fundamental aspects of a co-evolutionary-based conceptual framework

To understand what is required to jump onto a desirable SD pathway or, if already there, to remain on one, it is necessary to outline some of the fundamental aspects of a co-evolutionary-based conceptual framework. Doing so not only reinforces the importance of adopting a co-evolutionary approach, it highlights the impact that such an approach will have on the design of the policies and institutions required to move toward SD.

In the final section of the chapter, two things are very briefly considered. First, what impact does the co-evolutionary paradigm have for the SD concept itself? In the second instance, what does a co-evolutionary paradigm mean for both economic efficiency and the appropriate role and the domain of the market? This last aspect is important because the incessant change of a system's underlying parameters means the domain of the market is forever in a state of flux. So, therefore, are the relative prices it generates. Ensuring the market serves positively in the movement toward SD requires the appropriate evolution of its price-determining parameters, which, because of the importance of the market's institutional setting, demands the prior imposition of an appropriate set of institutions that deal adequately with inevitable and sometimes novel co-evolutionary change.

11.2.1 Macro stability and micro chaos — opposing aspects of the principle of self-organisation

Because co-evolution deals with the interdependent relationships between different systems and their constituent parts, it is necessary to understand the dynamic properties of systems. However, before such properties can be understood, one needs to know what constitutes a system. According to Weiss, a system can be operationally defined as:

> "...a rather circumscribed complex of relatively bounded phenomena which, within those bounds, retains a relatively stationary pattern of structure in space or of sequential configuration in time in spite of a high degree of variability in the details of distribution and interrelations among its constituent units of lower order" (Weiss, 1969, page 12).

In other words, a system is a discernible and stable entity that maintains its structure in spite of both the plasticity of its ongoing micro changes and the replacement of its components. Herein, of course, lies the observational paradox of systems. For as much as a system needs to exhibit well-defined regularities and behavioural patterns to in any way exist, the irregular and chaotic behaviour of its constituent parts and components will appear entirely at odds with the order and stability displayed by the system itself. Yet, both micro-level chaos and macro-level order coexist for a very logical reason. As opposite sides of the same coin, chaos and order are different aspects of the same dynamic principle, namely, the principle of *self-organisation* (Capra, 1982).

As a fundamental characteristic of the logos of nature, the principle of self-organisation is characterised, not by deterministic events, but by the continual presence of disequilibrium microstates (Clark and Juma, 1988). Consequently, what appears to be disorderly behaviour at the micro level is, at the macro level, an adherence to an orderly set of underlying patterns. Hence, chaos is, as Garfinkel describes it, "a higher form of order" (Rothschild, 1990).

The importance of the self-organisation principle lies in the fact that, with everything in a co-evolutionary world evolving together over time, the opportunity for further order and self-organisation at the systemic and global system level is, so long as the sun continues to shine, very real. In fact, the principle of self-organisation goes a long way toward explaining the importance of diversity and decentralisation in all spheres of life. It points to the importance of biodiversity in maintaining a planet fit for human habitation (something that humankind has probably already reduced through its more recent endeavours), and the importance of markets in being able to collect and communicate what would otherwise be a mass of scattered and useless information.

11.2.2 The principle of co-evolutionary balance

It has already been shown that, in a co-evolutionary world, each subsystem is not only related to all others but each is also changing and affecting evolution. As each subsystem puts selective pressure on each of the others, all subsystems ultimately reflect, reinforce, and accommodate each other. It is this reflection and reinforcement that gives rise to what can be loosely described as *co-evolutionary balance*. Co-evolutionary balance is a harmonious, if not short-lived relationship, existing between co-evolving systems. It is a relationship characterised by a high degree of reliance and dependency of one system on another. Co-evolutionary balance is an important aspect of the co-evolutionary paradigm because without it, the global system, or any other system, cannot possibly operate and reproduce itself through time (Hodgson, 1988).

With respect to any given higher-order system, co-evolutionary balance is a temporary phenomenon. Over the long term, the collapse of any form of co-evolutionary balance is inevitable because the underlying parameters of higher-order systems and, more particularly, those of its constituent subsystems and their relations, are continuously evolving. Because any form of co-evolutionary balance becomes progressively unstable over time, it is very useful to view the evolutionary dynamics of higher-order systems over

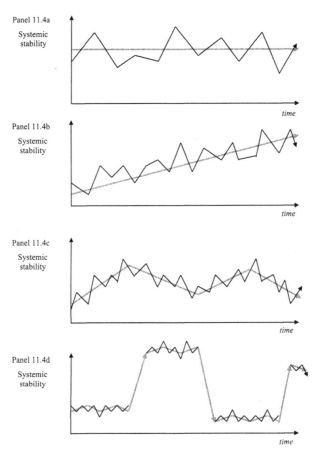

Figure 11.4   Representations of the evolutionary dynamics of higher-order systems in the very short-, short-, medium-, and long-term time durations. (Panel 11.4d has been adapted from Dury, 1981, p. 9.) With permission.)

definitive time periods. For instructive purposes, a system's evolutionary dynamics can be viewed over four different time frames — the very short-, short-, medium-, and long term time periods. Figure 11.4 represents these different views.

11.2.2.1 The very short term
Consider the state of a higher-order system in the very short term as depicted in Panel 11.4a. In the very short term, the underlying parameters of a higher-order system remain unchanged. Should the system be in any way perturbed, it is simply restored, not unlike a thermostat, to a "dynamic equilibrium" state. As such, the general features and characteristics of the system remain totally unaffected. Also remaining undisturbed during the very short

term is the underlying selection or fitness principle of the system, in effect, the necessary prerequisite for any system to maintain a state of homeostasis. Clearly, the very short term time frame elapses the moment the underlying parameters of a system change sufficiently enough to bring about an entirely new selection principle which, in doing so, precludes any possibility of a homeostatic equilibrium being maintained. Thus, by the very short term one is definitively speaking about a very limited time frame and, in some cases, a moment in time.

It is very important to note that when mention is made of the very short term one is not referring to a specific time duration for every system. What constitutes the very short term will vary significantly depending on the system in question. In the case of some systems, the very short term may be a matter of seconds, minutes, or hours, whereas for other systems it may be a matter of days, weeks, or months — perhaps even years.

11.2.2.2 The short term

As Panel 11.4b shows, a higher-order system no longer maintains a homeo-static equilibrium beyond the very short term. Instead, it begins to move along a particular evolutionary pathway (equivalent to moving from say, point B_2 to C_3 in Figure 11.3). Unlike the very short term, the underlying parameters of a higher-order system slowly evolve during the short term. Consequently, the system's general features and characteristics gradually change. Overall, however, the system's macrostate still remains relatively stable because within system variation is strictly confined to *adaptive* change (Capra, 1982).

Despite the system remaining relatively stable over a short term time frame, the system's constituent components and relations do vary considerably. A great number of constituent components and relations existing at the beginning of the time period lose their fitness status and no longer exist at the end of it. Conversely, many newly created components become established as they, in passing the ever changing fitness test, replace those considered unfit.

Notwithstanding the parametric change underlying the system, co-evolutionary balance is still largely maintained throughout a short term time duration. This is because the system's underlying selection or fitness principle changes only marginally from the beginning to the end of the period. Indeed, the short term period only ceases to exist the moment the underlying selection principle changes sufficiently enough to bring about a shift in the evolutionary pathway of the higher-order system.

11.2.2.3 The medium-term

Over the medium term, the underlying parameters of a higher-order system evolve considerably further. Within system variation transcends adaptive change to become something akin to a more complex form of *somatic* change. The general features and characteristics of a higher-order system now differ markedly to those exhibited at the beginning of the period, owing largely to the fact that, as depicted in Panel 11.4c, the direction of the system's evolutionary pathway is now shifting (equivalent to moving from point B_3 to C_6 and then on to D_{11} in Figure 11.3). Hence, the system is, overall, considerably less stable compared to the short term, if not altogether unstable.

In the medium term, the system's constituent components and relations are fundamentally altered. Only a small proportion of all initially existing components and relations remain because many will have failed a greatly modified selection or fitness principle. Because, over the medium term, the evolutionary pathway of systems change course, any semblance of co-evolutionary balance exists, but only tenuously. But exist it does because there is no discontinuity of component reflection, reinforcement, and accommodation over the period. Similarly, the selection principle also evolves without any discontinuity. It is important to recognise that a state of co-evolutionary balance can still be maintained despite considerable change in the system's component parts and relations.

11.2.2.4 The long term

In the long term, the underlying parameters of a higher-order system evolve to such an extent that they, along with the majority of its components, are irreconcilable with those in existence at the beginning of the period. Within system variation transcends somatic change to become something equivalent to *genotypic* change (mutation). Any notion of a previously existing form of co-evolutionary balance now reaches its ultimate conclusion such that the system itself experiences complete macro instability. As a consequence, the general features and characteristics of a higher-order system differ almost completely to those exhibited at the beginning of the period.

Panel 11.4d is a diagrammatical representation of a *step* function and illustrates what happens to a system in the long term. As can be seen from the figure itself, the long term is characterised by a series of medium-term time frames each punctuated by a "jump" between one form of co-evolutionary balance to another. In relation to Figure 11.3, this would amount to jumping from one evolutionary pathway to another (say jumping from B_1 to C_7, or C_5 to D_2 etc.). Each respective jumping event is known as a *bifurcation point* and occurs whenever the evolving parameters underlying a higher-order system exceeds a particular threshold level. As such, each bifurcation threshold is defined by the point at which a pre-existing form of co-evolutionary balance finally collapses. There is, at this point, a very rapid rate of evolutionary change which only moderates when a new state of co-evolutionary balance is reestablished. But, in moving to such a state, a couple of things ensue. First, because the characteristics of a higher-order system will no longer resemble those that previously existed, the system will now be comprised almost entirely of different parts, components, and relations. Only residues of the previously existing macrostate remain. Second, the newly emerging selection or fitness principle, as a function of the selective pressure that each of the new system components impose on each other, is completely modified. Most of the rules governing the relations between parts and components are now entirely reformed.

11.2.3 *The emergence of systemic instability and chaos through bifurcation*

As previously mentioned, bifurcation arises whenever the stability of a system is radically undermined by the continuing evolution of its underlying parameters. Once the parameters change sufficiently enough to cause the eventual cessation of any form of co-evolutionary balance, the system becomes sufficiently unstable to jump from one evolutionary pathway to another. As explained previously, it is at this threshold point where the bifurcation of systems takes place. But, in jumping from one pathway to another, the system merely moves to one of a number or range of potentially stable states. To understand this more clearly, consider Figure 11.5.

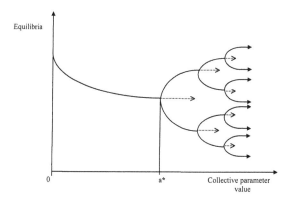

Figure 11.5 A diagrammatic representation of the bifurcation of systems. (From Faber and Proops, 1990, p. 86. With permission.)

Imagine that, over the short- and medium-term time frames, the underlying parameters of a higher-order system are evolving. As their numerical value increases (as measured on the horizontal axis), the overall structure of the system becomes progressively more unstable. Eventually, as a collective parameter value of a^* is reached (the value at which a bifurcation threshold exists), the system collapses. As a consequence, the system moves to one of a number or range of potential macrostates. Furthermore, as the value of the parameters continues to increase over time, the bifurcation process exhibits "cascading" characteristics that reflects the manner in which a system regularly jumps from one evolutionary pathway to another as it passes through successive bifurcation thresholds (Faber and Proops, 1990).

11.2.3.1 Implications

Because chaotic systems have the propensity to move to one of many potentially stable states, any ensuing evolutionary pathway of a system will depend very much on the nature of the boundary conditions and rules governing its macrostate. As Mokyr (1992) stresses, small differences in boundary conditions can lead to an entirely different course of events. This is particularly so in the neighbourhood of any bifurcation threshold where small and often random events significantly shape the evolutionary pathway of the system. Indeed, arbitrarily small disparities in the initial boundary conditions of nonlinear systems invariably lead to cumulatively growing disparities in their resulting pathways (Dosi and Metcalfe, 1991). It is for this reason that bifurcation has been associated with the *phenomenon of exponential divergence* — an exponentially increasing divergence between evolutionary pathways resulting from negligible disparities in the boundary conditions of nonlinear systems.

What makes the aforementioned so critical is not simply the fact that small and random events can greatly effect the pathway of the system. Growing evidence suggests that its eventual pathway may not necessarily be desirable, and may in fact be catastrophic. In other words, the selective process that determines the pathway a system eventually moves along, is by no means necessarily optimal (Mokyr, 1992). Given that the evolutionary pathway of the global system — the Great Economy — determines whether any society is operating commensurably with the SD objective, a number of very important questions emerge. All will be addressed in the remainder of the chapter. Suffice to say now, the answers to them have considerable policy implications. They include

- How rapidly and frequently do systems experience chaos-inducing bifurcation?
- To what extent can humankind predict the eventual pathway of the global system? — That is, to what extent can humankind predict both shifts in the direction of its present pathway (a medium-term effect) and the jumping of the system from one pathway to another (a long term bifurcating effect)?
- To what extent can humankind influence or direct the eventual pathway of the system by modifying its boundary conditions?
- To what extent can humankind modify the global system and its constituent parts and relations without prematurely promoting a chaos-inducing bifurcation of a catastrophic kind?
- Can humankind predict the impending onset of both a shift in the direction of the global system's pathway and a jump from one pathway to another? If so, to what degree and how precisely can it do so?

11.2.4 Rapidity and frequency of systemic chaos through bifurcation

In addressing the first of the previous questions, one needs to recognise that while far-from-thermodynamic-equilibrium systems eventually reach a bifurcation point of some kind, they do not all reach them with the same rapidity and frequency. This is very much dependent on a system's relative complexity. The greater is a system's complexity, the more self-assertive and self-autonomous it tends to be. This same correlation between systemic complexity and self-assertiveness has a similar bearing on how rapidly and how often a system is likely to reach a bifurcation threshold. Put simply, the more complex is any given system, the less often it passes through successive bifurcation thresholds, and the longer it maintains its present macrostate.

Take the global system for instance. As a highly complex system, it rarely passes through a chaos-inducing bifurcation threshold. Exactly how often it has done so is arguable in itself, but has probably only occurred on a handful of occasions. These would include, for instance, the moment life first began on Earth; the global emergence of oxygen bearing organisms; the extinction of the dinosaurs caused by the probable fallout from a massive meteorite explosion; the agricultural revolution, which facilitated the development of human civilisations; and, finally, the Industrial Revolution and the development of the modern market economy — labelled by Polanyi (1944) as the Great Transformation.

Exactly, what can one expect from the next global bifurcation, and when? This cannot be precisely known *a priori*, although some observers believe the emerging evidence of global warming, ozone depletion, acid rain, and biodiversity erosion, are the first signs of an impending catastrophic bifurcation of the global system. There are some who are of the opinion that the planet can deal with these "anomalies" by establishing suitable feedback mechanisms (something akin to Lovelock's Gaian hypothesis). In many ways, these observers are correct. But the establishment of Gaian feedback mechanisms does not, in itself, prevent any impending bifurcation of the global system. More importantly, while the establishment of a new "order" is likely to be desirable from the point of view of maintaining a planet fit for life, the new global macrostate may be completely inimical to the longevity of humankind.

The global system aside, what happens regards to the bifurcation of the lower-order systems of which it is comprised? Being of reduced complexity, lower-order systems experience instability from bifurcation much more rapidly and with greater frequently than the global system. Exactly how many bifurcation events there are and how often they occur, one cannot know, however, they occur within systems as large as the political systems of nation states (e.g., the demise of Communism) to others as small as the local football club or nearby mangrove swamp. However, what we do know is that all lower-order system will eventually disappear during the dynamic life of the global system, just as the global system will inevitably do as part of the greater universe.

11.2.4.1 Implications

Because one can expect the components and subsystems of the global system to undergo pathway changes and bifurcation more frequently than its parent system, the potential for impacts on human welfare and development are still quite marked even if the global system remains relatively stable. This is because the well-being of societies and its individual members is very much a function of the state of more immediate economic, social, and ecological systems. It is for this reason that human beings have a strong justifiable tendency to reduce or minimise the perturbation of those systems most critical to their immediate welfare interests. This, humankind has done in the past by establishing mechanisms to insulate itself from potentially destabilising effects. As will soon be explained, there is a limit to how much humankind can and should seek protection without concentrating, instead, on reducing undesirable human-induced perturbations. Prevention is always a more beneficial option than a cure.

11.2.5 Sources of surprise — risk, uncertainty, and human ignorance

As explained earlier in the chapter, an important element of the co-evolutionary paradigm is the notion of *surprise*. Surprising events occur because there is always a disparity between what humankind expects *ex ante* and what it experiences *ex post*. The notion of surprise has been given implicit attention by economists ever since the ground-breaking work of Knight (1921). Unfortunately, the treatment of surprise has been confined to the distinction between *risk* and *uncertainty*. As Faber and Proops (1990) point out, a co-evolutionary paradigm requires a third category of surprise, namely, human *ignorance*.

Because the existence of surprising events restricts humankind's ability to predict future outcomes, then, for two good reasons, it is necessary to gain a better understanding of their source. First, the precise nature and the source of a surprising event determines the degree to which humankind can make valid predictions regarding future events. Second, without a comprehensive knowledge of the sources of surprising events, humankind's ability to positively influence the evolutionary pathway of the global system is greatly reduced. To deal with surprise, Figure 11.6 is presented as a diagrammatic representation of its various sources. Also included is a simple taxonomy of ignorance.

11.2.5.1 Risk and uncertainty

Two kinds of surprising events are experienced by humankind as depicted in Figure 11.6. The first includes events where the range of all possible outcomes is known *a priori*. Here, humankind's understanding of the dynamic processes involved is sufficient to make useful, if not limited, predictions about the likely emergence of particular outcomes and events. Exactly how restricted humankind's predictive capacities are depends on its knowledge of the respective probabilities of each outcome emerging. Should all probabilities be known (e.g., it is known there is a 60, 30, and 10 per cent chance of X, Y, and Z occurring), future outcomes are predictable "in principle" (Faber, et al., 1992). Under these circumstances, one is dealing in the

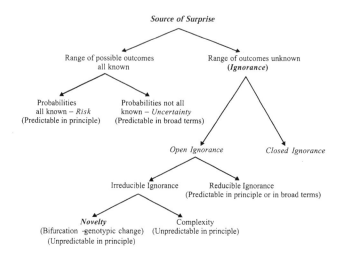

Figure 11.6 Sources of surprise and a taxonomy of ignorance. (Adapted from Faber et al., 1992, page 84. With permission.)

area of *risk* because, if X is the desired outcome, there is a 40 percent risk of it not occurring.

When the probabilities of a range of outcomes are not all known (e.g., it is known that X, Y, and Z may occur, but the probability of each emerging is not), one is dealing with *uncertainty*. On this occasion, future outcomes are only predictable "in broad terms" (Faber et al., 1992). That is, humankind is restricted to little more than saying something about the probable future behaviour of a system and a possible range of future events and outcomes (e.g., worst and best case scenarios). Clearly, when confronted with uncertainty, humankind's predictive powers are considerably weaker compared with instances involving risk.

11.2.5.2 Closed and open ignorance

The second category of surprising events involve those where the range of all possible outcomes are not known. It is here where humankind suffers from *ignorance* (Faber et al., 1992). As the taxonomy of ignorance in the right-hand side of Figure 11.6 shows, ignorance comes in two forms — *closed* and *open* ignorance.

When a society deliberately overlooks its ignorance, that is, it chooses to believe what it has yet to prove to be true, it is in a state of *closed ignorance*. Closed ignorance, particularly if it exists in the form of assumed omniscience (e.g., believing the macroeconomic subsystem can grow forever), is a significant barrier to humankind positively influencing the evolutionary pathway of the global system. In the event that a society is aware of its ignorance, that is, it chooses not to believe something until proven true, it is in a state of *open ignorance*. Only in a state of open ignorance is it possible for a society to fully experience novel and surprising events. Figure 11.6 indicates that open ignorance can be dichotomised into two forms — *reducible* and *irreducible* ignorance.

Reducible ignorance is ignorance that can be partially or fully overcome through learning and the application of the scientific method. Reducible ignorance exists because the stock of a society's knowledge is, at any moment, incapable of explaining and predicting the broadly explainable and predictable. Appropriate research makes it possible, at some future stage, to explain an event which has already taken place and/or to predict a greater range of future events.

The second form of open ignorance — irreducible ignorance — is never amenable to scientific tools of learning and research. In this instance, outcomes have the potential to emerge that one never could envisage *a priori*. As a consequence, irreducible ignorance involves a class of future events which are "unpredictable in principle." This means that one is unable to make even tentative predictions about the likely range of all possible outcomes. Irreducible ignorance severely restricts humankind's capacity to positively influence the evolutionary pathway of the global system.

Ignorance of the irreducible variety exists because of two ever present factors. The first is owing to *complexity*. Here, an outcome is unexpected because the complexity surrounding underlying processes of certain dynamic systems precludes the possibility of gaining a comprehensive understanding of them. In the second instance, irreducible ignorance stems from the emergence of *novelty*. Novelty arises because systemic parameters are forever evolving. This leads to adaptive and somatic change in the short and medium terms, and genotypic change (bifurcation) in the long term. Novelty results in irreducible ignorance because, in not knowing the initial boundary conditions governing the global system's evolutionary pathway, one cannot predict the future pathway of the global system in principle or in broad terms.

11.2.5.3 Implications

The inevitability of surprise, in all its previously described forms, obliges humankind to take note of the following. First, it cannot "control" the evolutionary pathway of the global system. The increasing ability of humankind to manipulate the ecosphere never translates to an equivalent increase in its ability to control the destiny of ecological and natural resource systems. Indeed, humankind can only hope to marginally increase its knowledge of the long term effects its own manipulative endeavours have on the global system. For this reason, humankind is strictly confined to positively "influencing" the pathway of the global system, which, moreover, it can only do in circumstances where it can make predictions about the implications of its endeavours in principle and in broad terms.

Second, because the logos of the global system is characterised by uncontrollable co-evolutionary processes decidedly more than by human teleology, humankind must obey the logos of the global system. As Laszlo (1972, page 75) puts it:

"There is freedom in choosing one's path of progress,
yet this freedom is always bounded by the limits of
compatibility with the dynamic structure of the whole
(global) system." (parentheses added)

It would appear, therefore, that only insofar as humankind learns to respect
and obey the logos of the global system can it, in Boulding's words, "move
away from the slavery of evolution to the freedom of teleology" (Boulding,
1970a, page 18). Clearly, for humankind to maximise its limited capacity to
positively influence the global system's pathway, it must recognise the circumstances under which it is a slave to the "rules" governing co-evolutionary processes (as opposed to being a slave to the process itself), and where
its actions are most likely to bring to bear catastrophic future macrostates of
the global system (i.e., where the impact of its own actions are unpredictable
in principle).

Third, humankind must abandon its futile attempts to bring more of the
ecosphere under its own manipulative control. In particular, it must recon-
sider its Faustian-like efforts to subdivide natural capital in order to make
it more amenable to private ownership. Last but not least, because novelty
leads to outcomes that are unpredictable in principle, nonprice rules and
institutional mechanisms must be suitably flexible to accommodate, for
instance, the need to revise the ultimate end in light of continuously conflicting signals; the need to reassess the maximum sustainable resource constraint as novel events alter the ecosphere's carrying capacity; and the need
to reestablish the market's functional prerequisites as they are periodically
undermined by the novel behaviour of market participants.

11.2.6 Feedback in co-evolutionary processes

Feedback is a process whereby a system regulates itself in response to the
recursive feeding back of the output generated by both its own within system
activities and the activities of other interdependent systems. Feedback is
omnipresent in a co-evolutionary world because, with everything interrelated, no individual system or system component can ever be entirely insu-
lated against the effects of perturbations originating external to it.

There are two types of feedback — *negative* and *positive* feedback. Traditionally, both have been viewed from a very mechanical perspective. For
instance, the term negative feedback has been used to explain how inbuilt
mechanisms, by minimising the effect of external perturbations, are able to
maintain the homeostatic equilibrium of a physical system. Negative feedback, therefore, conjures up images of a system maintaining a stable macrostate. On the other hand, positive feedback has been used to explain how
inbuilt mechanisms lead to the exponential amplification of external perturbations throughout the system in question. Because positive feedback is

ultimately a self-destructive process, it evokes an image of systemic instability. From a strictly mechanical viewpoint, negative feedback has been viewed as a "good" and positive feedback as a "bad."

The problem that emerges when considering feedback from a co-evolutionary world view is that, beyond a very short term time frame, dynamic systems are not characterised by negative feedback in the true homeostatic sense. Beyond the very short run, the overall structure of any given system becomes progressively more unstable as its underlying parameters evolve. Hence, from a mechanical interpretation of the feedback concept, co-evolution is a positive feedback phenomenon. While this suggests that co-evolution is a very destructive if not dialectical process, this is not necessarily so. In bringing about the end of a given macrostate, co-evolution merely generates the opportunities for the creation of new states which, in the case of biological evolution, is an unambiguously negentropic process. Indeed, by increasing the order and complexity of the global system, co-evolution has been anything but destructive, despite leaving in its wake numerous spent species and ecosystems.

In light of the dynamic nature of a co-evolutionary world, it is necessary to reinterpret the concepts of negative and positive feedback. Exactly how both concepts should be reinterpreted is open to conjecture. In the end, it depends on how and by what means one defines the end goal of the co-evolutionary process. Assume, for a moment, that it was defined purely in terms of maintaining the ecosphere's negentropic potential. In the face of constantly evolving systemic parameters, negative feedback would occur if the inbuilt feedback mechanisms of the ecosphere maintained the order and complexity of the global system. Feedback mechanisms of this sort are those which underpin the Gaian hypothesis previously introduced in chapter 4. Positive feedback would occur if, through the nature of the inbuilt feedback mechanisms or through their obliteration, the negentropic potential of the global system was systematically eroded.

As useful as it is to couch the feedback concept in negentropic terms, it has its obvious limitations. This is because feedback mechanisms that serve only to maintain the ecosphere's negentropic potential are of little anthropocentric value if the co-evolutionary process they promote does not advance the human condition — which it may not do given that sustainability, alone, does not guarantee SD. Clearly, one must extend any biophysical interpretation of feedback so that an instance of negative feedback is one where inbuilt feedback mechanisms direct the global system along an evolutionary pathway commensurate with the SD objective. Defined in this way, maintaining or increasing the negentropic potential of the global system is still important because, if eroded, it equates to a diminution in the ecosphere's carrying capacity. However, it is subsumed by the extended notion of negative feedback in recognition of the fact that sustainability is an insufficient condition to achieve SD.

11.2.6.1 Implications

Because the evolutionary pathway of the global system need not be commensurate with the SD objective, there is an overwhelming need to ensure appropriate negative feedback mechanisms are built into the global system (Boulding, 1990). This means humankind must, when and wherever possible, protect and preserve Gaian feedback mechanisms by adhering to the sustainability precepts developed in Chapter 4. It must also establish negative feedback mechanisms of its own to ensure, for instance, moral capital stays intact; that the distribution of income and wealth remains equitable; and that markets bring about the efficient allocation of the incoming resource flow. So too, must it weed out positive feedback mechanisms, in particular those it has implanted itself (e.g., mechanisms that institutionalise the GDP-poverty and GDP-unemployment links).

Three things are required to maximise the success of negative feedback mechanisms. First, they must be sufficiently rigid to ensure that, in the process of dealing with the recursive feeding back of the output generated by a variety of different systems, they themselves are not eroded. Should they be so, it is vital they be regenerated or re-installed. Second, as rigid as negative feedback mechanisms must be, they must remain adequately flexible to deal with novelty and surprise as they both emerge. Clearly, when it comes to designing negative feedback mechanisms, it is important to strike an appropriate balance between their rigidity and flexibility. As Hinterberger (1994, page 76) has described it,

> "Stability and flexibility are crucial for evolutionary processes; only a proper combination of both guarantees the conservation of reliable properties as well as the possibility for the progress of any system as a whole.... Both stability and flexibility are highly intertwined and depend on the workability of various natural, social, and economic feedback mechanisms."

Finally, in the process of establishing and maintaining feedback mechanisms, humankind must ensure they do not conflict with the logos of nature. It was earlier demonstrated that lower-order systems — many of which directly impinge on human welfare — are more chaotic than the global system. For this reason, many of the feedback mechanisms installed by humankind have been deliberate attempts to insulate itself from the potentially catastrophic consequences of external perturbations. Problems arise if the feedback mechanisms installed violate the rules governing co-evolutionary processes, because they instead become something equivalent to positive feedback mechanisms (e.g., feedback mechanisms that increase in the available supply of fossil fuels are leading to global warming). Of course, having to ensure feedback mechanisms obey the logos of nature does not mean that humankind is no longer free to reduce the frequency and extent to which lower-order systems experience chaotic fluctuations. However, its freedom is always confined to instances where outcomes are predictable in principle or in broad terms.

11.2.7 Human belief systems as a driving factor in global co-evolutionary processes

As already explained, human belief systems are an essential part of the co-evolutionary world view because the co-evolution of the global system's constituent systems and components are inextricably linked to the human belief systems underlying all their interdependent and dynamic manifestations. While the co-evolutionary world view also recognises that human belief systems are being constantly revised in light of signals emitted by the global co-evolutionary process itself, the extent to which human belief systems influence the evolutionary pathway of the global system is evidenced by the fact that a great number of systems now reflect many of the assumptions central to human belief systems and their corresponding world views.

A key question is: How much has, and does, the global system's characteristics reflect the beliefs, images, and values of human beings? Early on in humankind's development, very little. Given humankind's limited knowledge, its simple forms of human-made capital, and the lack of sophisticated institutional arrangements, the human capacity to manipulate and modify the natural environment was severely restricted. This rendered humankind almost entirely vulnerable to the often extreme and merciless whims of the natural environment (in particular, flood, drought, disease, etc.). Because, in its early stages, humankind's main concern was that of its survival, the knowledge, technologies, and the simple institutional structures of different societies and cultures were largely a reflection of the surrounding environment.

Over time, humankind has been able to free itself, to some degree, from the tyranny of co-evolutionary processes. It has done so in two main ways. In the first instance, it has significantly augmented its technological capacity to greatly modify all corners of the planet. Second, it has developed the whereabouts to correct naturally occuring and human-induced positive feedbacks. This has prolonged the co-evolutionary balance of the global system's constituent parts and relations and, in the process, has kept at bay — at its own great expense — the catastrophic bifurcation of critical subsystems. Hence, in contrast to humankind's beginnings, the global system has come to increasingly resemble the assumed beliefs, images, and values of human belief systems.

11.2.7.1 Implications

Because humankind must obey the logos of nature, there is a limit to how far the global system can increasingly resemble human beliefs, images, and values. This has a number of implications. First, at some point, humankind must acknowledge the need for a greater degree of "symmetry" in the co-evolutionary process. This means having to take greater notice of the signals being emitted by the global system as well as having to revise belief systems, when necessary, to ensure they comply with the logos of nature. Second, where existing human belief systems transgress the logos of nature, a beneficial "jump" to a SD pathway is unlikely to occur without a radical modification of currently existing human belief systems — something that technological, institutional, and politically induced change cannot, by themselves, redress. It is the required change in human belief systems that underlies the ultimate need for moral growth to initiate a movement toward SD.

11.2.8 Institutions as the biological equivalent to the gene

The importance of human belief systems in co-evolutionary processes highlights an aspect of human behaviour that is largely absent in the biological domain — the presence of conscious goal-seeking behaviour (Hodgson, 1991). Despite a considerable degree of randomness and indeterminacy associated with human activity, human endeavours are shaped and influenced by a host of stabilising forces. These include norms, customs, habits, and other nonprice rules that are characteristically embodied in a range of institutional structures and arrangements. All such stabilising forces have an inert quality inasmuch as they are consistently communicated over time as they are passed on from one institutional form to another. It is by promoting regular and predictable human behaviour in a world characterised by indeterminacy, novelty, and surprise, that institutions are able to keep the evolutionary pathway of the global system commensurate with the SD objective. For this reason, institutions serve as important feedback mechanisms and, in many ways, play a similar evolutionary role to that of the gene in the biological world (Hodgson, 1988). But, in playing such a role, it is important to recognise that institutions can also act as genes in the Lamarkian evolutionary sense.

To explain what is meant here, it needs to be understood that the norms, customs, and human knowledge transmitted by institutions collectively constitute the macroeconomy's *genotype*. The genotype defines the potentialities of any system. The realisation of these potentialities — the macroeconomy's *phenotype* — emerges as the genotype interacts with other interdependent systems. While, in biological terms, the genotype directly affects the development of the phenotype (e.g., the genes of a human being allow it to develop the language skills to communicate with another human being), the genotype cannot inherit the characteristics acquired during the life of the phenotype (e.g., a human being is not born with the language skills acquired by its parents). That is, Lamarkian evolution cannot occur in the biological world. In the anthropocentric world, however, Lamarkian evolution is possible because, as explained, institutions and other organisational forms are able to communicate and transmit their acquired characteristics onto tomorrow's institutions. For instance, the nonprice rules established in response to a revision of human belief systems or the production techniques developed by a firm wishing to operate more efficiently can be readily passed onto any newly established institutional arrangement or any newly established firm without either having to be acquired by the institutions themselves.

11.2.9 History matters — chreodic evolutionary pathways

Because the gene-like function of institutions leads to certain patterns of human behaviour, past events are rendered important insofar as they significantly influence and structure the present. Indeed, once a set of events has taken place, the range of evolutionary pathways a particular system can take is, even though many in number, largely predetermined. Thus, for some time at least, a system can be expected to travel along a relatively stable trajectory or *chreod* — better known as a necessary or "fated" path (Waddington, 1977). Of course, a chreodic pathway is never indefinitely stable because all systems eventually collapse. Nonetheless, the past remains significant because, as the phenomenon of exponential divergence suggests, two seemingly unrelated pathways can eventuate even from the smallest variation in a system's initial boundary conditions.

11.2.9.1 Implications
The existence of chreodic pathways has a number of important SD implications. First, because of the stabilising impact of institutions and other organisational forms, one can justifiably conclude that chreodic-type pathways are more evocative of evolutionary processes than the traditional Darwinian position of natural selection would indicate (Hodgson, 1991). Second, the pathways of both the global system and its constituent subsystems will be as much determined by their past as by their unique ability to fit the present. Third, while the inertial qualities of institutions can serve a valuable negative feedback role, they can also lead to the "locking in" of a range of undesirable pathways from which humankind may have considerable difficulty extricating itself.

11.2.10 Path dependency in co-evolutionary processes

The final aspect of the co-evolutionary paradigm concerns the *path-dependent* nature of many co-evolutionary processes. Path-dependency occurs because, in the presence of chreodic-like pathways, the choice between alternative macrostates is often restricted to a limited range of substitution possibilities — all of which may be suboptimal. A path-dependent sequence of events will frequently occur whenever important restrictive influences on the evolutionary pathway of systems are "exerted by temporally remote events, including chance elements as well as systemic forces" (David, 1985, page 332).

According to David, there are two major factors contributing to path-dependency. The first involves *structural interrelatedness*. To survive in a co-evolutionary world, subsystems, components, and the relations between them typically embody the assumed values underlying the global co-evolutionary process. Because only the technologies, institutions, routines, etc., that adopt the assumed values generally succeed, there is a tendency for some to become locked in and difficult to remove. The second involves system scale economies or what is sometimes referred to as the *quasi-irreversibility of investment*. This occurs when already adopted technologies, institutions, and routines are preferred to alternatives because, as a consequence of them already being in place, the long-run discounted costs of continuing to employ them is less than the cost of their replacement. It is because of system scale economies that a macroeconomic system can be gradually locked into a restricted range of undesirable macrostates, such as the non-SD pathways in Figure 11.3. All too often, these are not easily alterable (i.e., it is difficult to jump onto a SD pathway) and, in most cases, the range of alternative macrostates is impossible to predict in advance (Arthur, 1989).

It is important to recognise that path dependency is not just a function of structural factors. Because a comparison of the discounted net benefits of alternative pathways is the basis upon which a particular pathway is invariably chosen, human beliefs, images, and values are particularly influential. After all, it is these values which, along with structural, social, and ecological factors, shape the market's price-determining parameters and thus determine the discounted net benefits upon which many collective and resource allocation decisions are made. Clearly, therefore, a path-dependent sequence of events is inextricably "value laden" and perceptually influenced as much as structurally determined (O'Connor, 1989a). Hence, a simple change in beliefs, values, and preferences, by altering discounted net benefits, has the capacity to alter a society's perception of the range of pathways available to it at a given point in time. That having been said, it would be a grave mistake to believe that a change in human beliefs, images, and values can completely override the structural factors underlying a path-dependent sequence of events. For example, a fossil fuel-reliant macroeconomic system will, for some time, continue to be dependent on fossil fuels despite any concerted effort to make the transition toward renewable energy sources.

11.2.10.1   *Implications*

Path dependency has a number of considerable SD implications. To begin with, once locked in, particular technologies, institutions, and routines may become immune from threats posed by alternatives despite the fact that the alternatives could be notionally superior (Dosi and Metcalfe, 1991). Second, path dependency implies that humankind is not always free to do what it wants. Constituent components of the global system and the relations between them cannot be freely altered at will because the remainder of the global system can often react in a conservative, identity-preserving way (Silverberg, 1988). This suggests humankind will, at times, have difficulty initiating such things as institutional modifications; variations in nonprice rules, imposing a previously nonexistent resource constraint, and/or imposing a Pigouvian tax in the hope of internalising a worrisome externality.

Of course, path dependency does not altogether preclude policy measures to bring about a more desirable evolutionary pathway of the global system. But it does mean that whatever changes are required can only be implemented slowly and gradually. Moreover, and largely because of the phenomenon of exponential divergence, the longer humankind waits to implement the required changes, the more perceptively prohibitive does the necessary shift in the current pathway become.

Finally, the path-dependent nature of co-evolutionary processes indicates that if the market mechanism is solely relied on to bring about a more desirable pathway of the global system, it is possible to never reach one. Why? Because, as shown in Part II, the market is only capable of differentiating between efficient and inefficient pathways. The relative prices generated by markets cannot be relied upon to select pathways that are also sustainable and equitable. Thus, if the discounted net benefits of continuing on with a non-SD pathway forever exceed those associated with a SD pathway, it is possible for a nation's macroeconomy to grow beyond its optimal scale, and indeed, its maximum sustainable scale.

11.3 Concluding remarks

Now that a co-evolutionary-based conceptual framework has been outlined, including many of its broader implications, the chapter is rounded off by making a couple of brief observations. The first deals with the SD concept itself. While a co-evolutionary world view has no major implications for the SD concept developed and defined in Part I, the same cannot be said in terms of what is required to satisfy the SD goal. It should now be clear that, in accordance with a co-evolutionary paradigm, the parameters of the global system, along with those of its constituent subsystems and components, are forever evolving. Although the set of SD-based parameters, guidelines, and minimum positions outlined in Chapter 5 are entirely valid, satisfying them requires humankind to make, in many instances, considerable institutional and policy adjustments over time.

For instance, because the rules governing the co-evolutionary process perpetually evolve, what is perceived today as fair, just, and equitable, may not seem so tomorrow. Even a society's delineation between wealth and illth will change over time. In addition, because the ecosphere's negentropic potential is in a constant state of flux, there will always be the need to vary the quantitative restriction on the incoming resource flow to ensure the macroeconomy operates within the ecosphere's changing carrying capacity (which, depending on whether the ecosphere's negentropic potential increases or decreases, allows for either a larger or smaller maximum sustainable scale).

In the second instance, the following needs to be considered. Where does a co-evolutionary-based conceptual framework leave the two crucial aspects discussed at length in Part II — namely, the importance of economic efficiency and the market in moving toward SD? In other words, what role does economic efficiency and the market play in a co-evolutionary world? While the need to apply a co-evolutionary approach should in no way diminish their importance, it is necessary to show how both can be meaningfully incorporated into the co-evolutionary-based conceptual framework. If not, it will be impossible to claim success at bridging the crucial gap between economic theory and concrete reality.

chapter twelve

The market: A co-evolutionary feedback mechanism

12.1 The role of the market in the co-evolutionary process

One of the critical messages to emerge from the previous chapter was the importance of having in place appropriate feedback mechanisms to ensure the evolutionary pathway of the global system is in keeping with the SD objective. Of course, one these mechanisms happens to be the market. By reflecting relative scarcities, the market's role is to ensure that, as the incoming resource flow is allocated among alternative product uses, the co-evolutionary "selection" process is one that facilitates improvements in allocative efficiency. That is, the market's role is to diminish or eliminate reducible forms of "allocative" ignorance through the promotion of beneficial forms of novelty. However, the market may or may not be a desirable negative feedback mechanism. Whether it is depends on whether the market selection process promotes evolutionary pathways of the ecological economic efficiency kind. This depends entirely upon the qualitative nature of the market's price-determining parameters.

In order to demonstrate the importance of allocative efficiency and the market place in a co-evolutionary world, we now turn our attention to the market by employing an evolutionary economic schema set within the conceptual framework developed in the previous chapter. While the schema draws heavily from the pioneering work of Faber and Proops (1990), it is adapted, where necessary, to be commensurate with the SD concept established in Part I. The evolutionary economic schema enables one to conceive the co-evolutionary process from the perspective of the market. Conceiving the co-evolutionary process in this way permits two things. First, it permits one to describe the co-evolutionary interdependencies between the macro-economy and other components or subsystems of the global system. Second, it enables one to better appreciate the impact these interdependencies have on the market, the price signals it generates, and ultimately, the type of human behaviour it encourages. This last point is important because should the selection process go wrong, it is possible for a nation to become locked into undesirable chreodic pathways.

12.1.1 Economic genotypes and phenotypes

12.1.1.1 The economic macrogenotype

The evolutionary economic schema to be employed is based on the important notions of *genotypic* and *phenotypic* evolution. As explained in Chapter 11, the genotype refers to the various individual elements which collectively define the "potentialities" of any given economic system. The phenotype, on the other hand, is the "realisation" of a macroeconomy's genotypic potential. The difficulty one has in distinguishing between economic genotypes and phenotypes is that, unlike the biological world, there is no immediate or obvious evidence of an economic genotype. For this reason, there is no immediate way of recognising genotypic changes (Faber and Proops, 1990). Notwithstanding this apparent shortcoming, it is possible to observe the emergence of new structures and elements that, in most instances, lead to the complete reorganisation of the macrophenotype of any macroeconomic system. It is these new structures and elements that constitute suitable candidates for the economic genotype. They include

- Technological know-how (human knowledge) — in effect, the set of all known and available resource exaction, waste insertion, and production techniques.
- Social, economic, political, legal, and religious institutions, and the nonprice rules they embody.
- The objective values pertaining to human belief systems.
- Society's moral capital — the social virtues that maintain not only a shared sense of and dogmatic belief in objective value, but also mutual standards of honesty, trust, and co-operation.
- Preference orderings (tastes) of individual consumers. These are, to a large degree, a product of the previous two candidates.

The nature of these elements, most of which are non-economic, define the potentialities of a macroeconomic system. A change in any one of these elements corresponds to the genotypic change of a macroeconomic system and, thus, a change in its potentialities.

Before moving onto economic phenotypes, there is a need to differentiate between two possible interpretations of the genotypic potential of macroeconomic systems. In the first instance, one can view the macrogenotype in terms of its potential to maximise the transformation of natural to human-made capital — the macroeconomy's physical growth potential. Conversely, one can view the macrogenotype in terms of its potential to achieve an optimal macroeconomic scale. One must bear in mind that should they focus entirely on the growth potential of the macroeconomy, the potential to meet critical considerations necessary to achieve an optimal macroeconomic scale, such as distributional equity and a sustainable rate of resource throughput, maybe overlooked. Clearly, an economic macrogenotype required to maximise the physical scale of the macroeconomy will differ considerably from a macrogenotype aimed at delivering an optimal macroeconomic scale. The latter, for example, would include fewer throughput-increasing but considerably more efficiency-increasing techniques than the former.

12.1.1.2 The economic macrophenotype

The economic phenotype results from the interplay of the economic genotype and the greater ecosphere (Faber and Proops, 1990). The ecosphere is a critical phenotypic realisational factor because the quality of its instrumental services greatly enhances or constrains the macroeconomy's capacity to realise its own potential. Hence, even if the macrogenotype were to remain unchanged, the capacity for phenotypic realisation can drastically alter, for better or worse, simply because of changes in the ecosphere's source, sink, and life support functions. Because the economic macrophenotype constitutes the realisation of an macroeconomy's genotypic potential, its macrophenotype will contain some of the following elements.

- The resource exaction, waste insertion, and production techniques actually employed (as opposed to all known and available techniques).
- The different types and quantities of producer goods used in the production of new commodities.
- The range and quantity of newly produced human-made capital.
- The relative distribution of both the incoming resource flow as embodied in human-made capital — income, and the accumulated stock of human-made capital — wealth.
- The different market structures in existence throughout the macroeconomy (e.g., from monopolistic through to competitive markets).
- Although not a desired realisation, the generation of high entropy waste and its insertion back into the accommodating ecosphere.

In the same way that it is important to distinguish between the genotypic potential of the macroeconomy on pure quantitative and qualitative grounds, so must one do the same with the realisation of the macroeconomy's genotypic potential — its macrophenotype. Specifically, the macrophenotype can be interpreted in terms of whether it has sufficiently met a predetermined growth target or an optimal macroeconomic scale. Again, an economic macrophenotype reflecting the successful maximisation of a macroeconomy's growth potential will differ considerably to one reflecting the successful realisation of an optimal macroeconomic scale. The latter, for instance, would more than likely have a smaller but qualitatively superior and more equitably distributed stock of human-made capital.

12.1.2 The price-influencing effect of genotypic and phenotypic evolution

To understand the economic role played by genotypic and phenotypic evo-lution, it is necessary to consider how both affect a market's price-determin-ing parameters. Price-determining parameters are the binding constraints that, in the absence of externalities, manifest themselves as market prices. From a co-evolutionary perspective, a market's price-determining parame-ters are the product of the combined forces of the economic macrogenotype, the accommodating ecosphere, and the economic macrophenotype emerging from their interaction. As any one or all of these factors or forces vary, so also do the market's price-determining parameters. This leads to relative price changes and, as a consequence, to a change in the quantity and the mix of physical commodities bought and sold in the market.

12.1.2.1 The price-influencing effect of phenotypic evolution

Phenotypic evolution involves the changing nature of and relationships between mutually interacting phenotypes in the presence of a stable genotype. The concept of phenotypic evolution presupposes the absence of any recursive feedback that might otherwise alter the macroeconomy's genotypic potential. An instance of phenotypic evolution is potentially observable when

- Currently employed techniques are rejected and replaced by alterna-tive but already existing techniques.
- When labour and producer goods are either relocated or reorganised to produce a different but preferred macrophenotype.
- As the ecosphere's source, sink, and life support functions vary over time.

Because the concept of phenotypic evolution assumes a stable macrogeno-type, it presupposes the attainment of equilibrium macrostates and the poten-tial for equilibrium prices and quantities. For example, consider Figure 12.1 which, apart from the obvious absence of demand and supply curves, is a typical diagram in price/quantity phase space. Figure 12.1 illustrates an initial equilibrium state (point A) with a corresponding equilibrium market price and quantity (P_1, Q_1). Imagine that a new economic phenotype is introduced — the result of employing a previously known but disused production tech-nique. By perturbing the economic system, the new phenotype causes supply and/or demand factors to shift until a new equilibrium market price and quantity (P_2, Q_2) is established at point B. Between the initial and final equi-librium states, the equilibrium market position travels along the phase path between points A and B. As it does, the relative abundancies of the various economic phenotypes adjust. As is customary with phenotypic evolution, there is no corresponding alteration of the underlying genotypic potential of the macroeconomy. The new equilibrium position at point B is maintained until such time as there is a further instance of phenotypic evolution.

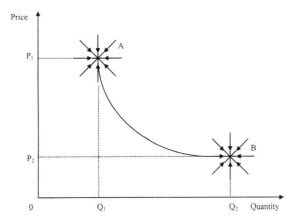

Figure 12.1 Price and quantity adjustments in the presence of phenotypic evolution in a market economy. (Adapted from Faber and Proops, 1990, page 32. With permission.)

12.1.2.2 *The price-influencing effect of genotypic evolution*

Genotypic evolution involves the modification of the macroeconomy's genotypic potential and, consequently, its progeny over time. As explained in Chapter 11, the presence of feedback loops and the concomitant evolution of underlying systemic parameters means that genotypic change is part and parcel of a co-evolutionary world. Indeed, in the case of economic genotypes, genotypic evolution is often as rapid as phenotypic evolution (Faber and Proops, 1990). Moreover, genotypic evolution is manifestly unpredictable "in principle." The unpredictability of genotypic evolution arises because any change in the macroeconomy's genotypic potential is equivalent to the emergence of novelty, and underlying systemic processes are so complex that, in most instances, the mutation of economic genotypes cannot be specified beforehand.

Figure 12.2 shows how genotypic evolution moves a macroeconomy's phenotypic equilibrium from point *A* to *B* and subsequently to points *C*, *D*, and beyond. Characteristic of economic systems, the rate of genotypic evolution is approximately the same as the rate of phenotypic evolution toward each of these successive equilibria. This results in phenotypic evolution proceeds along the ray, beginning at point *A*, toward the new stable equilibrium at point *B*. However, because the genotype evolves before a phenotypic equilibrium is reached, the phase path shifts to one converging on point *C*, the new equilibrium of the system. The new phase path heading for the new equilibrium replaces the old phase path beginning at point *b*. Again, because of further genotypic changes, a new phase path, starting from point *c*, begins its way toward point *D*, and then from point *d* onward. So long as genotypic evolution continues at a rate similar to phenotypic evolution, a phenotypic equilibrium is never attained. Clearly, in the case of rapid and

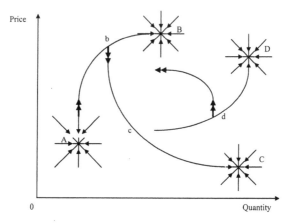

Figure 12.2   Price and quantity adjustments in the presence of combined genotypic and phenotypic evolution in a market economy. (Adapted from Faber and Proops, 1990, p. 47. With permission.)

ongoing genotypic change, relative prices become, not a product of independent and comparatively static factors, but a reflection of a multitude of interdependent and incessantly co-evolving market forces. Hence, one is left to conclude that, in a co-evolutionary world, market disequilibrium is the normal state of affairs. Quite literally, pecuniary economies and diseconomies arising from perturbations originating external to the market are, as O'Connor (1989a, page 8) points out, "the tangible economic evidence of the dynamism of the market system and of the incessant shifting towards new structures and patterns of supply and demand." Thus, in keeping with the conclusions of Chapter 8, the resultant change in relative prices is likely to reflect a changing efficiency context as often as it does a variation in allocative efficiencies. What matters, therefore, are two things. First, and most importantly, is the prevailing efficiency context still one of the ecological economic efficiency kind? That is, are market outcomes, even if allocatively efficient, still commensurate with a set of SD-based parameters, guidelines, and minimum positions? Second, is the changing efficiency context qualitatively improving over time — in other words, is there a gradual evolution of a more appropriate market domain?

12.1.3 A schematic representation of the co-evolutionary process

It is important, at this stage, to gain a clearer understanding of the association between economic genotypic and phenotypic evolution and how this association, mediated via the market selection process, influences the evolutionary pathway of the global system. It is through this understanding that the critical role of both the market and its price-determining parameters become increasingly apparent. The association between genotypic and phenotypic evolution is best understood in terms of an interaction between *invention* and *innovation*. By defining these two terms as well as another — *technology* — one is better placed to describe the evolutionary process of economic systems. According to Faber and Proops (1990),

- The human *technology* of a macroeconomy is the set of all known and available exaction, insertion, and production techniques.
- *Invention* is the emergence of a novel technique that augments the set of all known techniques. The generation and emergence of an invention can be classed as an example of genotypic evolution.
- *Innovation* is the process of introducing a novel technique and is, therefore, an example of phenotypic evolution.

The interplay of genotypic and phenotypic evolution can be described in the following manner. First, consider a macroeconomy which, via the market mechanism, has established a predictable phenotypic equilibrium state. In Figure 12.1, this would be equivalent to an equilibrium at point A. Suppose the stock of natural capital is being steadily depleted. One would expect marginal adjustments in the relative prices and quantities of both low entropy resources and human-made capital to reflect the increase in resource scarcity. For example, one would expect, in the first instance, an increase in low entropy prices to reflect the higher marginal opportunity costs of maintaining intact the stock of human-made capital. Second, one could expect the increase in low entropy resource prices to induce a novel efficiency-increasing invention. Should this occur, the emergence of the new invention would give rise to a new and unpredictable macrogenotypic which, in turn, would give rise to a new macrophenotype. Via the market selection process, the new macrophenotype would then be required to establish a new and predictable equilibrium state. Once achieved, relative prices and quantities would remain stable until such time as there was further genotypic evolution.

Figure 12.3 serves as a generic representation of the co-evolutionary process as viewed from the market perspective. The genotype (G) and phenotype (P) represent the macrogenotype and macrophenotype for a particular market economy. Because the macrophenotype results from the interplay of the macrogenotype and the ecosphere then, for convenience and diagrammatic simplicity, it will be assumed that the macrogenotype — and the symbol (G) — includes the ecosphere's source, sink, and life-support functions. The evolutionary economic schema represented in Figure 12.3 is divided into three phases. The beginning of Phase 1 and the conclusion of Phase 3 reflect initial and final equilibrium phenotypic states. In between, there are two major events. The first is a perturbed macrogenotype arising out of a newly developed technique (invention). The second is the market selection process (phenotypic evolution) which leads, ultimately, to a new phenotypic equilibrium. At this stage of proceedings, it is assumed that there is no recursive feeding back of the resultant phenotype on the macroeconomy's genotypic potential.

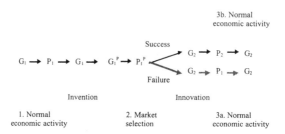

Figure 12.3   The evolutionary economic schema for the macroeconomy with no recursive feedback. (Adapted from Faber and Proops, 1990, p. 52. With permission.)

Phase 1 — The macroeconomy is initially exhibiting "normal economic activity." This may be supposed as a steady state equilibrium with a stable equilibrium price/quantity combination. The genotypic potential of the macroeconomy or the economic macrogenotype (G_1) is being realised in economic activity through the macrophenotype (P_1). As with biological systems, the genotype and phenotype are in an interdependent, self-maintaining relationship.

Phase 2 — As the second phase begins, an invention emerges in response to the relative prices generated by the market. The new invention augments the macrogenotype such that we now have (G_1^P) the perturbed macrogenotype. This, in turn, produces a corresponding new macrophenotype (P_1^P). Because of the perturbation of both the macrogenotype and macrophenotype, a new set of relative prices is established in the market to reflect the change in the market's price-determining parameters. The market now acts as a "filter" to determine whether the invention leads to increased allocative efficiency and, therefore, whether it should be adopted.

Phases 3a and 3b — Irrespective of the success or failure of the novel invention, the augmentation of the macroeconomy's macrogenotype (in this case, an increase in the set of all known and available techniques) is retained. (G_1), the original macrogenotype, becomes a subset of the new macrogenotype (G_2). Where the invention fails (Phase 3a), the realisation of the macroeconomy (P_1) does not alter despite the increase in the system's genotypic potential. Conversely, where it succeeds (Phase 3b), both the new macrogenotype (G_2) and the new macrophenotype (P_2) become established. In both cases, a state of normal economic activity is restored.

12.2 The market as a co-evolutionary feedback mechanism

Phase 2 in Figure 12.3 demonstrates the important role of the market in a co-evolutionary world. By generating relative prices and acting as a mechanism to filter genotypic changes, the market influences the evolutionary pathway of the macroeconomy. As such, it is the co-evolutionary equivalent of a feedback mechanism. However, as mentioned at the beginning of the chapter, the market may or may not be a desirable negative feedback mechanism. This will depend largely on the qualitative nature of the market's price-determining parameters and, therefore, on the collective price-influencing decisions made prior to the allocation process. Only when the prices generated are commensurate with an ecological economic efficiency context is the market best able to induce appropriate forms of human behaviour and promote an evolutionary pathway of the global system in keeping with the SD objective.

12.2.1 *Incorporating recursive feedback into the evolutionary economic schema*

In the generic representation of the evolutionary economic schema depicted in Figure 12.3, the notion of recursive feedback was deliberately ignored. To continue on with schematic representations of the evolutionary economic process in a manner consistent with reality, it is necessary to incorporate feedback into the conceptual approach. To recall, recursive feedback occurs when the realisation of a system's potentialities — its phenotype — feeds back on itself and other interdependent systems such that the system's very own genotypic potential is eventually perturbed.

As already mentioned, the evolution of the macrophenotype can indirectly or directly affect the macrogenotype in two ways, indirectly, by offering opportunities by way of niche establishment (Darwinian evolutionary feedback), and directly, as institutions and other organisational structures communicate and transmit acquired characteristics into the future (Lamarkian evolutionary feedback). Also, given the systemic interdependencies characteristic of co-evolutionary processes, a changing macrophenotype will always impact, in some way, shape, or form on the source, sink, and life-support functions of the ecosphere. Hence, the two major factors which collectively determine the potentialities of the macroeconomy — namely, the macrogenotype and the accommodating ecosphere — will always be directly or indirectly influenced by the macrophenotype itself. A recursive feedback effect is illustrated in Figure 12.4.

> *Phase 1* — As shown in Figure 12.3, the co-evolutionary process begins with the macroeconomy exhibiting normal economic activity. The genotypic potential of the macroeconomy (G_1) is being realised through the macrophenotype (P_1) which, in turn, allows for the maintenance of the macrogenotype (G_1).
>
> *Phase 2* — As the second phase begins, an invention augments the macrogenotype resulting in the perturbed macrogenotype (G_1^P). This leads to the creation of a new macrophenotype (P_1^P) and, ultimately, to a new set of relative prices to reflect the change in the market's

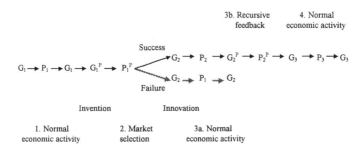

Figure 12.4   The evolutionary economic schema for the macroeconomy in the presence of recursive feedback — a generic representation.

price-determining parameters. The market acts as a filter or feedback mechanism to determine whether the invention is successful.

Phase 3a — In the event that the invention fails, the original macrogenotype (G_1) is subsumed by the new macrogenotype (G_2). Because the initial macrophenotype (P_1) is retained, there is no recursive feedback and a state of normal economic activity returns.

Phase 3b — Where the invention succeeds, a new macrophenotype (P_2) prevails. The new macrophenotype (P_2) initiates a fourth phase by recursively feeding back on the new macrogenotype (G_2) in a typically random and unforeseen manner. This modifies or perturbs the macrogenotype leading to both (G_2^P) and a corresponding new macrophenotype (P_2^P). Ultimately, a new macrogenotype (G_3) subsumes the previously existing macrogenotype (G_2).

Phase 4 — At this point, one could continue the evolutionary economic schema to represent the continuation of the co-evolutionary feedback process into the future. For illustrative purposes, it is only necessary to demonstrate the impact once. Hence, it is assumed that, following on from Phase 3b, a state of normal economic activity is restored in Phase 4.

Whether the feedback impact of the macrophenotype (P_2) on the macrogenotype is of the negative or positive feedback kind depends on the comparison between the macrogenotype at the end of Phase 3b (G_3), and the macrogenotype at the completion of Phase 3a (G_2). Should the genotypic potential of the macroeconomy, as the potential to achieve an optimal macroeconomic scale, be greater at the completion of Phase 3b than at the end of Phase 3a (i.e., $G_3 > G_2$), then the market can be said to be a negative feedback mechanism. In effect, the market's filtering process, by inducing an efficiency-increasing invention or a beneficial institutional change, will have increased the potential to move toward an optimal macroeconomic scale (and to increase a nation's sustainable net benefits). On the other hand, should the market's filtering process undermine the macroeconomy's potential to achieve an optimal scale (i.e., $G_3 < G_2$), the market can be considered a positive feedback mechanism.

It is important to recognise that, despite the best efforts of any society to ensure the evolution of an appropriate market domain, there will always be instances where the genotypic potential of the macroeconomy will suffer deleteriously. Irrespective of how vigilant a society might be in the establishment and subsequent maintenance of negative feedback mechanisms, it is because the byproduct of co-evolutionary feedback processes are random and novel that no society can guarantee that the perturbed macrogenotype (G_2^P) and the newly established macrogenotype (G_3) will be always be notionally superior to that which previously existed (G_2). Given, therefore, that a market will always make some undesirable selections, the key to having in place markets of the negative feedback variety is to establish them in ways that minimise undesirable feedbacks while maximising the possibility of beneficial novelty. This reinforces the need for the market's functional prerequisites to be suitably resistant to the harmful perturbating effects of a newly ordered macrophenotype but sufficiently malleable to be readily modified in order to respond to mutations in the underlying parameters of the global system.

12.2.2 The market as a negative feedback mechanism

In order to gain a much clearer picture of how the market acts as a negative feedback mechanism consider Figure 12.5.

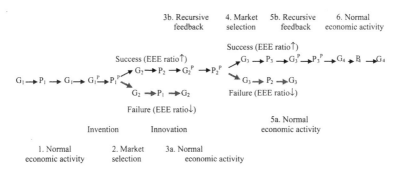

Figure 12.5 The evolutionary economic schema for the macroeconomy with the market as a negative feedback mechanism.

Phases 1 and 2 — Both phases are the same as the previous example. Note that for an invention to be successful, it must increase the ecological economic efficiency (EEE) ratio. The invention is rejected, at least for now, if its adoption reduces the EEE ratio.

Phase 3a — Phase 3a is also the same as in the previous example. Where the invention fails, the original macrogenotype (G_1) is subsumed by the new macrogenotype (G_2). There is no recursive feedback and a state of normal economic activity returns.

Phase 3b — Where the invention succeeds, a new macrophenotype (P_2) prevails. The new macrophenotype (P_2) recursively feeds back on the new macrogenotype (G_2). This perturbs the macrogenotype leading to both (G_2^P) and a corresponding new macrophenotype (P_2^P). A new macrogenotype (G_3) subsumes the previously existing macrogenotype (G_2). Because the market is, in this instance, a negative feedback mechanism, the new macrogenotype (G_3) is greater than that immediately following the emergence of the original invention (G_2) (i.e., $G_3 > G_2$). The recursive feeding back of the macrophenotype on the macrogenotype can increase the macroeconomy's genotypic potential by inducing the emergence of another novel, yet beneficial, invention, institution, or non-price rule.

Phase 4 — The market once again acts as a filter to determine whether the newly emerged addition to the macrogenotype is successful or not. Despite a mutationally improved macrogenotype, it need not necessarily pass the market selection test. There is still the possibility that should the new and novel economic genotype be adopted, the subsequent feeding back of the perturbed macrophenotype (P_3^P) on the macrogenotype (G_3) could be harmful (i.e., reduce the EEE ratio). Should this be the case, the market will reject the new genotype and the previous macrophenotype (P_2) will be retained. In the event that the new economic genotype succeeds, a new macrophenotype (P_3) will prevail.

Phase 5a — Where the invention fails, the macrogenotype (G_2) is subsumed by the latest macrogenotype (G_3). Because the macrophenotype (P_2) is retained, a state of normal economic activity returns.

Phase 5b — Where the invention passes the market selection test (i.e., increases the EEE ratio), a new macrogenotype (G_4) emerges from the recursive feeding back of the macrophenotype (P_3) on the previous macrogenotype (G_3). Because the market is a negative feedback mechanism, the genotypic potential of the macroeconomy increases (i.e., $G_4 > G_3$).

Phase 6 — While the evolutionary economic process would continue until a new addition to the macrogenotype failed the market selection test, it is once again assumed that, for illustrative purposes, a state of normal economic activity is restored in Phase 6.

12.2.3 *The market as a positive feedback mechanism*

Next, in Figure 12.6, consider how the market can instead act as a positive feedback mechanism.

Phases 1 and 2 — Both phases are the same as the previous example except, to pass the market selection process, the new outcome need only lower the cost of resource allocation regardless of the impact on the rate of resource throughput or the initial distribution of income

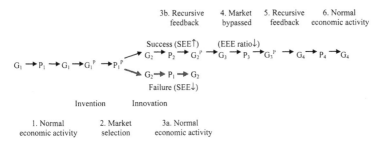

Figure 12.6 The evolutionary economic schema for the macroeconomy with the market as a positive feedback mechanism.

and wealth (i.e., can lead to an increase in standard economic efficiency even at the expense of lowering the EEE ratio).

Phase 3a — Where the invention fails the market selection test, the original macrogenotype (G_1) is subsumed by the new macrogenotype (G_2). There is no recursive feedback and a state of normal economic activity returns.

Phase 3b — Should the invention be successful, the new macrophenotype (P_2) recursively feeds back on the macrogenotype (G_2). Via the perturbed macrogenotype and macrophenotype (G_2^P) and (P_2^P), a new macrogenotype (G_3) emerges. Because the market is a positive feedback mechanism, the new macrogenotype (G_3) is less than the initial macrogenotype (G_1) such that, overall, the genotypic potential of the ecosphere has been severely undermined (i.e., $G_1 < G_2 < G_3$).

The recursive feeding back of the macrophenotype on the macrogenotype can diminish the macroeconomy's genotypic potential in any one or more of the following ways. First, it can lead to the depletion of society's moral capital. Second, by reinforcing and rejecting certain beliefs and images, it can lead to the irrecoverable loss or breakdown of important nonprice rules, norms, and conventions. Last, should it violate one or more of the sustainability precepts established in Chapter 4, it can erode the ecosphere's sustainable carrying capacity.

Phase 4 — Unlike the fourth phase in the negative feedback example, here the market selection process is completely bypassed. Following on from Phase 3b, the new macrophenotype (P_3) emerges, not out of choice, but simply because the macroeconomy no longer has the same genotypic potential.

Phase 5 — The new macrophenotype (P_3) recursively feeds back on the macrogenotype (G_3). This results in the perturbation of the macrogenotype (G_3^P) and thus a new macrogenotype (G_4). Whether G_4 is less than G_3 depends largely on whether some action is taken, in the meantime, to correct the problems associated with the previous positive feedback effect, that is, whether the market's functional prerequisites and/or other negative feedback mechanisms have been reinforced or reinstalled. If they have, there is the possibility of a beneficial genotypic change emerging to offset the initial harmful effect. If not, one can expect a further depreciation of the macrogenotype.

Phase 6 — It is assumed that, for illustrative purposes, a state of normal economic activity is restored in Phase 6.

Figure 12.6 demonstrates what can happen when a market is characterised by an inappropriate or a misleading set of price-determining parameters. With the market acting as a positive feedback mechanism, the ensuing outcome simply becomes a case of making the best of a worsening set of circumstances, even though it must be Pareto improving to pass the market selection test. Indeed, given the chreodic nature of evolutionary pathways, it is possible for the misleading price signals to result in what Mokyr (1992) refers to as "myopia" and what Costanza (1987) calls a "social trap." Both are merely different descriptions of the same phenomenon and occur when the so-called "road signs" that humankind rely on are both inaccurate and misleading. In this case the misleading road signs are the relative prices generated by an inappropriate set of price-determining parameters. Should the road signs be followed, a society can become trapped because, with the macroeconomy having been directed along an errant pathway, its identity-preserving nature makes the switch to a more desirable pathway difficult, if not impossible, without a collective desire of the present generation to bear the large structural adjustment costs necessary to resume a SD pathway.

12.3 Further applications of the evolutionary economic schema

Describing the co-evolutionary interdependencies between the macroeconomy and other components of the global system is particularly useful inasmuch as it brings to light a number of important policy-related areas that would otherwise go unrecognised and untreated within the confines of a neoclassical economic approach to SD. In what follows, the evolutionary economic schema will be used to highlight three policy-related areas. While the policy prescriptions will themselves be left to Chapter 15, the aim here is to examine the issues in question in order to ensure the most appropriate policy responses are eventually invoked.

12.3.1 Human responses to positive feedback effects

Humankind has devoted considerable effort correcting for impacts incurred as a result of naturally occuring and human-induced positive feedback events. In what are essentially efforts to insulate the more immediate systems critical to its welfare interests from the potentially chaotic consequences of systemic perturbations, humankind has continuously searched for ways to maintain, if not increase, the genotypic potential of the macroeconomy. All too often, however, such attempts have led, not to the long-term maintenance of the macroeconomy's SD potential, but to short-term increases in the macroeconomy's growth potential. As a consequence, many positive feedback effects have been brought on by the divestment of natural capital, by growing disparities in rich and poor, and by a decline in the restraining influence of society's moral capital. However, instances of undesirable positive feedback can and ought to be avoided. To reveal how, a diagrammatic interpretation of a positive feedback response is portrayed in Figure 12.7.

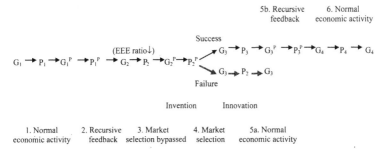

Figure 12.7   The evolutionary economic schema for the macroeconomy — correcting for positive feedback.

> *Phase 1* — In the initial phase, the macroeconomy is exhibiting normal economic activity with the macrogenotype (G_1) realised via the macrophenotype (P_1).
>
> *Phase 2* — It is assumed in Phase 2 that the macrophenotype (P_1) eventually feeds back on the macrogenotype (G_1) in a typically novel fashion. This leads to a perturbed macrogenotype and macrophenotype (G_1^P) and (P_1^P).
>
> *Phase 3* — The market selection process is bypassed. Being a positive feedback effect, the new macrogenotype (G_2) and macrophenotype (P_2) emerge, not out of choice, but simply because the macroeconomy no longer has the same genotypic potential. It is plausible to no longer be on a SD pathway. The emergence of the new macrogenotype and macrophenotype lead to a change in the market's price-determining parameters and, hence, a new set of relative prices.
>
> *Phase 4* — The changed set of relative prices induces a corrective response which modifies or perturbs the macrogenotype (G_2^P). This, in turn, produces a corresponding new macrophenotype (P_2^P). Two types of responses are likely at this stage. Whether the corrective measure passes the market selection test depends on the objective value criteria embodied in the market's price-determining parameters. For instance, a throughput-increasing invention, if it were to lower the relative prices of some forms of low entropy resources in a conventional economic

efficiency context, would pass the market selection test. However, this would not occur in an ecological economic efficiency context if the incoming resource flow was already at the maximum sustainable rate.*

Phase 5a — If the corrective measure fails the market selection test the macrogenotype (G_2) is subsumed by the new macrogenotype (G_3). The macrophenotype (P_2) is retained and a state of normal economic activity returns.

Phase 5b — If the corrective measure survives the market selection test, both the new macrogenotype (G_3) and a new macrophenotype (P_3) are established. In typical co-evolutionary fashion, the new macrophenotype now feeds back on and modifies the macrogenotype (G_3). This leads to a perturbed macrogenotype and phenotype of G_3^P and P_3^P respectively. Depending on whether the genotypic potential of the macroeconomy is increased or impaired, the perturbed macrophenotype will either be subjected to the market selection process or bypass it altogether. In both instances, an entirely new macrogenotype (G_4) is established.

Phase 6 — A state of normal economic activity returns until such time as there is further genotypic change. The genotypic potential of the macroeconomy (G_4) is realised in economic activity through the macrophenotype (P_4).

Whether the genotypic potential of the macroeconomy is ultimately restored or increased depends upon the comparison one draws between the eventual new macrogenotype (G_4) and the original macrogenotype (G_1). In comparing the macrogenotypes G_1 and G_4, one of three possibilities will emerge. These are as follows.**

1. The first is where the corrective measure more than offsets the reduction in the genotypic potential of the macroeconomy brought about the initial positive feedback effect. Here G_4 is greater than G_1 such that the macroeconomy's overall genotypic potential is enhanced. This constitutes a movement toward SD.

2. The second is where the corrective measure is sufficient enough to exactly offset the reduction in the macroeconomy's genotypic potential. In this instance, G_4 equals G_1 and the genotypic potential of the macroeconomy is maintained.

3. The final possibility is where the corrective measure insufficiently compensates for the reduction in the macroeconomy's genotypic potential. Here G_4 is less than G_1 such that the macroeconomy's genotypic potential is undermined. Under these circumstances, a nation

---

* In an ecological economic efficiency context, the invention can only pass the market selection test if the rate of low entropy resource throughput is no greater than the maximum sustainable rate (i.e., if $h = h_R + h_{NR} \leq s_R$).

** Note that, in all three cases, a comparison between macrogenotypes G_1 and G_3 must allow for the opportunity cost of the resources utilised in carrying out the positive feedback correcting response.

is forced to accept, for the meantime at least, a less desirable evolutionary pathway of its macroeconomy.

Of course, when comparing G_1 and G_4 it is necessary to assess the aggregate effect on the macrogenotype in terms of the long-run sustainable capacity of the macroeconomy. A corrective measure that merely permits the macroeconomy to maintain its short-term growth potential (e.g., via a throughput-increasing invention) cannot be considered to have genuinely maintained the macrogenotype intact. Clearly, it is insufficient to judge the effectiveness of the corrective measure by comparing G_1 and G_3 because it is demonstrably possible for the recursive feeding back of P_3 on G_3 to undermine the genotypic potential of the macroeconomy (i.e., to have $G_4 < G_1$). All told, this simply means that a positive feedback resulting from the imposition of a corrective measure can more than offset any gains made by the corrective measure itself. As such, the detrimental impact of positive feedback events can be magnified by ill-conceived attempts to maintain the macroeconomy's growth potential rather than its long-run sustainable capacity.

12.3.2 Path-dependency implications for policy implementation

To recall, path dependency occurs when, as a result of the chreodic-like nature of evolutionary pathways, the choice between alternative macrostates is often restricted to a limited range of substitution possibilities. Path dependency implies the possibility of particular elements of a macroeconomy's genotypic potential becoming immune from threats posed by new inventions or any other novel form of genotypic change.

One of the potential implications of a path-dependent set of events is the stultifying effect it can have on policy implementation. For example, human responses to instances of positive feedback cannot always be freely executed. Frequently, the measures which offer the most preferable long-term solution (e.g., quantitative restrictions on the rate of resource throughput) involve a high degree of structural change. Because their implementation conflicts with the identity-preserving nature of the co-evolutionary process, they are invariably difficult to implement in the first place. It is for this reason that undesirable and, in many instances, positive feedback-exacerbating corrective measures (e.g., throughput-increasing innovations) are often implemented at the expense of more desirable measures.

To illustrate the problems that characterise the path-dependent nature of co-evolutionary processes, consider a situation where a "carbon" tax has been proposed to internalise the spillover costs of fossil fuel consumption. Also, consider, Figure 12.8.

Phase 1 — The macroeconomy is exhibiting normal economic activity with the macrogenotype (G_1) being realised through the macrophenotype (P_1).

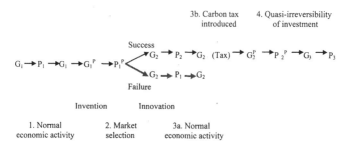

Figure 12.8 The evolutionary economic schema for the macroeconomy — a representation of a path-dependency implication.

Phase 2 — An invention permitting the increased throughput of hydro-carbons (say a technique that enables lower grade and previously inaccessible oil deposits to be exploited) augments the macrogenotype. This results in a perturbed macrogenotype (G_1^P) and, in turn, a perturbed macrophenotype (P_1^P). A new set of relative market prices emerges to reflect the change in the market's price-determining parameters. However, because of the open access resource characteristic of air, there is a disparity between the marginal opportunity costs reflected by the market and that incurred on society as a whole (hence the reason for the negative externality). The market now acts as a filter to determine whether the invention is successful.

Phase 3a — Where the throughput-increasing invention fails the market selection test, the macrogenotype (G_2) subsumes the original macro-genotype (G_1), and the original macrophenotype (P_1) is retained. A state of normal economic activity returns. Failure of the invention is most likely in an ecological economic efficiency context, particularly if the total exaction rate of low entropy is greater than the regenerative capacity of the stock of renewable natural capital (i.e., if $h = h_R + h_{NR} > s_R$). As per sustainability precept No. 1b, it is still possible for the throughput-increasing invention to pass the market selection test should h be greater than s_R but only if the excess quantity of low entropy exacted is offset by the cultivation of a suitable renewable resource substitute and/or by an increase in the renewable natural capital stock/low entropy surplus ratio.

Phase 3b — Where the invention succeeds, the ensuing amount of oil-based products used exceeds the Pareto optimal level.* Having discovered the existence of a negative externality, a carbon tax is introduced to equilibrate the marginal private and social cost of fossil

---

* It is possible for the throughput-increasing invention to pass the market selection test even in an ecological economic efficiency context. To recall from Chapter 7, the mere fact that the rate of resource throughput is limited to the maximum sustainable level does not rule out the possibility of negative externalities in an ecological economic efficiency context nor the ineffi-ciencies associated with them. It merely rules out the potential for an unsustainable rate of resource use.

fuel consumption. This leads to a new set of relative market prices. The new market prices induce a reallocation of resources throughout the macroeconomy and, as a consequence, leads to novel genotypic changes. The macrogenotype is modified and perturbed along with the macrophenotype, such that there now is (G_2^P) and (P_2^P), respectively.

Phase 4 — Because of the quasi-irreversibility of investment and the extent to which macroeconomic systems are locked into petroleum-based energy systems, a long-term preferable pathway is not established. The macrogenotype and macrophenotype (G_3) and (P_3) overwhelmingly contain residuals of (G_2) and (P_2). Because the majority of benefits associated with an inevitable reduction in hydrocarbon use are, in all probability, enjoyed by future people while the structural adjustments costs are largely borne by the present, the establishment of the new macrophenotype (P_3) comes at considerable expense to the present generation. Unless there is a willingness of the present generation to absorb such an expense, there is likely to be considerable opposition to the introduction of a carbon tax. Hence, the tax is unlikely to be imposed, as a recent Australian experience can testify.

In all, this particular example demonstrates the difficulties that often arise whenever policy initiatives are being considered. While, in this instance, a carbon tax scenario has been used to demonstrate a point concerning path-dependency implications, a similar conclusion can be reached by considering, for example, the need for institutional modifications; for variations in nonprice rules; or for a quantitative restriction on the incoming resource flow. Because the path-dependent nature of macroeconomic systems must be considered before any policy initiatives are implemented, its implications for policy setting are further dealt with in Chapter 15.

12.3.3 Externalities from a co-evolutionary perspective

The last important area requiring consideration is that concerning externalities. In a co-evolutionary world, the concept of an externality takes on an additional dimension that is not revealed in a conventional, A-M-based, neoclassical framework. An externality is an indirect benefit or cost conferred to, or imposed upon, one or more parties by the activity of another which is not fully reflected by the market, nor paid to the benefactor (if a benefit) or by the perpetrator (if a cost).

From a co-evolutionary perspective, the shortcomings associated with the mainstream conception and treatment of an externality stem from the fact that externalities are viewed within the confines of a stable set of underlying conditions and parameters consistent with a static, Newtonian perspective. As shown in Figure 12.2, market prices are forever in a state of flux because the constant evolution of price-determining market parameters — brought on by incessant genotypic as well as phenotypic changes —

precludes equilibrium market outcomes. As a number of observers have pointed out (Swaney, 1981; Dragun, 1983; O'Connor, 1989a, 1989b), it is often the evolution of underlying price-determining parameters that result in spill-over benefits and costs. While the market is not necessarily incapable of reflecting spillover benefits and costs as they emerge, market participating agents are generally unable to specify beforehand how and in what way future spillover costs and benefits will result from present human activities. Hence, in the present, markets cannot internalise the future benefits and costs of current actions. Nor, furthermore, can the policy maker always act to internalise these costs and benefits because they, themselves, are as igno-rant of future spillover effects as are market participants. Hence, the bene-factors of spillover benefits and the perpetrators of spillover costs cannot always be rewarded or held responsible for their present actions.

From the point of view of spillover benefits, a positive externality is not so much of a problem. The beneficial activity will take place irrespective of whether a benefactor receives adequate payment for the spillover benefits that subsequently arise — although, from an equity perspective, it is unfair for the benefactor to go unrewarded. It is with respect to costs that, for three main reasons, problems arise. First, unlike benefits, a cost-incurring activity is one which is preferably prevented but, unless the impact is known *a priori*, is still likely to eventuate. Second, because the spillover costs are imposed unilaterally on unsuspecting parties, their imposition is, in most instances, unjust. Injured parties not only bear a cost they would prefer to avoid, they are not suitably compensated for their losses. This is accentuated by the fact that prevailing institutional arrangements — including the nature of prop-erty rights associated with such arrangements — readily facilitate the pro-curement of benefits but, conversely, the off-loading of costs onto other parties and society as a whole (Dragun, 1983). Third, and as the co-evolu-tionary paradigm informs us, cost-inducing activities have the potential to set in motion undesirable evolutionary pathways that a nation may have difficulty extricating itself from (social traps as described previously).

Clearly, in these circumstances, the more usual means of ameliorating externalities, such as the imposition of Pigouvian taxes and subsidies, are rendered impotent. Indeed, externalities of this nature are likely to be diffi-cult if not impossible to avoid. Hence, designing appropriate solutions will require, as O'Connor (1989a, page 34) suggests, "thinking creatively about the manifest *reciprocities* that exist amongst humans in their respective soci-eties, and between human and non-human components of the real world." That is, solutions will need to involve the institutionalisation of solidarity amongst members of society and, as such, the necessary introduction of the "interests of others" into the individual's decision-making process. Even though such solutions will fail to eliminate all negative externalities of the co-evolutionary kind, they will at least reduce the extent to which costs are undesirably off-loaded onto other parties.

In many ways, the institutionalisation of solidarity to acknowledge the existence of reciprocities again reflects the importance of maintaining the market's functional prerequisites (particularly its moral capital foundations), as well as the need to ensure appropriate negative feedback mechanisms are installed so as to bring forward potential future costs into the decision-making domain of the present. Exactly how externalities of this nature might best be ameliorated is discussed and put forward in Chapter 15.

12.4 Concluding remarks

This chapter has attempted to demonstrate the important role played by the market in a co-evolutionary world. By employing an evolutionary economic schema, it has been shown that the market is an important co-evolutionary feedback mechanism. Failure to see the market in this way can lead to a number of potential oversights. These include an inability to come to grips with the dynamic and inherent disequilibrium properties of markets; a failure to recognise the important role played by the market's price-determining parameters in shaping the evolutionary pathway of macroeconomic systems and the global system of which they are a part; an inability to recognise the problem of externalities in all its potential manifestations; and a general failure to see how price-determined outcomes, through the process of recursive feedback, can dramatically influence a nation's ability to attain an optimal macroeconomic scale, particularly in view of the fact that the chreodic-like evolutionary pathways of macroeconomic systems can greatly reduce the range of solutions available to policy makers.

IV

Toward sustainable development

chapter thirteen

Toward sustainable development: a framework for policy setting and national accounting reform

13.1 Policy setting, national accounting, and the optimal macroeconomic scale

So far, this book has focussed on what needs to be considered to move toward SD. Part IV of the book focuses on how to get there. In order to lend structural support to Part IV, this chapter outlines a policy and indicator framework to achieve SD. It does this by refocussing on what we previously described as the primary macroeconomic objective, namely, the attainment of an optimal macroeconomic scale. It is through a reconsideration of an optimal scale that one is best able to prescribe the policies and institutional mechanisms to achieve SD. It also provides the basis for more suitable SD indicators, in particular, an indicator framework to reform the current system of national accounts.

13.1.1 The failure of policy — a failure to embrace the notion of an optimal macroeconomic scale

As explained many times throughout the book, the formulation and implementation of policy has failed to move humankind toward SD for the reason that it is overwhelmingly devoted to maximising the growth of macroeconomic systems. This does not mean that social and ecological policy objectives have been overlooked. But, for two main reasons, they have been largely subsumed by growth-based policies. First, policies devoted to social and ecological objectives run the risk of conflicting with the growth objective. For example, nature conservation is seen as a threat to the access of natural resources and the growth of many resource-based industries, while a strong social welfare system is often regarded as a drain on the resources that could otherwise be used to prevent social problems from emerging in the first place. Second, growth is seen as the panacea for all social and ecological ills — we are forever told that growth makes us richer which, apart from alleviating poverty, enables more resources to be devoted to environmental protection and rehabilitation. Unfortunately, this policy approach fails to recognise that growth and the institutions that facilitate it are the major cause of the social and ecological problems that now plague us.

Because the majority of policies put forward by policy makers are inimical to the concept of an optimal macroeconomic scale, it is necessary to understand some of the co-evolutionary explanations behind the assumed need for continued growth. Knowing what underlies the so-called "growth imperative" brings forth, for example, policies to sever the GDP-employment and GDP-poverty links that presently serve as an "escape hatch" for growth advocates who insist that a call for an eventual nongrowing macroeconomy will lead to intolerably high levels of unemployment and absolute poverty. It also leads to a reassessment of current international trade policy. The co-evolutionary emergence of the World Trade Organisation and the institutionalisation of free trade in the pursuit of global economic growth greatly disadvantages nation states that wish to introduce or maintain an efficient policy of standards setting and spillover cost internalisation. This has made it exceedingly difficult for nations to implement the domestic policies required to move toward SD.

13.1.2 Co-evolutionary explanations for the perceived growth imperative

Why, in the face of increasing evidence that growth is rendering many nations poorer rather than richer (Daly and Cobb, 1989; Max-Neef, 1995; Jackson and Stymne, 1996; Castaneda, 1999; Lawn and Sanders, 1999), does a growth imperative exist? A number of explanations have already been outlined in earlier chapters, for example, the process of "establishment appropriation" that led to the initial emergence of the SD concept, the linear throughput representation of the macroeconomy, and the failure of economists to extend the theme of optimal scale from micro- to macroeconomics. However, it would seem the main reason for the assumed need for growth is that macroeconomic systems have co-evolved along with social and ecological systems over a period in which growth has been the primary social and political objective. In a number of different ways, this has structurally biased macroeconomic systems toward the need for their continued expansion.

First, it has led to a preoccupation with flow concepts such as production, consumption, and GDP, and a minimal emphasis on such stock concepts as capital intactness and the durability and recyclability of physical wealth. It is because of this "flow fetishism" that a considerable incentive exists for the producers of physical commodities to produce illth and cajole people into desiring it, to minimise the durability of human-made capital (because this speeds up the rate of consumption and the need for additional production), and to program the obsolescence and self-destruction of physical commodities. This incentive has been heightened by the fact that domestic markets have becoming increasingly concentrated by the proliferation of corporate mergers. Not only have many markets become supply- rather then demand-driven (with demand being created through persuasive advertising), but the competitive pressure to produce durable, high quality goods has diminished. Of course, it is true that the globalisation of markets through international trade has increased the competitive pressure on domestic producers. However, because international trade is now governed by the law of absolute advantage and the means by which an internationally traded good is produced need only be subject to the workplace and environmental standards of the production location, there is growing pressure to reduce costs in ways that do not reflect improvements in economic efficiency of the ecological economic kind (e.g., lower wages, increased part-time employment, and less stringent environmental standards). This, unfortunately, has only accentuated the perverse growth-based incentives that have long existed in domestic markets.

Second, most contemporary societies fail to be organised on the basis of the performance of duties and obligations. Consequently, private property rights are largely unconditional, that is, they fail to be defined in a way that ensures one's right to use, employ, or exchange either low entropy resources or physical commodities conditional upon the ensuing final goods yielding service to the eventual possessor. This not only promotes quantitative expansion (growth) rather than qualitative improvement (development), it increases the incidence of negative externalities arising from the off-loading of costs onto unsuspecting parties. Indeed, it would seem that the existence of unconditional or "functionless" property rights has long promoted the Lauderdale Paradox — the incentive for property owners to increase their private riches at the expense of public wealth (Lauderdale, 1819).

Third, because humankind has invested so heavily into divesting the stock of nonrenewable natural capital instead of ways of making better use of the renewable resource flow, it has become increasingly reliant or path dependent on nonrenewable resources and throughput-increasing technological progress. As Norgaard (1992) points out, this has led to the co-evolution of two distinct categories of institutional structures and arrangements, both of which reinforce the perceived need for continued growth. The first are those that have evolved to fit the growth-related opportunities provided by the exploitation of a one-off geological bonanza of nonrenewable low entropy. The second category includes those that have more recently evolved to minimise all growth-induced social and ecological damage (i.e., positive feedback correcting mechanisms).

Fourth, in the modern world of finance, money — as both debt and the abstract symbol of exchange value — continues to grow exponentially at the rate of compound interest. Unlike the human-made capital that money is designed to symbolise, debts do not wear out because they are not subject to the Entropy Law. Herein lies one of the factors contributing to the growth imperative. Unless the macroeconomy expands to keep up with the exponential growth of a nation's money supply, hyperinflation and/or widespread debt repudiation follows. Thus, growth is maintained to avoid the potentially catastrophic impact that a nongrowing macroeconomy would have on national and global financial markets.

Fifth, while beyond the optimal scale, the marginal costs of a growing macroeconomy exceed the marginal benefits, the procurement of benefits and the incursion of costs fall disproportionately across all members of a given society. For privileged members, the marginal benefits personally enjoyed can often exceed the marginal costs incurred despite the opposite occurring in the aggregate. For the underprivileged, marginal costs can exceed the marginal benefits even before the macroeconomy reaches the optimum, that is, while the sustainable net benefits from a growing macroeconomy are still rising. Because decision makers are those which predominantly fall into the former category, growth appears unambiguously good and, as a consequence, the policies they esteem reflect their distorted panoramic view of the aggregate.

Last but not least, the national accounting measure of GDP includes, as benefits, such costs as natural capital consumption and social and ecological defence and rehabilitation expenditures. This not only distorts society's view of the net benefits derived from a growing macroeconomic scale, it indicates nothing about whether the resource throughput required to maintain the macroeconomy at its present physical scale can be ecologically sustained. What is more, for the decision makers who see growth as a desirable policy objective, increases in GDP reinforce the policies they promote and prescribe.

13.2 A framework for policy setting and national accounting reform

Table 13.1 reveals a policy framework to achieve SD as well as a list of national accounting reforms to indicate whether a nation is moving closer to the SD goal. At the top of Table 13.1 is the primary macroeconomic objective in achieving SD — an optimal macroeconomic scale. The table is then divided into separate policy setting and indicator sections. On the second level of the policy section are the four intermediate macropolicy goals to attain an optimal scale. These are, of course, a sustainable rate of resource throughput, distributional equity, allocative efficiency, and moral growth. In brackets are the policy instruments pertaining to each of the policy goals. At the next level are the specific policy prescriptions and the institutional mechanisms embodying the respective policy instruments.

Because a sustainable rate of resource throughput requires an adherence to the sustainability precepts established in Chapter 4, the prescribed policy measures are designed to ensure a number of things. The first is to keep the rate of low entropy resource input and high entropy waste output within

Table 13.1 Policy and National Accounting Framework

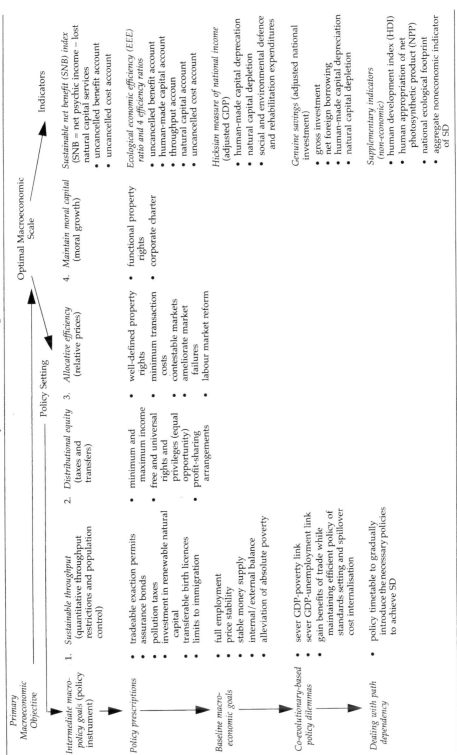

Primary Macroeconomic Objective	1. Sustainable throughput (quantitative throughput restrictions and population control)	*Policy Setting* →		*Optimal Macroeconomic Scale* →	Indicators →
		2. Distributional equity (taxes and transfers)	3. Allocative efficiency (relative prices)	4. Maintain moral capital (moral growth)	
Intermediate macro-policy goals (policy instrument)	Sustainable throughput (quantitative throughput restrictions and population control)	Distributional equity (taxes and transfers)	Allocative efficiency (relative prices)	Maintain moral capital (moral growth)	*Sustainable net benefit (SNB) index* (SNB = net psychic income – lost natural capital services) • uncancelled benefit account • uncancelled cost account
Policy prescriptions	• tradeable exaction permits • assurance bonds • pollution taxes • investment in renewable natural capital • transferable birth licences • limits to immigration	• minimum and maximum income • free and universal rights and privileges (equal opportunity) • profit-sharing arrangements	• well-defined property rights • minimum transaction costs • contestable markets • ameliorate market failures • labour market reform	• functional property rights • corporate charter	*Ecological economic efficiency (EEE) ratio and 4 efficiency ratios* • uncancelled benefit account • human-made capital account • throughput accoun • natural capital account • uncancelled cost account
Baseline macro-economic goals	• full employment • price stability • stable money supply • internal/external balance • alleviation of absolute poverty				*Hicksian measure of national income* (adjusted GDP) • human-made capital depreciation • natural capital depletion • social and environmental defence and rehabilitation expenditures
Co-evolutionary-based policy dilemmas	• sever GDP-poverty link • sever GDP-unemployment link • gain benefits of trade while maintaining efficient policy of standards setting and spillover cost internalisation				*Genuine savings* (adjusted national investment) • gross investment • net foreign borrowing • human-made capital depreciation • natural capital depletion
Dealing with path dependency	• policy timetable to gradually introduce the necessary policies to achieve SD				*Supplementary indicators* (*non-economic*) • human development index (HDI) • human appropriation of net photosynthetic product (NPP) • national ecological footprint • aggregate noneconomic indicator of SD

the ecosphere's regenerative and waste assimilative capacities (tradeable exaction permits, assurance bonds, and pollution taxes). The second is to keep the stock of natural capital intact by replacing depleted nonrenewable resource stocks with cultivated renewable resource substitutes. Finally, the aim of the transferable birth licences and immigration restrictions is to limit a nation's population level to one that can be provided sustainably with a sufficient per capita stock of human-made capital.

To resolve the intermediate goal of distributional equity, policies are prescribed to ensure an ethically permissible range of inequality and to guarantee the free and universal distribution of basic rights, entitlements, and privileges. The latter is required because although taxes and transfers are able to determine minimum and maximum income levels, they do not prevent unjust forms of exploitation and abuse.

The policy measures to achieve allocative efficiency are aimed at improving the effectiveness with which markets allocate the incoming resource flow. While the measures listed are familiar to most economists (well-defined property rights, minimising transaction costs, ameliorating instances of market failure, and making markets as contestable as possible), care must be taken in their design to ensure other macropolicy goals are not undermined. For example, property rights must do more than simply determine who has exclusive ownership and access to property. It is important that they be defined on the basis of the performance of duties and obligations. This, in part, ensures the institutionalisation of a SD domain necessary to facilitate efficient outcomes of the ecological economic kind.

The final macro goal of moral growth is dealt with in Chapter 16. As important as moral growth is, so is the maintenance of any moral capital that is generated. For this reason, two policy initiatives (functional property rights and a corporate charter) are put forward to assist in maintaining the market's moral capital foundations — the very foundations that, as explained in Chapter 9, the market has a propensity to erode.

On the next level of the policy section are a number of baseline macroeconomic goals such as full employment, price stability, internal and external balances, a stable money supply, and the alleviation of absolute poverty. Because SD is promoted by growth only in the early phase of a nation's development process, it is no longer possible to rely on growth-based policies to achieve full employment and to alleviate poverty. Moreover, and given the fact that an exponential increase of the money supply "compels" a nation to expand the macroeconomy to avoid hyperinflation and/or debt repudiation, a greater sense of urgency is required on the part of governments to control the growth of a nation's money supply. For these two reasons alone, a less-than-conventional approach will be required in Chapter 15 to address these concerns.

At the next level are some of the policy dilemmas that have arisen because of the co-evolutionary factors underlying the growth imperative. Two of these policy dilemmas concern the institutionalised GDP-employment and GDP-poverty links that presently render "prohibitive" the transition toward SD through the eventual nongrowth of macroeconomic systems. In the minds of many ecological economists, the severing of these two links constitutes one of the most critical policy-related issues requiring resolution to achieve SD. The other major policy dilemma confronting nations is how, in a globalised economic environment, the benefits of international trade can be gained without compromising the ability to set and maintain an efficient policy of standards setting and cost internalisation. Because this and the fact that problems such as global warming, ozone depletion, and acid rain have an international dimension, the success of policies at the national level are clearly conditional upon nations establishing a consensus on many SD-related issues. Once again, these concerns need to be adequately addressed in Chapter 15.

The bottom level of the policy section serves as a reminder of the need to deal with the conservative, identity-preserving nature of the global system and the path dependency of macroeconomic systems. For reasons explained in Chapter 12, a path-dependent set of events can have a stultifying effect on policy implementation. There is, therefore, a need to gradually introduce the policies required to achieve SD, although the speed of their implementation depends largely on the extent to which moral growth leads to the legitimation of painful policy measures. To deal with the path dependency of macroeconomic systems, a tentatively suggested policy timetable is put forward in Chapter 15.

On the right-hand side of Table 13.1 are a number of suggested indicators to reform the current system of national accounts. Given that SD requires the transition toward an optimal macroeconomic scale, the essential aim of the indicators put forward in Chapter 14 is to replace GDP as the principal policy-guiding index. In order to do this, it is necessary for the economic indicators to reflect a number of things. The first is a measure of sustainable net benefits which, in order to calculate, requires the compilation of individual "benefit" and "cost" accounts. The second is a "strong" sustainability measure of Hicksian income and genuine savings (the latter of which keeps track of a nation's income-generating capital). The last is a measure of the efficiency with which natural capital is transformed into human-made capital, including separate measures to determine how successfully a nation is dealing with each of the ecological economic subproblems.

Of course, economic indicators reveal only so much about the SD performance of a nation. It is, therefore, necessary that they be supplemented with noneconomic indicators. From a development perspective, a number of noneconomic indicators are recommended to reflect the quality of life of a nation's citizens. As for determining whether a nation is operating sustainably, the suggested noneconomic indicators focus on biophysical assessments of natural capital and the ecological impact of a nation's economic activity (e.g., the appropriation of the net primary photosynthetic product and a measure of the national "ecological footprint"). Finally, a method is put forward to combine the development and sustainability indicators in order to assess and rank the SD performance of different countries.

chapter fourteen

Toward sustainable development: sustainable development indicators

14.1 The need for national accounting reform

In order to provide policy guidance and to ascertain whether a nation is moving toward SD, the national accounting system requires urgent reform. In line with the indicator framework put forward in the previous chapter, four major reform measures are proposed. These are

- The construction of a sustainable net benefit (SNB) index; this requires the compilation of currently nonexistent "benefit" and "cost" accounts.
- A "strong" sustainability-based measure of Hicksian income and genuine savings to take account of the glaring deficiencies of GDP; given the difficulties associated with the construction of any measure of sustainable economic welfare, Hicksian income and genuine savings serve as useful backup indicators to the SNB index.
- The calculation of the ecological economic efficiency (EEE) ratio and its four component ratios; this enables one to ascertain the efficiency with which natural capital, and the low entropy matter–energy it provides, is transformed into human-made capital. It also reveals how well a nation is dealing with each of the subproblems that make up the larger ecological economic problem. To calculate all four component ratios, three additional accounts are required, namely, a human-made capital account, a throughput account, and a natural capital account.
- A set of supplementary sustainability and development indicators; supplementary indicators based on noneconomic factors are required given that monetary measures are questionable indicators of sustainable human progress, in particular, the ecological sustainability of economic activity.

Of these four reform measures, the SNB index is calculated for Australia for the period 1966–1967 to 1994–1995 along with the EEE ratio and the four component ratios. The results are revealed and interpreted later in the chapter.

Before discussing the four reform measures, it is appropriate to say something about the policy-guiding value of each of the performance indicators. In terms of economic indicators, it is always better to have a somewhat inaccurate but conceptually sound indicator than a performance indicator that is relatively accurate but conceptually flawed. GDP, the most commonly used economic indicator, belongs to the latter category. A measure of sustainable net benefits belongs to the former. Hicksian income is somewhere in between, although the strong sustainability measure leans more heavily to the former category than its weak sustainability counterpart. However, as a back up to the more contentious SNB index, the strong sustainability measure of Hicksian income is the next best alternative economic indicator of SD.

Noneconomic indicators, particularly physical indicators based on biophysical assessments, are of greatest value to policy makers in ascertaining the ecological impact of human activities. Unlike economic indicators, which, at best, merely count the benefits and costs of economic activity, noneconomic indicators can more reliably determine whether a nation's macroeconomy is approaching or has exceeded its maximum sustainable scale. However, because noneconomic indicators do not count the benefits and costs of growth, they are of little value in determining whether a nation has surpassed its optimal macroeconomic scale. Given that an economic limit to growth is likely to precede a biophysical limit (see again Figure 6.1 where S_* is smaller than S_S), this has important implications for the economic valuation of ecosystem services. Contrary to what some observers believe, it means the economic valuation of natural capital and sacrificed ecosystem services has a crucial role to play in the development of any set of SD indicators.

Finally, it should be pointed out that any aggregate measure or index, whether it be economic or noneconomic, must be treated with caution. Given the so-called "fallacy of composition", the possibility of an undesirable event having a perversely positive influence on any index is always present (Daly, 1979). For this reason alone, policy setting never should be made on the evidence provided by any one particular SD indicator. A range of plausible SD factors should always be taken into account since only then can a relatively accurate assessment be made of a nation's SD performance.

14.2 Reform no.1: A strong sustainability measure of Hicksian income

Reform measure No. 1 involves the establishment of a strong sustainability measure of Hicksian income. To recall, Hicksian income is defined as the monetary value of the goods and services a nation can consume over a given accounting period without undermining its ability to consume the same amount over future accounting periods. The strong sustainability measure, as one of three Hicksian income measures outlined in Chapter 1, requires the maintenance of both natural and human-made capital. While a Hicksian measure of income can never be considered a genuine measure of human development nor, as will be later explained, a genuine measure of income in the Fisherian sense, there are many good reasons for including it as one of the reform measures. First, the correlation between Hicksian income and human welfare is an overwhelmingly positive one (Pearce and Atkinson, 1993). Second, a considerable amount of work has been undertaken by both academic and governmental economists in establishing measures of Hicksian income. Third, the calculation of Hicksian income does not interfere with the current structure of national accounting systems and procedures. Nor does it result in any loss of historical continuity and/or comparability (Daly, 1996).

The means by which a strong measure of Hicksian income can be calculated is best achieved by closely examining the conventional measure of national income, namely, GDP. GDP is the monetary value of the goods and services annually produced by domestically located factors of production (i.e., domestically located human-made and natural capital). GDP is measured in both nominal and real terms. Nominal GDP is the monetary value of the goods and services measured in current prices. Annual variations in nominal GDP result from changes in both prices and the quantity of goods and services produced. Real GDP eliminates the effect that inflation (deflation) has on the nominal value of GDP by measuring the annual production of goods and services in terms of a constant price level.* Consequently, real GDP is a more accurate approximation of the annual variation in the *physical* volume of goods and services produced. It is for this reason that real GDP is used as a guide to a nation's income and its economic welfare.

Given the definition of Hicksian income, the question that needs to be asked is this: Can a nation consume its entire GDP without undermining its capacity to produce and consume the same amount in future years? If the answer is yes, then one can safely use GDP as a measure of Hicksian income. Of course, the answer is no. At least some of the nation's annual GDP must be set aside to replace depreciating and worn out producer goods such as buildings, roads, machinery, and equipment. In addition, a fund must be established in order to respond to undesirable positive feedback events induced by a growing macroeconomy. The fund, which would otherwise finance the consumption of goods and services, is required to

---

* Real GDP is calculated with the use of a GDP deflator. The GDP deflator is a price index equal to the aggregate price level in any year divided by the aggregate price level in a chosen base year. For example, assume that the financial year 1990–1991 is the chosen base year. It would be assigned an index value of 100. Should the aggregate price level be 20 percent higher in 1995–1996, the financial year 1995–1996 would be assigned an index value of 120. The GDP deflator for 1995–1996 would be equal to $100 \div 120 = 0.83$. Hence, if the nominal GDP for 1995–1996 was $100 billion, the real GDP for 1995–1996 measured in 1990–1991 prices would be $83 billion.

finance just some of the following defensive and rehabilitation expenditures (Leipert, 1986):

- Expenditures related to environmental protection activities and environmental damage compensation.
- Expenditures induced by urbanisation and spatial concentration, such as increasing commuting and housing costs.
- Expenditures related to crime prevention and protection against industrial accidents, sabotage, and technical failure.
- Expenditures induced by traffic accidents (e.g., vehicle repairs and medical expenses).
- Expenditures arising from unhealthy living patterns and working conditions.

Finally, in most instances, the production of a nation's GDP involves the unsustainable exploitation of the natural capital that is required to fuel future production and consumption. This depreciation of natural capital does not constitute income but the liquidation of an income-generating capital asset. Yet GDP counts the entire proceeds derived from the depletion of natural capital as if it were income.

14.2.1 Establishing a strong sustainability measure of Hicksian income

To calculate Hicksian income, the following needs to be subtracted from the conventional measure of GDP:

- The depreciation of producer goods (plant, machinery, and equipment). This converts GDP to a measure of Net Domestic Product (NDP).
- The depletion and consumption of natural capital.
- Social and ecological defence and rehabilitation expenditures.

Thus, in its most basic form, Hicksian income is equal to the following (Daly and Cobb, 1989, page 71):

Hicksian income = GDP ± depreciation of producer goods

$$\pm \text{ depletion of natural capital} \pm \text{defence and rehabilitation expenditures}$$

$$(14.1)$$

While the largest possible value for Hicksian income is intuitively desirable, Hicksian income, alone, does not indicate whether the productive capacity of a nation's macroeconomy is being maintained — even if it is increasing over time. This is because Hicksian income approximates the maximum value of the annual product that a country can consume should it be forced to keep its capital stock intact. But what if it chooses not to? Certainly its sustainable productive capacity would eventually decline, as would its Hicksian income. However, this is not always immediately obvious from a current measure of Hicksian income because factors unrelated to the state of its natural and human-made capital, and therefore its sustainable productive capacity, can also cause Hicksian income to fall.

To appreciate whether a nation is maintaining its productive capacity, it is necessary to ascertain measures of the capital stock which, at least in the short run, must be kept intact. To what extent a final Hicksian measure of income can be considered "sustainable" depends on whether a strong or weak sustainability measure is sought. In both instances, the final measure of Hicksian income will be the same. However, it is because of the assumed substitutability of human-made for natural capital in the weak sustainability case that the depletion of natural capital beyond some critical level is of little sustainability concern. So long as the proceeds from natural capital depletion have been sufficiently invested in human-made capital to maintain intact a combined stock of capital, it is assumed that sustainability is being achieved (Hartwick, 1977, 1978). Of course, if a strong sustainability measure of Hicksian income is considered conceptually superior to its weak sustainability counterpart, as ecological economists believe to be the case, it is essential that both human-made and natural capital be kept intact.

Given that a Hicksian measure of income does not keep track of the capital stock, it therefore stands as an incomplete reform measure. Indeed, according to Hamilton (1994), the inability of Hicksian income to reflect the sustainable productive capacity of a nation's macroeconomy means that adjusted or "green" measures of national income do not readily translate into, or suggest, any SD-related policy consequences. To bestow a Hicksian measure of income with a sense of policy relevance, it needs to be accompanied by a measure of *genuine savings*. A measure of genuine savings provides a better indication of the productive capacity of a nation's macroeconomy by keeping account of the capital stock. For instance, a nonnegative measure of genuine savings indicates a constant capital stock and, in theory, the maintenance of a nation's productive capacity.

There are many formulas available for calculating genuine savings (Pearce and Atkinson, 1993; Hamilton, 1994). To capture the full impact of capital stock variations on a nation's productive capacity, a measure of genuine savings must incorporate changes in the waste assimilative capacity and life-support function of natural capital. This is owing to the fact that diminutions in these two functions can severely erode the source capacity of natural capital and, in turn, the productive capacity of a nation's macroeconomy. Hence, as a variation on a genuine savings formula suggested by Pearce et al. (1996, page 90), the following formula is presented:

$$\text{Genuine savings} = Inv \pm NFB \pm d \cdot Kh \pm r \cdot \left(h \pm s_R \right) \pm p \cdot \left(W \pm A \right) + q \cdot \Delta HI \quad (14.2)$$

- *Inv* equals gross investment in human-made producer goods (e.g., plant, machinery, equipment, and labour).
- *NFB* equals net foreign borrowing.
- *d.Kh* equals the depreciation of producer goods (where *d* equals the depreciation rate; and *Kh* equals the stock of producer goods).
- $r \cdot (h - s_R)$ equals the cost of natural capital depletion (where *r* equals the net resource rent; *h* equals the harvest or exaction rate of all low entropy, that is, $h = h_R + h_{NR}$; and s_R equals the low entropy surplus or regeneration rate of renewable natural capital).
- $p \cdot (W - A)$ equals the cost of high entropy waste (where *p* equals the marginal cost of pollution; *W* equals high entropy waste; and *A* equals the ecosphere's waste assimilative capacity).
- $q \cdot \Delta HI$ equals the value of increasing/decreasing life-support services (where *q* equals the marginal value of biodiversity; and ΔHI equals a change in the ecosystem health index, where a negative change in the index denotes a reduction in the ecosphere's life-support services).

Beginning with the weak sustainability perspective, consider the information that can now be provided by a measure of Hicksian income and genuine savings. If genuine savings is nonnegative, it implies that the stock of income-generating capital has been kept intact. Whether a nation's sustainable productive capacity is increasing or diminishing depends on what is happening to Hicksian income. If Hicksian income is rising, the sustainable productive capacity of the country in question is increasing. If Hicksian income is falling, it means that, in keeping the capital stock intact, a country has been forced to reduce the value of the annual product that it had previously been consuming. It also indicates that its sustainable productive capacity is in decline. Naturally, if the measure of genuine savings is negative then even the productive capacity implied by a measure of Hicksian income is unsustainable.

From the strong sustainability perspective, where complementarity between human-made and natural capital is assumed, little of the preceding is particularly relevant. Hicksian income and genuine savings remain useful indicators only if natural capital remains at or above a critical or minimum level. For as soon as the critical level is reached, any further natural capital depletion implies an unsustainable outcome irrespective of the final adjusted measure of Hicksian income or genuine savings (note that Hicksian income can be nonnegative even when natural capital is falling).

In view of the ecological economic preference for a strong sustainability measure of Hicksian income, the following is abundantly clear. Apart from having to ascertain national measures of Hicksian income and genuine savings, it is also necessary to determine the critical or minimum stock of natural capital (i.e., the ecosphere's sustainable carrying capacity), and to keep abreast of the changes in the stock of natural capital to ensure a nation's sustainable carrying capacity has not been undermined. In the light of this, a strong sustainability measure of Hicksian income and genuine savings requires a satellite account to keep track of changes in the stock of natural capital. Of course, to be truly effective, the satellite account must not only include physical measures of natural capital stocks, but also biophysical assessments of the qualitative state or "health" of vital ecosystems.*

14.2.2 A previous attempt at a national measure of Hicksian income

One of the best known and publicised attempts at a Hicksian or green measure of national income was that undertaken by Repetto et al. (1989). In this particular example, adjustments were made to Indonesia's real GDP for the period 1971 to 1984. The extent of the adjustments made by Repetto et al. were not nearly as comprehensive as those recommended in Eq. (14.1). Repetto et al. merely deducted the depreciation of three major elements of Indonesia's natural capital stock — namely, its reserves of oil, timber, and soil assets. Overlooked in relation to Eq. (14.1) were allowances for the depreciation of producer goods and defence and rehabilitation expenditures.

Repetto et al. began their study by first constructing a set of physical resource accounts. These accounts kept track of the physical changes in the three selected natural capital assets and were designed to incorporate and reflect any changes arising from depletion, resource discoveries, and natural regeneration. Having achieved this, Repetto et al. then applied monetary valuations to each asset. These were based on the market or export price for each resource less their respective extraction of harvesting costs. The combined change in the real monetary value of the three resource assets was then subtracted (or added) from real GDP to obtain the adjusted income measures.

Despite the lack of comprehensiveness involved in the adjusted measures of Indonesia's GDP, the final figures provoked considerable interest and debate. Repetto et al. found that while GDP had grown at an annual average of 7.1 percent over the study period, Net Domestic Product (NDP) — defined as a measure of GDP less the depletion of the three natural capital assets — grew by only 4 percent per annum. Although the difference is not particularly large, it is significant in that NDP declined in some years despite GDP increasing over the entire study period. Moreover, early boosts to NDP were largely the result of oil discoveries rather than a deliberate policy of natural capital stock maintenance.

The study of Repetto et al. has not escaped its fair share of criticism (see, for instance, Common, 1990). The criticisms have ranged from the lack of a comprehensive range of GDP adjustments to the valuation techniques adopted. Perhaps the study's most glaring deficiency is its failure to determine whether the stock of natural capital at any stage fell below a critical level. Indeed, no attempt was made to estimate a sustainability threshold or construct a satellite account to monitor the stock of natural capital relative

---

* See Costanza et al. (1992) on the different approaches to assessing the health of ecosystems.

to it. The lack of such an attempt renders this Hicksian income attempt, at best, one of the weak sustainability kind.

Also overlooked by Repetto et al. was a measure of genuine savings to accompany the adjusted measures of national income. Consequently, one cannot ascertain, even from a weak sustainability perspective, whether the sustainable productive capacity of the Indonesian macroeconomy was maintained over the study period. Given that an adjusted measure of national income is likely to be motivated by its potential policy guiding implications, the lack of an accompanying savings measure is what Common (1990) considers to be the study's greatest single weakness.

14.2.3 Problems with Hicksian income — the need for a sustainable net benefit index

While Hicksian income has the advantage of excluding many costs that GDP counts as benefits, it has a number of shortcomings. First, it is arguable whether it is a true measure of income. Like GDP, Hicksian income counts any additions to the stock of human-made capital as current income. Throughout this book, capital has been defined in the Fisherian sense as any physical commodity subject to ownership that is capable of satisfying human needs and wants. Income, on the other hand, has by and large been treated as the service — the net psychic income — enjoyed by the final user or consumer of human-made capital. The significance of Fisher's view of income and capital is that, when applied to national accounting, it avoids the conflation of the value of all services (income) with the value of the human-made capital that yields the service (capital). For example, the value of a car produced this year is not counted as part of this year's income but as an addition to the stock of human-made capital. Only as a car depreciates and wears out through use can the annual value of its depreciation be regarded as income. Consider, therefore, a $20,000 car that takes ten years to "wear out." In the year of the car's manufacture, a $20,000 addition is made to the stock of human-made capital. No income is derived. In the years that follow, and until the car wears out, there is an annual service value or income of $2,000 per year ($20,000 ÷ 10 years). Unfortunately, and contrary to this position, Hicksian income counts the initial $20,000 addition to the stock of human-made capital as current income and subtracts the annual depreciation value of $2,000 in subsequent years.

Admittedly, because Hicksian income involves an adjustment to GDP, it includes as income the annual service value of all consumer goods privately rented. However, it wrongly excludes the services derived from privately owned consumer goods (e.g., household furniture, televisions, and refrigerators) and the stock of publicly provided assets such as art galleries, museums, libraries, and roads and highways. Only the imputed service value of owner-occupied houses is included. Overlooked, also, are the welfare benefits of nonpaid household and volunteer work, and the annual value of natural capital services that appear to have greater welfare significance than any human-made capital (Costanza et al., 1997a).

Second, Fisher's concept of income and capital recognises that the production required to replace worn out human-made capital is a cost. This is because production requires the throughput of matter–energy and, as previously explained, throughput is a cost-inducing physical flow (i.e., it requires natural capital exploitation and leads to the inevitable loss of natural capital services). Thus, in many ways, Hicksian income is a measure of sustainable cost — the maximum cost that can be sustained to keep the stock of human-made capital intact. As useful as this is, it indicates little about the sustainable economic welfare enjoyed by a nation's citizens.

Third, the calculation of Hicksian income assumes that an extra dollar of income received by the rich adds as much to the nation's economic welfare as an extra dollar of income to the poor. As alluded to in Chapter 6, this assumption runs counter to the law of diminishing marginal utility. Of course, a nation's economic welfare would not be maximised by equalising the incomes of its citizens. Nevertheless, an adjustment of some description should be made to reflect the welfare implications of any changes over time in the distribution of income.

Finally, like GDP, Hicksian income fails to include the social cost of unemployment, in particular, long-term unemployment, as well as the potential sustainability implications of a growing foreign debt. Would, for instance, a steady rise in Hicksian income represent an increase in sustainable economic welfare if a growing percentage of the labour force was unemployed or if, in servicing a growing foreign debt, a nation was forced to liquidate its natural capital? The answer is an unambiguous no although, again, one could not rely on Hicksian income to reflect these impacts on the welfare of a nation's citizens.

It is with these shortcomings in mind that the need for a conceptually superior indicator of SD arises. Hence, and without wanting to downplay the usefulness of a strong sustainability measure of Hicksian income, the sustainable net benefit (SNB) index is presented in the following section as a superior single measure of sustainable economic welfare and, thus, of SD.

14.3 Reform no.2: a sustainable net benefit index

There have been various attempts at constructing a single index of sustainable economic welfare. While each specific index has its own peculiar deficiencies — some more so than others — they all have something to contribute to the construction of the SNB index. It is, therefore, a useful exercise to examine some of the better known indexes and the methods used in their calculation.

14.3.1 Nordhaus and Tobin's measure of economic welfare

The first of these indexes is Nordhaus and Tobin's (1972) well known Measure of Economic Welfare (MEW). The MEW was the first genuine attempt at deriving an index of sustainable economic welfare (Singer, 1973). To calculate the MEW, Nordhaus and Tobin made a number of adjustments to the United States' measure of GDP for the period 1929 to 1965. Because economic welfare is more closely correlated to consumption than production, Nordhaus and Tobin began by isolating consumption expenditure from capital investment and intermediate expenditures. Adjustments were then made to account for both capital depreciation and the fact that capital reinvestment is required for per capita consumption to be sustained. The second category of adjustments by Nordhaus and Tobin involved the imputation and addition of the following welfare-related factors capital stock services, leisure activities, and nonmarket productive activity. Finally, Nordhaus and Tobin made some deductions on the basis that urban life involves a number of negative amenities that are normally, albeit falsely, counted as additions to economic welfare (e.g., "regrettable necessities" such as commuting costs, police services, sanitation costs, road and highway maintenance, and national defence expenditures).

Without wishing to downplay the usefulness of the MEW, Daly and Cobb (1989) question whether the list of regrettable expenditures deducted from GDP by Nordhaus and Tobin was sufficiently inclusive. They also criticise Nordhaus and Tobin's interpretation of their own results. For the record, the MEW grew by 1.0 percent over the period 1929 to 1965, whereas NDP — as GDP less capital depreciation — grew at a rate of 1.7 percent over the same period (Nordhaus and Tobin, 1972, Table 18, page 56). This led Nordhaus and Tobin to conclude that existing measures of GDP and NDP correlate sufficiently well with economic welfare to make unnecessary the permanent establishment of a MEW or any alternative indicator.

Daly and Cobb (1989) drew vastly different conclusions from Nordhaus and Tobin's results. By comparing MEW and NDP over specific periods, Daly and Cobb point out that the disparity between the two measures is too large to be ignored. They show that, between 1935 and 1945, per capita GDP increased by almost 90 percent whereas per capita sustainable MEW grew by just 13 percent. Also, between 1947 and 1965 per capita GDP grew by 2.2 percent per year as opposed to a growth rate of 0.4 percent for the per capita sustainable MEW. Significantly, when a number of the subjective assumptions used to impute certain items are altered, per capita sustainable MEW varies markedly over specific periods between 1929 and 1965. Indeed, in some instances, it declines when, at the same time, NDP is clearly rising. This leaves Daly and Cobb believing that the MEW has considerable merit despite its apparent shortcomings. Moreover, Daly and Cobb's interpretation of the study's results suggest that the economic welfare of Americans was not accurately reflected by changes in either GDP or NDP.

14.3.2 Zolotas's index of economic aspects of welfare

The next index worth consideration is an Index of the Economic Aspects of Welfare (IEAW) propounded by Zolotas (1981). The IEAW was calculated for the United States for the years 1950 to 1977. Unlike Nordhaus and Tobin, Zolotas elected to pay little attention to the affects of capital accumulation and none at all to sustainability concerns, although deductions were made for resource depletion and the cost of pollution. This aside, the remainder of the IEAW resembles the MEW quite closely. For instance, imputed values for leisure and household services were added while commuting to work was deducted as a regrettable necessity. In addition, a subtraction was made for expenditures on consumer durables and public buildings to reflect that they represent additions to the stock of human-made capital, not income. In keeping with the Fisherian concept of income and capital, additions were made only for the imputed annual services provided by human-made capital.

As expected, a comparison between the MEW and the IEAW for the United States reveals only a minor disparity between the two indexes (Daly and Cobb, 1989). From 1950 to 1965, the per capita growth of the IEAW was approximately 9 percent (0.57 percent per year). For the MEW, it was 7.5 percent (0.4 percent per year). These compare with an average annual 2.2 percent per capita increase in real GDP over the same period. Over the entire study period, 1950 to 1977, the IEAW consistently rose around one-third as rapidly as GDP. Like the MEW, the IEAW suggests that the economic welfare of Americans did not rise as rapidly as indicated by increases in GDP.

14.3.3 Daly and Cobb's index of sustainable economic welfare

Daly and Cobb's (1989) construction of the Index of Sustainable Economic Welfare (ISEW) was motivated by two main concerns. The first was their belief that many policies required to achieve SD do not get a fair hearing because, although they might increase the sustainable economic welfare of a nation's citizens, they are likely to reduce GDP, and it is GDP that is currently used as a nation's welfare barometer. Their second concern was with the state of economic welfare indexes, including the two just mentioned. These deficiencies left Daly and Cobb firmly believing in the need for a new index, indeed, one built on the strengths of pre-existing measures of economic welfare.

The ISEW that Daly and Cobb subsequently developed was first calculated for the United States for the period 1950 to 1986. As a foundation to the index, Daly and Cobb began with private consumption expenditure rather than GDP or NDP. This figure was first weighted by an index of distributional inequality — the index being a variant on the well known Gini Coefficient. This adjusted measure was further modified by the addition of the following "goods" or benefits

- Unpaid housework (measured at the cost of hiring someone else to do the housework).
- Services provided by consumer durables and publicly funded infrastructure (the latter of which included streets and highways).

- The portion of public health and education expenditures that are non-defensive and, therefore, equivalent to consumption expenditures.
- Net additions to the capital stock above the minimum sustainable requirement.

The ISEW was finally calculated by subtracting a range of social, economic, and environmental "bads" or costs.

- Defensive expenditures (such as medical and other repair bills, commuting costs, expenditures on national advertising, and household expenditures on pollution control devices).
- Costs of urbanisation and noise pollution.
- Water and air pollution expenditures.
- Resource depletion (including the depletion of farmland, wetlands, and nonrenewable resources).
- Long-term environmental damage.
- Automobile accident costs.
- Increases in net international indebtedness.

Daly and Cobb show that in spite of per capita GDP continuing to increase for the majority of the period 1950 to 1986, the per capita ISEW levelled off in 1969, enjoyed a slight peak in 1976, and effectively fell thereafter. A more recent measure of the Genuine Progress Indicator (GPI) — a modified version of the ISEW — shows the downward trend has continued through to 1994 (Redefining Progress, 1995).

14.3.4 *How useful is the index of sustainable economic welfare?*

In the pursuit of policy-guiding indicators, the following needs to be asked. How useful is the ISEW as an indicator of SD? It would seem, if only because of its comprehensiveness, that the ISEW supersedes the MEW and IEAW as the best indicator of sustainable economic welfare. Nonetheless, like all indicators, it does have many shortcomings. Daly and Cobb have highlighted a number of these themselves. The most important of these limitations is its reliance on private consumption expenditure as the reference point for measuring economic welfare. While there is little doubt that the quantity of human-made capital consumed is a better indicator of human well-being than the overall level of production, it is a far from perfect reference point. This is because consumption is, in most cases, a "bad" and should, therefore, be minimised. As indicated previously, the majority of consumer goods are purchased, not with the idea of consumption in mind, but largely for the services they yield. Moreover, for two reasons, an increase in private consumption expenditure cannot be directly equated with a proportionate increase in psychic income. The first is owing to the law of diminishing marginal utility which suggests that as one increases their consumption of physical commodities, the service they enjoy increases at a diminishing rate.

The second is owing to the fact that an increase in the rate at which some individual goods are consumed may not increase the service one enjoys at all. Consider, for example, the lighting of a room by a single light bulb. Is more service experienced if three light bulbs are worn out or "consumed" over one year compared to just one light bulb because the latter is more durable? No, because the total service provided by the three fragile light bulbs is the same as that provided by the more durable light bulb.

Despite this, consumption expenditure may still prove the best available reference point in the estimation of economic welfare. As explained in Chapter 7, people will generally pay a higher price for a commodity embodying superior service-yielding qualities. Consequently, a measure of psychic income can be approximated with the use of market prices. For instance, the rental value of a car, a house, a TV, or a refrigerator — that is, the amount paid to rent durable commodities for a one year period — can be used as a proxy measure of the annual services they yield. The service yielded by commodities entirely consumed during the accounting period in which they are purchased can be valued at their actual market prices (Daly, 1991b).

Unfortunately, the use of market prices to value commodity services is an easier task said than done. Market prices and the rental values of physical commodities are, of course, readily available. However, prices vary for reasons other than from changes in their service-yielding qualities. The price of a commodity is also affected by the relative prices of the different forms of low entropy available to produce it, the actual quantity or supply of the commodity itself, and changes in taxes, the nominal money supply, and the opportunity cost of holding money. Clearly, for prices to remain a proxy indicator of psychic income, it is necessary to eliminate all price-influencing factors other than those related to a commodity's service-yielding qualities. Given that this is an impossible task, there are two choices available. The first option is to leave prices as they are, that is, to rely on current prices. The second is to deflate the nominal annual value of private consumption expenditure by an aggregate price index, such as the Consumer Price Index (CPI). If the former option is chosen, the nominal value of private consumption expenditure will embody unwanted price influences over and above any use value-related influences. If the latter is chosen, one obtains a real value of private consumption expenditure. But, in so doing, one also eliminates the price-influencing effect of varying commodity use values — the very influence that one wants to maintain in order for prices to be used as an approximate measure of psychic income. What, then, is one left to do? The most desirable option is to follow the lead of Daly and Cobb and use, as a reference point, the real value of private consumption expenditure.

This second option is desirable for the following reason. While the law of diminishing marginal utility suggests that an increase in psychic income will be proportionately less than any increase in the quantity of physical commodities consumed, the law is based on the assumption that there is no change in their service-yielding qualities. However, it is reasonable to assume that, through technological progress, the service-yielding qualities of most commodities will continue to increase for some time to come (i.e., increase the service efficiency of human-made capital). If so, this will largely offset the effect of the law of diminishing marginal utility. To what extent it does so, one cannot ascertain, however, it should be sufficient to ensure that any positive impact on psychic income over time is probably well reflected by an increase in real private consumption expenditure.

Another important limitation of the ISEW rests with the fact that, in constructing the index, Daly and Cobb were forced to measure the seemingly unmeasurable and, in their own words, "to make some heroic assumptions" (Daly and Cobb, 1989, page 416). This has resulted in considerable disagreement on how virtually every aspect of the index has been calculated (Lintott, 1996). Accordingly, Ekins and Max-Neef (1992) doubt whether the ISEW will ever gain the widespread support necessary for it or any other similar index to be officially recognised. While, for the moment, such an assessment is probably correct, this may not be so in the long run. Assuming that humankind takes the necessary steps to move toward SD then, for a couple of reasons, something akin to the ISEW should eventually surface as a principal policy-guiding indicator. First, and as mentioned at the beginning of the chapter, a conceptually sound indicator is intuitively preferable to a conceptually flawed performance indicator, such as GDP. Second, the fact that alternative indexes are currently out of favour indicates more about humankind's present dependence on GDP than of the state of alternative indexes. To recall, in a co-evolutionary world, the components of the global system co-evolve such that each component reflects all others. With an economic growth imperative currently prevailing, it is only natural that GDP should predominate as the principal policy-guiding index. But it only does so because it conveniently measures that which the global system is "designed" to accomplish. Clearly, for now, the ISEW will have to wait its turn. But one can be fairly certain that GDP, as a performance indicator, will struggle to maintain a sense of co-evolutionary fitness should a SD pathway ever be established. Conversely, the ISEW or something like it will.

14.3.5 Introducing the sustainable net benefit index

There seems little doubt that the ISEW or its close cousin, the GPI, is the best single economic indicator of SD so far established. However, an SNB index is now presented as a variant of the ISEW. The rationale for its construction is as follows. In the compilation of the ISEW, Daly and Cobb include a range of benefits and costs in a single account and, as such, make no explicit attempt to separate benefits from costs (see Daly and Cobb, 1989, Table A.1, pages 418–419). This, it would seem, is a mistake, even if it reflects no more than an oversight on the part of its proponents. To overcome this weakness, the SNB index is constructed following the compilation of two clearly distinct uncancelled benefit and uncancelled cost accounts. While this approach does not, in the end, alter the final measure of sustainable economic welfare, it is preferable in the sense that it is more consistent with Fisher's concept of income and capital.

14.3.6 Account no.1 — the uncancelled benefit account

The primary aim of the uncancelled benefit account is to calculate a pecuniary measure of net psychic income. The construction of the uncancelled benefit account involves a number of steps. First, there is a need to identify and value the psychic income derived from the use and/or consumption of human-made capital. Second, the disservices experienced as a consequence of transforming natural to human-made capital — the psychic outgo of economic activity — must also be identified and valued. Having achieved both, the total of the latter is subtracted from the former by adhering to the following simple identity:

$$\text{Net psychic income} = \text{psychic income} - \text{psychic outgo} \quad (14.3)$$

The identification of the psychic income and psychic outgo factors of economic activity is made easier by classifying services and disservices according to four modes of human experience (Ekins, 1990). The four modes of experience are as follows:

1. *Having.* The "having" category of human experience includes the services enjoyed from the use and/or consumption of human-made capital produced either at the workplace, on the farm, or in the home.
2. *Being.* This category of human experience includes the impact on one's psychic income of the immediate built and natural environment and of the quality of human capital (i.e., one's education and general skill and aptitude levels). Unlike the "having" mode of experience, the "being" mode involves gains (services) as well as losses (disservices). Services, for example, are gained directly from the life-support amenities provided by the ecosphere, and indirectly in terms of the aesthetic, recreational, and existence values of natural capital. Disservices arise from a polluted and unpleasant built and natural environment.
3. *Doing.* The "doing" category of human experience includes the services and disservices derived from being engaged in production and leisure activities. Services are gained from having made a purposeful contribution to society and from having satisfied individual esteem and self-actualisation needs. Disservices result from the disutility of some forms of work and, as a result of being engaged in the production process, having to forego service-yielding pursuits.
4. *Relating.* The final mode of human experience includes the impact on one's psychic income of a society's institutional and organisational structures and arrangements. The "relating" mode of experience includes the enabling role of the family unit, community-based organisations, and relationships formed at the workplace.

By using the preceding classification, the following psychic income-related items (Items A to L) are used to ascertain a measure of net psychic income for Australia for the period 1966–1967 to 1994–1995. The means by which a monetary valuation is estimated for each item is outlined and discussed in Appendix 1: the Uncancelled Benefit Account for Australia, 1966–1967 to 1994–1995. Their respective values are based on 1989-90 prices.

- Item A — *private consumption (having)*. Private consumption expenditure serves as the initial reference point for the uncancelled benefit account.
- Item B — *index of distributional inequality (having)*. As previously explained, the psychic income enjoyed by a nation as a whole depends very much on the distribution of income. To accommodate this, an index of distributional inequality is used to weight private consumption expenditure.
- Item C — *distributional weighting of private consumption expenditure*. Once the index of distributional inequality is calculated for each year, Item A is weighted to reflect annual changes in the distribution of income.
- Item D — *services yielded by the stock of consumer durables (having)*. Item D imputes the annual services derived from the wearing out or part "consumption" of consumer durables over each accounting period in which they are used (calculated on the basis of an annual depreciation rate of 15 percent per year).
- Item E — *services yielded by the stock of public dwellings (having)*. The services provided by government are, to a large degree, defensive in nature and do not add to a nation's psychic income (Daly and Cobb, 1989). However, some clearly do. These include, for example, the services provided by art galleries, museums, and libraries. The imputed rental (service) rate of public dwellings is assumed to be equal to that of owner-occupied private dwellings.
- Item F — *services yielded by roads and highways (having)*. As per Item E.
- Item G — *services provided by non-paid household labour (having)*. Significant benefits are derived from nonpaid household labour activities such as household cleaning, cooking, child care, household maintenance and renovation, and gardening.
- Item H — *services provided by volunteer labour (having)*. The psychic income enjoyed by a nation also includes the invaluable services provided by volunteer workers.
- Item I — *public expenditure on health and education counted as consumption (having)*. Not all public expenditure on health and education directly benefits a nation's citizens. Much of it can be classed as defensive expenditure. This item includes the estimated portion of public expenditure on health and education that contributes to the improvement of a nation's health and vitality.

- Item J — *net producer goods growth (having)*. A strong measure of sustainability requires the stock of human-made capital to be kept intact as well as the stock of natural capital. Thus, to sustain the net benefits of economic activity, the quantity of producer goods per worker must be maintained. Net producer goods growth constitutes any increase in producer goods above the necessary minimum requirement.
- Item K — *change in net international position (having)*. Item K is included because a nation's long-term capacity to sustain net benefits depends very much on whether capital is domestically or foreign owned.
- Item L — *imputed value of leisure time (doing)*. Leisure is an important component of any nation's well-being. It also one of the more difficult items to calculate. The number of leisure hours was estimated by calculating the average daily number of nonleisure hours (Australians aged fifteen years and above) and deducting it from the number of hours in a day. It was assumed that the real value of leisure time remained constant over the study period.
- Item M — *total psychic income*. Equal to the sum of Items C to L.
- Item M1 — *Australian population*. Included to calculate per capita psychic income.
- Item M2 — *per capita psychic income*. Item M divided by Item M1.

Items N to X constitute the psychic outgo items which, when deducted from Item M, provide a measure of uncancelled benefits (net psychic income) for Australia for the period 1966–1967 to 1994–1995. The means by which a monetary valuation is estimated for each item are explained in Appendix 1: The Uncancelled Benefit Account for Australia, 1966–1967 to 1994–1995. They, too, are based on 1989–1990 prices.

- Item N — *expenditure on consumer durables (having)*. In line with Fisher's concept of income and capital, the portion of private consumption expenditure allocated to the acquisition of consumer durables in any given year must be counted as an addition to the stock of human-made capital in that year. It should not be counted as the value of the psychic income enjoyed in that year which, instead, is experienced in future years as the consumer durables wear out through use. As such, it is necessary to subtract all current expenditure on consumer durables.
- Item O — *defensive private health and education expenditure (having)*. There is a need to subtract the portion of all private education and health expenditures that are clearly defensive and do not add to the psychic income of a nation's citizens.
- Item P — *cost of private vehicle accidents (being)*. The cost of vehicle accidents is an obvious form of expenditure that does not contribute directly to the nation's psychic income. However, the entire cost of

all vehicle accidents is not deducted because much of it is incurred by the business sector. While vehicle damage raises the cost of production and draws resources away from productive activities, it does not directly lower the psychic income of a nation's citizens. It can only lower it indirectly by reducing, for example, the quantity of producer goods and the physical commodities available for consumption. This, however, is captured by other items, such as lower values for Items A and J. Hence, only the cost of private vehicle accidents is deducted.

- Item Q — *cost of noise pollution (being).* Because the "being" mode of experience involves the impact that the quality of the immediate built and natural environment has on one's psychic income, a number of environmental costs need to be deducted. These constitute, in a sense, the disservices or psychic outgo-related costs attributable to the production process. The first of these involves the cost of noise pollution.
- Item R — *direct disamenity cost of air pollution (being).* Along with many other forms of pollution, air pollution costs reflect the loss of the ecosphere's sink function. However, some portion of all air pollution costs includes the direct impact that pollution has on a nation's psychic income. These include the impact of air pollution on urban property values and wages, and urban aesthetics (Freeman, 1982). It is assumed that these two impacts constitute 40 percent of the total cost of air pollution.
- Item S — *cost of unemployment (doing).* In the same way leisure time constitutes a psychic income-related benefit, so unemployment constitutes a psychic outgo-related cost.
- Item T — *cost of underemployment (doing).* As per Item S, underemployment occurs when the number of hours worked by those employed is less than the amount desired.
- Item U — *cost of commuting (doing).* As cities and towns grow, the length of time spent commuting increases thereby reducing the nation's net psychic income.
- Item V — *cost of crime (relating).* As mentioned previously, a number of cost-related expenditures reflect a breakdown in the quality of social, organisational, and institutional structures. The first of these is the cost of crime. The cost of crime is assumed to be the sum of the theft of privately owned property (excluding the cost of crime inflicted upon the business and public sector for reasons explained in the calculation of Item P), and the cost of being confined indoors or to particular places owing to the fear of being a victim of crime.
- Item W — *cost of family breakdown (relating).* Whereas the family contributes positively to the net psychic income of society, a breakdown in the family unit incurs a significant psychic outgo-related cost.
- Item X — *total psychic outgo.* Equal to the sum of Items N to X.
- Item X1 — *Australian population.* Included to calculate per capita psychic outgo.
- Item X2 — *per capita psychic outgo.* Item X divided by Item X1.

- Item AA — *net psychic income.* Equal to total psychic income less total psychic outgo (Item M less Item X).
- Item BB — *Australian population.* Included to calculate per capita net psychic income.
- Item CC — *per capita net psychic income.* Item AA divided by Item BB.

The net psychic income calculated for Australia for the period 1966–1967 to 1994–1995 appears in Table 14.1. Australia's net psychic income, psychic income, and psychic outgo is indicated by Figure 14.1. Both Table 14.1 (column AA) and Figure 14.1 show that Australia's net psychic income generally increased over the study period ($262,604 million in 1966–1967 to $479,482 million in 1994–1995). This, of course, suggests that psychic income (column M) rose more than psychic outgo (column X). The steepest rise in net psychic income occurred during the period 1966–1967 to 1973–1974. Between 1973–1974 and 1986–1987, Australia's net psychic income increased only marginally during which a small decline was experienced between 1975–1976 and 1979–1980. Apart from the recessionary period of 1988–1989 to 1990–1991, net psychic income again increased beyond 1986–1987. However, the rate of such an increase was considerably smaller than that experienced between the years of 1966–1967 to 1973–1974. The lower rate of increase in net psychic income beyond 1973–1974 can be explained by the deterioration, at times, of net producer goods growth, the distribution of income, the net international position, and to three short periods of rapid unemployment growth. Interestingly, apart from the growth in producer goods, neither of these factors has any affect on conventional measures of GDP.

Of greater importance however, was the change in per capita net psychic income (column CC). Following a steady rise after 1966–1967, per capita net psychic income peaked in 1973–1974 at $29,239 per Australian. It then fell quite significantly through to 1979–1980 ($24,986 per Australian). A momentary rise following 1979–1980 was followed by a decline in per capita net psychic income in the 1982–1983 recessionary year. By 1986–1987, per capita net psychic income had recovered to mid 1970s levels where it remained largely unchanged apart from a small decline during the 1989 to 1991 recession. All told, there was only a small increase in per capita net psychic income over the entire study period ($22,256 per Australian to $26,558).

14.3.7 Account no.2 — the uncancelled cost account

The principal objective of the uncancelled cost account is to estimate a pecuniary measure of the natural capital services sacrificed in the process of producing and maintaining the stock of human-made capital. To construct the uncancelled cost account, it is first necessary to ascertain the extent to which the ecosphere's source, sink, and life-support functions have been lost during each accounting period. This involves a number of steps. The first relates to the source function of natural capital. As explained in Chapter 4,

Table 14.1 Uncancelled Benefit Account for Australia, 1966–1967 to 1994–1995

Year	Private Consumption Expenditure A	Index of Distributional Inequality B	Weighted Consumption (A/B × 100) C	Services from Consumer Durables D	Services from Public Dwellings E	Services from Roads and Highways F	Services from Household Labour G	Services from Volunteer Labour H	Public Expend. on Health & Educ. Counted as Consumption I	Net Producer Goods Growth J
1966–1967	93,129	100.0	93,129	6,753	373	1,959	119,111	1,197	2,649	8,973
1967–1968	98,259	100.0	98,259	7,014	396	2,057	133,504	1,353	2,934	8,741
1968–1969	103,414	100.0	103,414	7,357	413	2,164	137,150	1,403	3,390	10,644
1969–1970	109,551	98.8	110,882	7,614	436	2,276	138,300	1,430	4,010	11,178
1970–1971	114,117	97.6	116,923	8,010	469	2,365	150,297	1,572	4,848	13,751
1971–1972	118,793	96.4	123,229	8,303	476	2,491	154,411	1,638	5,606	15,250
1972–1973	125,420	95.2	131,744	8,441	442	2,712	163,423	1,759	5,954	18,411
1973–1974	132,252	93.9	140,843	8,909	415	3,076	179,055	2,139	7,202	24,652
1974–1975	136,920	98.8	138,583	9,463	454	3,254	184,879	2,055	9,492	25,831
1975–1976	141,446	103.6	136,531	9,899	499	3,178	193,485	2,207	10,965	22,076
1976–1977	144,983	108.5	133,625	10,262	551	3,126	197,055	2,303	10,796	18,447
1977–1978	147,548	113.3	130,228	10,453	621	3,151	201,544	2,414	10,916	16,345
1878–1979	152,757	118.2	129,236	10,402	674	3,274	199,508	2,452	10,909	10,933
1979–1980	156,867	119.1	131,719	10,425	692	3,425	199,381	2,514	10,516	6,897
1980–1981	163,277	120.3	135,725	10,504	697	3,613	205,461	2,661	11,071	14,456
1981–1982	170,000	121.2	140,264	10,800	712	3,734	207,278	2,759	11,316	17,403
1982–1983	172,117	122.1	140,964	10,892	773	3,772	208,997	2,918	11,599	13,884
1983–1984	176,500	122.7	143,847	10,861	828	3,759	211,829	3,105	12,644	9,642
1984–1985	182,749	123.6	147,855	11,243	883	3,779	214,577	3,301	14,169	18,343
1985–1986	189,709	124.2	152,745	12,158	959	3,779	214,859	3,464	14,618	12,493
1986–1987	191,420	125.8	152,162	12,768	1,056	3,759	229,602	3,883	14,898	9,258
1987–1988	199,228	127.3	156,503	12,776	1,082	3,726	228,422	4,044	14,928	11,023
1988–1989	208,100	128.8	161,568	12,655	1,079	3,717	227,285	4,214	15,543	14,006
1989–1990	217,428	130.3	166,867	12,874	1,132	3,617	228,809	4,445	15,857	694
1990–1991	218,890	131.5	166,456	13,037	1,222	3,797	232,672	4,771	17,073	1,159
1991–1992	224,704	132.7	169,332	13,186	1,292	3,824	238,871	5,104	17,921	–658
1992–1993	230,762	133.9	172,339	13,644	1,344	3,821	243,207	5,454	18,542	–1,862
1993–1994	236,642	135.2	175,031	13,117	1,347	3,826	248,357	5,847	19,054	–4,521
1993–1995	248,722	136.4	182,348	14,589	1,367	3,833	249,720	6,174	19,537	–1,403

Table 14.1 Uncancelled Benefit Account for Australia, 1966–1967 to 1994–1995 (Continued)

Year	Change in Net International Position K	Imputed Value of Leisure Time L	Total Psychic Income (C to L) M	Australian Population (Thousands) M1	Per Capita Psychic Income (M/M1) M2	Expenditure on Consumer Durables N	Private Defense Expend. on Health and Educ. O	Cost of Private Vehicle Accidents P	Cost of Noise Pollution Q	Direct Disamenity Cost of Air Pollution R	Cost of Unemploym. S
1966–1967	0	50,295	284,439	11,799	24,107	-7,208	-832	-1,177	-694	-1,676	-6,666
1967–1968	-1,713	52,445	304,990	12,009	25,397	-7,581	-1,069	-1,160	-719	-1,738	-5,919
1968–1969	-2,336	54,355	317,954	12,263	25,928	-7,982	-1,267	-1,176	-782	-1,892	-5,671
1969–1970	-489	55,084	330,721	12,507	26,443	-8,424	-1,470	-1,210	-826	-1,996	-5,606
1970–1971	-234	60,792	358,793	13,067	27,458	-8,721	-1,791	-1,224	-866	-2,094	-6,816
1971–1972	4,704	63,424	379,532	13,304	28,528	-9,277	-2,169	-1,217	-901	-2,177	-10,435
1972–1973	6,928	65,251	405,065	13,505	29,994	-10,121	-2,333	-1,201	-935	-2,260	-6,848
1973–1974	1,181	68,102	435,574	13,723	31,740	-11,780	-2,256	-1,213	-949	-2,285	-9,901
1974–1975	-325	70,653	444,339	13,893	31,983	-12,707	-2,348	-1,223	-955	-2,297	-19,153
1975–1976	1,735	72,636	453,211	14,033	32,296	-13,388	-2,542	-1,219	-964	-2,315	-20,493
1976–1977	-3,286	74,436	447,315	14,192	31,519	-13,300	-2,639	-1,201	-793	-2,330	-24,550
1977–1978	-4,977	76,857	447,552	14,359	31,169	-12,433	-2,724	-1,188	-977	-2,336	-28,368
1978–1979	-4,411	78,785	441,762	14,516	30,433	-11,983	-3,035	-1,230	-995	-2,369	-26,954
1979–1980	-21,665	80,216	424,111	14,695	28,861	-11,894	-2,896	-1,324	-1,001	-2,381	-28,464
1980–1981	-7,140	83,032	460,080	14,923	30,830	-12,634	-3,206	-1,443	-1,014	-2,357	-27,358
1981–1982	-4,395	87,396	477,267	15,184	31,432	-13,178	-3,607	-1,588	-1,022	-2,357	-32,834
1982–1983	-9,332	90,854	475,321	15,394	30,877	-12,903	-4,095	-1,730	-1,017	-2,357	-49,855
1983–1984	-4,972	92,400	483,943	15,579	31,064	-13,337	-4,394	-1,854	-1,037	-2,357	-43,462
1984–1985	-27,663	95,517	482,004	15,788	30,530	-13,708	-4,448	-1,967	-1,054	-2,357	-40,245
1985–1986	-18,973	97,284	493,386	16,018	30,802	-14,096	-4,710	-2,040	-1,067	-2,356	-41,409
1986–1987	-16,210	100,061	511,237	16,264	31,434	-14,702	-5,252	-2,051	-1,077	-2,333	-41,878
1987–1988	79	101,256	533,839	16,532	32,291	-14,702	-5,698	-2,015	-1,094	-2,309	-38,163
1988–1989	-15,739	102,537	526,865	16,814	31,335	-15,026	-6,285	-2,028	-1,111	-2,287	-32,738
1989–1990	-9,871	105,985	530,409	17,065	31,082	-15,373	-6,710	-2,083	-1,117	-2,264	-41,953
1990–1991	-15,532	110,810	535,645	17,284	30,991	-14,550	-7,428	-2,072	-1,116	-2,242	-57,655
1991–1992	-4,961	113,042	556,953	17,489	31,846	-14,877	-8,174	-2,053	-1,117	-2,219	-65,029
1992–1993	-14,189	113,981	556,281	17,656	31,507	-15,301	-8,638	-2,064	-1,123	-2,197	-66,699
1993–1994	-17,660	113,638	559,036	17,838	31,340	-15,727	-9,110	-2,063	-1,132	-2,176	-57,856
1994–1995	-13,011	114,715	577,869	18,054	32,008	-16,276	-9,519	-2,109	-1,140	-2,154	-51,139

Table 14.1 Uncancelled Benefit Account for Australia, 1966-1967 to 1994-1995 (Continued)

Year	Cost of Under-Employment T	Cost of Commuting U	Cost of Crime V	Cost of Family Breakdown W	Total Psychic Outgo (N to W) X	Australian Population (Thousands) X1	Per Capita Psychic Outgo (X/X1) X2	Net Psychic Income (M − X) ($m at 1989–1990 Prices) AA	Australian Population (Thousands) BB	Per Capita Net Psychic Income ($ at 1989–1990 Prices) (AA/BB) CC
1966–1967	−48	−1,458	−1,821	−255	−21,835	11,799	−1,851	262,604	11,799	22,256
1967–1968	−44	−1,563	−1,975	−267	−22,035	12,009	−1,835	282,955	12,009	23,562
1968–1969	−42	−1,702	−2,175	−283	−22,972	12,263	−1,873	294,982	12,263	24,055
1969–1970	−45	−1,811	−2,405	−302	−24,095	12,507	−1,927	306,626	12,507	24,516
1970–1971	−50	−1,930	−2,720	−328	−26,540	13,067	−2,031	332,253	13,067	25,427
1971–1972	−57	−2,017	−3,121	−365	−31,736	13,304	−2,385	347,796	13,304	26,142
1972–1973	−45	−2,008	−3,185	−408	−29,344	13,505	−2,173	375,721	13,505	27,821
1973–1974	−67	−2,937	−3,406	−433	−34,327	13,723	−2,501	401,247	13,723	29,239
1974–1975	−81	−2,035	−3,860	−535	−45,194	13,893	−3,253	399,145	13,893	28,730
1975–1976	−98	−2,153	−3,995	−897	−48,064	14,033	−3,425	405,147	14,033	28,871
1976–1977	−126	−2,095	−4,195	−1,158	−52,567	14,192	−3,704	394,748	14,192	27,815
1977–1978	−179	−2,146	−4,638	−1,072	−56,061	14,359	−3,904	391,491	14,359	27,264
1978–1979	−175	−2,271	−4,947	−981	−54,940	14,516	−3,785	386,822	14,516	26,648
1979–1980	−203	−2,528	−5,280	−964	−56,935	14,695	−3,874	367,176	14,695	24,986
1980–1981	−186	−2,622	−5,676	−988	−57,484	14,923	−3,852	402,596	14,923	26,978
1981–1982	−241	−2,673	−6,079	−1,047	−64,626	15,184	−4,256	412,641	15,184	27,176
1982–1983	−277	−2,860	−6,648	−1,073	−82,815	15,394	−5,380	392,506	15,394	25,497
1983–1984	−253	−2,895	−6,879	−1,040	−77,508	15,589	−4,975	406,435	15,589	26,089
1984–1985	−242	−3,047	−7,096	−995	−75,159	15,788	−4,761	406,845	15,788	25,769
1985–1986	−283	−3,114	−7,751	−951	−77,840	16,018	−4,860	415,546	16,018	25,942
1986–1987	−327	−3,242	−8,160	−950	−79,366	16,264	−4,880	431,871	16,264	26,554
1987–1988	−293	−3,348	−8,050	−969	−76,641	16,532	−4,636	457,198	16,532	27,655
1988–1989	−334	−3,379	−8,453	−989	−72,630	16,814	−4,320	454,235	16,814	27,015
1989–1990	−388	−3,594	−8,813	−1,008	−83,303	17,065	−4,882	447,106	17,065	26,200
1880–1991	−533	−3,957	−9,056	−1,059	−99,668	17,284	−5,766	435,977	17,284	25,224
1991–1992	−628	−4,020	−9,311	−1,096	−108,524	17,489	−6,205	448,429	17,489	25,641
1992–1993	−610	−3,976	−9,358	−1,128	−111,094	17,656	−6,292	445,187	17,656	25,214
1993–1994	−584	−4,047	−9,544	−1,159	−103,398	17,838	−5,797	455,638	17,838	25,543
1994–1995	−603	−4,142	−10,130	−1,175	−98,387	18,054	−5,450	479,482	18,054	26,558

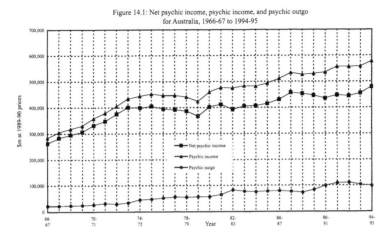

Figure 14.1 Net psychic income, psychic income, and psychic outgo for Australia, 1966–1967 to 1994–1995.

the ecosphere's source function depends on the stock of renewable and nonrenewable natural capital. However, the contribution of each differs markedly. While renewable low entropy can be harvested with minimal effect on both the renewable natural capital stock itself or the low entropy surplus it generates, any exaction of nonrenewable low entropy diminishes the source function by an amount equal to the quantity of low entropy harvested.

In keeping with sustainability precept No. 1, the ecosphere's source function is sacrificed, to some degree, if the total harvest rate of low entropy ($h = h_R + h_{NR}$) is greater than the natural regenerative capacity of the stock of renewable natural capital (s_R), that is, if $h = h_R + h_{NR} > s_R$. However, as outlined in sustainability precept No. 1b, even if the ecosphere's regenerative capacity has been exceeded, the source function can still be maintained provided nonrenewable low entropy is exploited at a rate equal to the creation of suitable renewable resource substitutes and/or at a rate equal to an increase in the renewable natural capital/low entropy surplus ratio equal to the difference between the rate at which all natural capital is harvested and the regenerative capacity of renewable natural capital.

Thus, in order to deduce the extent to which the source function has been sacrificed, it is necessary to determine

a. The rate at which both renewable and nonrenewable low entropy has been harvested, that is, $h = h_R + h_{NR}$.
b. The renewable low entropy surplus (s_R).
c. The extent to which renewable resources have been created or cultivated to offset the depletion of nonrenewable low entropy.
d. Movements in the renewable natural capital/low entropy surplus ratio.

Having achieved this, the extent to which the ecosphere's source function has been lost requires a comparison between (a) and (b) with allowances made for variations in (c) and (d).

With this in mind, the items listed next are used to ascertain a measure of the sacrificed source function of natural capital. The means by which a monetary valuation is estimated for each item are discussed in Appendix 2: the Uncancelled Cost Account for Australia, 1966–1967 to 1994–1995. The values of all items are based on 1989–1990 prices.

- Item 1 — *user (depletion) cost of metallic mineral stocks.* As previously explained, the exaction of nonrenewable resources constitutes a direct decline in the source function of natural capital. The method used to calculate the user cost of nonrenewable resources is a variation on that suggested by El Serafy (Lawn, 1998). The first of these resources concerns the cost of metallic mineral stock depletion.
- Item 2 — *user (depletion) cost of coal stocks. As per Item 1.*
- Item 3 — *user (depletion) cost of oil and gas stocks. As per Items 1 and 2.*
- Item 4 — *user (depletion) cost of nonmetallic mineral stocks.* As per Items 1, 2, and 3.
- Item 4A — *user (depletion) cost of nonrenewable resources.* Item 4A merely sums the total user cost of depleting the stock of all nonrenewable natural capital.
- Item 5 — *cost of lost agricultural land.* As Daly and Cobb (1989) show, the productivity of farmland is essentially reduced in two ways. The first is through urban expansion where, in some cases, the most productive arable farmland is converted into outlying city or town suburbs. The second is through productivity losses resulting from soil erosion, compaction, acidification, increasing salinity, and soil structure decline.
- Item 6 — *user cost of timber resources.* As a renewable resource, the impact of timber harvesting on the ecosphere's source function depends critically on the rate at which it is harvested (h_{timber}); its regenerative rate (s_{timber}); and the extent to which additional timber stocks have been cultivated.
- Item 7 — *user cost of fishery resources.* As per Item 6, fishery stocks included the stock of prawns, rock lobster, abalone, scallops, oysters, fish (excluding tuna), and tuna.
- Item 8 — *cost of lost or degraded wetlands, mangroves and saltmarshes.* While wetlands serve as a vital component of the ecosphere's capacity to assimilative waste, the cost of sacrificed wetlands is borne disproportionately in terms of its indirect impact on a variety of source-related functions. For this reason, the cost of lost wetlands is included as a sacrificed source item.
- Item 9 — *cost of sacrificed source function of natural capital. Equal to the sum of Items 4A to 8.*

The second major step involved in the compilation of the uncancelled cost account relates to the sink function of natural capital. As per sustainability precept No. 2, the ecosphere's sink function is sacrificed, to some extent, when the quantity and qualitative nature of high entropy waste (W) exceeds the ecosphere's waste assimilative capacity (A). With a good deal of reliability, one can claim that the ecosphere's waste assimilative capacity has been exceeded whenever the emission of pollution imposes discernible environmental costs. Hence, the following items have been chosen to estimate the loss of the ecosphere's sink function:

- Item 10 — *cost of water pollution.* The cost arising from deteriorating water quality is generally the result of one of the following: damages attributed to point source discharges from the household, the factory, and the farm, damages owing to siltation, and damages owing to rising groundwater salinity (Daly and Cobb, 1989, pages 425–429).
- Item 11 — *cost of air pollution.* Air pollution costs pertaining to the loss of the ecosphere's sink function include damage to agricultural vegetation, materials damage, the cost of cleaning soiled articles, and damage from acid rain (Freeman, 1982). These four categories of air pollution are assumed to constitute 60 percent of the total cost of air pollution. The remaining 40 percent of the total is counted as a psychic outgo-related cost (Item R).
- Item 12 — *cost of solid waste pollution.* Because the production of human-made capital involves the generation of hazardous and non-biodegradable substances, solid waste pollution has a tendency to contaminate soil, watercourses, and groundwater aquifers.
- Item 13 — *cost of ozone depletion.* The cost of ozone depletion is equivalent to the amount needed to compensate future generations for the accumulated effects of ozone-depleting substances in the Earth's atmosphere.
- Item 14 — *cost of sacrificed sink function of natural capital.* Equal to the sum of Items 10 to 13.

The completion of the uncancelled cost account requires the estimation of two important aspects related to the ecosphere's life-support function. The first of these relates to the cost of long-term environmental damage. The second involves the calculation of an Ecosystem Health Index. The index is used to weight the running total of uncancelled costs. Consequently, the uncancelled cost account is finalised by the following items.

- Item 15 — *cost of long-term environmental damage.* To a considerable degree, long-term environmental damage occurs as a result of the diminishing source and sink functions of natural capital. However, it also results from diminishing life-support services brought on by the excessive use of fossil fuels. To calculate Item 13, Australia's annual consumption of energy was converted to a crude oil barrel equivalent and multiplied by a cost value of $2.50 per barrel. Like Item 12, the cumulative figure is equivalent to the amount needed to compensate future generations for the excessive use of energy resources.

- Item 16 — *total unweighted uncancelled costs.*   Sum of Items 9, 14, and 15.
- Item 17 — *ecosystem health index.*   Many sacrificed life-support services are exceedingly difficult to measure. To overcome this, the preceding running total of uncancelled costs (Item 16) is weighted by an index reflecting the health of Australia's ecosystems. This is necessary given that the items included to account for lost source and sink services do not account for impacts on the life-support function of natural capital. For example, mining not only diminishes the source function of natural capital, it also reduces its life-support function by degrading ecosystems.
- Item 18 — *total weighted uncancelled costs.*   This item is simply Item 16 weighted by the ecosystem health index (Item 17).
- Item 19 — *Australian population.*   Included to calculate per capita uncancelled costs.
- Item 20 — *per capita uncancelled costs.*   Item 18 divided by Item 19.

The estimation of the total weighted uncancelled costs for Australia for the period 1966–1967 to 1994–1995 appears in Table 14.2 (column 18). Figure 14.2 shows Australia's total weighted uncancelled costs and the breakup of the total into the costs arising from the sacrificed source and sink functions of natural capital, and the cost of long-term environmental damage. Overall, the total weighted uncancelled costs for Australia rose throughout the entire study period ($108,941 million in 1966–1967 to $247,456 million in 1994–1995). Similarly, per capita uncancelled costs (column 20) increased in all but the financial year 1970–1971 ($9,233 in 1966–1967 to $13,706 in 1994–1995 per Australian). Both Table 14.2 and Figure 14.2 indicate that the transformation of natural to human-made capital is coming at the growing expense of the sacrificed source, sink, and life-support services of natural capital.

Of the items included in the uncancelled cost account, the biggest rises occurred in the user cost of nonrenewable resources, the loss of agricultural land, and in the cost of ozone depletion and long-term environmental damage. While pollution costs generally increased over the study period, such increases were moderated by improvements in waste treatment, pollution abatement technologies, and tighter legislative controls on waste emissions. Owing largely to extensive native vegetation clearance in the states of Queensland and New South Wales (about one-third of the Australian continent), the health of Australia's ecosystems deteriorated over the study period despite a small net increase in timber stocks. The ecosystem health index constantly fell from a value of 100.0 in 1966–1967 to a value of 90.0 by the year 1994–1995.

Table 14.2 Uncancelled Cost Account for Australia, 1966–1967 to 1994–1995

Year	User Cost of Metallic Minerals 1	User Cost of Coal 2	User Cost of Oil and Gas 3	User Cost of Non-Metallic Minerals 4	User Cost of Non-Renewable Resources 4A	Loss of Agricultural Land 5	User Cost of Timber Resources 6	User Cost of Fishery Resources 7	Loss and/or Degradation of Wetlands 8	Sacrificed Source Function 9	Cost of Water Pollution 10
1966-1967	1,377.0	671.2	79.0	459.4	2,587	57,515	317	88	24,047	84,554	3,801
1967-1968	1,680.6	755.0	141.1	480.1	3,057	58,678	317	96	24,431	86,579	3,943
1968-1969	1,954.7	762.4	140.7	545.3	3,403	60,088	317	103	24,816	88,727	4,293
1969-1970	2,530.9	874.0	289.0	615.3	4,309	61,444	317	111	25,201	91,382	4,528
1970-1971	2,651.3	981.3	679.6	650.9	4,963	63,002	-575	121	25,585	93,096	4,750
1971-1972	2,694.2	1,062.7	795.0	681.9	5,234	64,450	-575	129	25,969	95,207	4,939
1972-1973	2,681.3	1,144.4	840.2	717.6	5,384	65,768	-575	139	26,354	97,070	5,127
1973-1974	3,065.0	1,167.2	905.7	747.7	5,886	67,158	29	136	26,739	99,948	5,213
1974-1975	3,153.8	1,858.1	894.7	718.6	6,625	68,587	29	115	27,123	102,479	5,253
1975-1976	2,881.9	2,186.0	839.7	697.5	6,605	70,063	28	111	27,508	104,315	5,313
1976-1977	3,070.6	2,332.5	836.6	712.3	6,952	71,565	73	120	27,892	106,602	5,368
1977-1978	2,964.4	2,386.7	966.1	738.8	7,056	72,928	-187	125	28,276	108,198	5,388
1978-1979	3,195.2	2,324.2	1,220.7	754.4	7,495	74,620	537	132	28,662	111,446	5,502
1979-1980	4,074.3	2,228.2	1,413.1	840.8	8,556	76,228	354	141	29,046	114,325	5,541
1980-1981	3,459.7	2,739.6	1,740.0	854.1	8,793	77,719	71	159	29,431	116,173	5,615
1981-1982	3,245.2	3,007.8	1,933.8	869.1	9,056	79,387	122	169	29,815	118,549	5,664
1982-1983	3,410.5	3,282.2	2,035.9	810.6	9,542	80,821	-13	186	30,045	120,581	5,630
1983-1984	3,184.5	3,100.4	2,523.2	876.7	9,685	82,576	20	177	30,275	122,733	5,715
1984-1985	3,684.7	3,621.0	3,208.1	927.8	11,442	84,360	-3	152	30,507	126,458	5,786
1985-1986	3,910.8	4,013.1	2,879.2	1,040.4	11,844	86,107	107	137	30,737	128,932	5,840
1986-1987	4,408.4	3,719.5	2,550.2	1,071.2	11,749	87,926	156	134	30,967	130,932	5,877
1987-1988	5,045.3	3,118.2	2,665.3	1,001.7	11,831	89,725	95	138	31,198	132,987	5,949
1988-1989	n.a.	n.a.	n.a.	n.a.	11,795	91,543	62	143	31,428	134,971	6,020
1989-1990	n.a.	n.a.	n.a.	n.a.	14,634	93,370	10	138	31,658	139,810	6,066
1990-1991	n.a.	n.a.	n.a.	n.a.	14,724	95,270	28	169	31,889	142,080	6,054
1991-1992	n.a.	n.a.	n.a.	n.a.	14,552	97,068	-61	189	21,119	143,867	6,060
1992-1993	n.a.	n.a.	n.a.	n.a.	16,347	98,940	54	184	32,349	147,874	6,087
1993-1994	n.a.	n.a.	n.a.	n.a.	15,180	100,884	-19	159	32,579	148,783	6,123
1994-1995	n.a.	n.a.	n.a.	n.a.	15,099	102,592	3	150	32,810	150,654	6,154

n.a. denotes not available.

Table 14.2 Uncancelled Cost Account for Australia, 1966–1967 to 1994–1995 (Continued)

Year	Cost of Air Pollution 11	Cost of Solid Waste Pollution 12	Cost of Ozone Depletion 13	Sacrificed Sink Function (10 to 13) 14	Cost of Long-Term Environmental Damage 15	Total Uncancelled Costs (Unweighted) 16	Ecosystem Health Index 17	Total Uncancelled Costs (Weighted) 18	Australian Population (Thousands) 19	Per Capita Uncancelled Costs (Weighted) 20
1966–1967	2,514	2,322	2,160	10,797	13,590	108,941	100.0	108,941	11,799	9,233
1967–1968	2,607	2,389	2,490	11,429	14,211	112,219	99.5	112,783	12,009	9,392
1968–1969	2,838	2,453	2,850	12,434	14,873	116,034	98.9	117,325	12,263	9,567
1969–1970	2,904	2,539	3,263	13,324	15,572	120,278	98.4	122,234	12,507	9,773
1970–1971	3,141	2,658	3,660	14,209	16,294	123,599	97.8	126,379	13,067	9,672
1971–1972	3,266	2,679	4,208	15,092	17,056	127,355	97.3	130,889	13,304	9,838
1972–1973	3,390	2,740	4,755	16,012	17,856	130,938	96.7	135,406	13,505	10,026
1973–1974	3,428	2,801	5,378	16,820	18,711	135,479	96.2	140,830	13,723	10,262
1974–1975	3,446	2,908	6,038	17,645	19,592	139,716	95.7	145,994	13,893	10,508
1975–1976	3,472	2,908	6,600	18,293	20,484	143,092	95.1	150,465	14,033	10,722
1976–1977	3,496	2,963	7,200	19,027	21,434	147,063	94.6	155,458	14,192	10,954
1977–1978	3,504	3,018	7,770	19,680	22,409	150,287	94.0	159,879	14,359	11,134
1978–1979	3,554	3,076	8,318	20,450	23,406	155,301	93.5	166,098	14,516	11,442
1979–1980	3,571	3,137	8,843	21,092	24,429	159,846	93.0	171,878	14,695	11,696
1980–1981	3,535	3,204	9,353	21,707	25,457	163,337	92.4	176,772	14,923	11,846
1981–1982	3,535	3,289	9,870	22,358	26,515	167,422	91.9	182,179	15,184	11,998
1982–1983	3,535	3,356	10,350	22,871	27,536	170,989	91.7	186,465	15,394	12,113
1983–1984	3,535	3,424	10,875	23,549	28,588	174,869	91.6	190,905	15,579	12,254
1984–1985	3,535	3,494	11,438	24,253	29,689	180,399	91.5	197,158	15,788	12,488
1985–1986	3,534	3,573	12,000	14,947	30,801	184,680	91.3	202,278	16,018	12,628
1986–1987	3,499	3,655	12,608	25,639	31,950	188,522	91.2	206,712	16,264	12,710
1987–1988	3,464	3,741	13,260	26,414	33,134	192,534	91.0	211,576	16,532	12,798
1988–1989	3,430	3,832	13,898	27,180	34,386	196,537	90.9	216,213	16,814	12,859
1989–1990	3,396	3,918	14,453	27,833	35,675	203,318	90.7	224,165	17,065	13,136
1990–1991	3,362	3,997	14,828	28,241.	36,965	207,286	90.6	228,793	17,284	13,237
1991–1992	3,329	4,070	15,180	28,639.	38,273	210,779	90.4	233,163	17,489	13,332
1992–1993	3,296.	4,144	15,503	29,030	39,606	216,510	90.3	239,767	17,656	13,580
1993–1994	3,263	4,214	15,795	29,395.	40,970	219,148	90.1	243,227	17,838	13,635
1994–1995	3,231	4,272	16,035	29,692	42,365	222,711	90.0	247,456	18,054	13,706

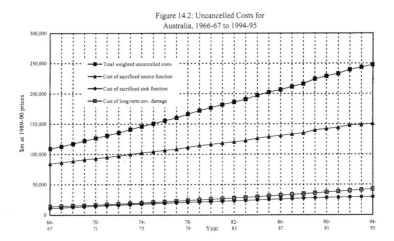

Figure 14.2   Uncancelled costs for Australia, 1966–1967 to 1994–1995.

14.3.8 *Calculating the sustainable net benefit index*

Having estimated the pecuniary value of uncancelled benefits and costs, the SNB index for Australia is calculated by subtracting the latter from the former. Both an aggregate and per capita measure of the SNB index for the period 1966–1967 to 1994–1995 appear in Table 14.3. In Figure 14.3, a comparison is drawn between Australia's per capita SNB and per capita real GDP.

Table 14.3 shows that the SNB index increased markedly from its 1966–1967 value of $153,663 million before peaking in 1973–1974 at $260,417 million (column c). It then steadily declined to a value of $195,298 million by 1979–1980. The index recovered slightly in 1980–1981 but continued to fluctuate up to 1986–1987. In 1987–1988 there was a sharp increase in the SNB index. The index then proceeded to gradually but marginally decline to 1992–1993. There was a very small increase in the SNB index in the last two years of the study period. By 1994–1995, the SNB index was less than its 1973–1974 peak ($232,026 million in 1994–1995 as opposed to $260,417 million in 1973–1974).

As for the per capita measure (column e), the trend is similar to the aggregate index but, given the importance of per capita measures, is of greater interest. The per capita SNB index began in 1966–1967 at $13,023. While the per capita SNB index rose sharply before peaking in 1973–1974 at $18,977, it eventually declined to $12,852 by 1994–1995 — less than its initial value. Both the aggregate and per capita SNB index indicate the strong likelihood of the Australian macroeconomy having exceeded its optimal scale, if not, given the length of time the SNB index has been in decline, its maximum sustainable scale.

Table 14.3   Sustainable Net Benefit (SNB) index for Australia, 1966–1967 to 1994–1995

Year	Uncancelled Benefits ($m at 1989–1990 Prices) a	Uncancelled Costs ($m at 1989–1990 Prices) b	SNB Index (a − b) ($m at 1989–1990 Prices) c	Australian Population (Thousands) d	Per Capita SNB Index (c/d) ($ at 1989–1990 Prices) e
1966–1967	262,604	108,941	153,663	11,799	13,023
1967–1968	282,955	112,783	170,172	12,009	14,170
1968–1969	294,982	117,325	177,657	12,263	14,487
1969–1970	306,626	122,234	184,392	12,507	14,743
1970–1971	332,253	126,379	205,874	13,067	15,755
1971–1972	347,796	130,889	216,907	13,304	16,304
1972–1973	375,721	135,406	240,315	13,505	17,795
1973–1974	401,247	140,830	260,417	13,723	18,977
1974–1975	399,145	145,994	253,151	13,893	18,221
1975–1976	405,147	150,465	254,682	14,033	18,149
1976–1977	394,748	155,458	239,290	14,192	16,861
1977–1978	391,491	159,879	231,612	14,359	16,130
1978–1979	386,822	166,098	220,724	14,516	15,206
1979–1980	367,176	171,878	195,298	14,695	13,290
1980–1981	402,596	176,772	225,824	14,923	15,133
1981–1982	412,641	182,179	230,462	15,184	15,178
1982–1983	392,506	186,465	206,041	15,394	13,385
1983–1984	406,435	190,905	215,530	15,579	13,835
1984–1985	406,845	197,158	209,687	15,788	13,281
1985–1986	415,546	202,278	213,268	16,018	13,314
1986–1987	431,871	206,712	225,159	16,264	13,844
1987–1988	457,198	211,576	245,622	16,532	14,857
1988–1989	454,235	216,213	238,022	16,814	14,156
1989–1990	447,106	224,165	222,941	17,065	13,064
1990–1991	435,977	228,793	207,184	17,284	11,987
1991–1992	448,429	233,163	215,266	17,489	12,309
1992–1993	445,187	239,767	205,420	17,656	11,635
1993–1994	455,638	243,227	212,411	17,838	11,908
1994–1995	479,482	247,456	232,026	18,054	12,852

14.3.9   Cross-country comparison of the sustainable net benefit index and indexes of sustainable economic welfare

How does the evidence provided by the SNB index for Australia compare with the evidence provided by the ISEW calculated for other "well-developed" countries? Figure 14.4 shows the ISEW and GDP of six nations from 1950 to the early 1990s.* While Figure 14.4 reveals the ISEW increasing and decreasing at different rates and beginning its decline at different points of time during the study period (the late 1960s for the United States, the mid

* The following are the sources of the Indexes of Sustainable Economic Welfare calculated for the six countries represented in Figure 14.2: (a) United States — Daly and Cobb (1989), (b) Germany — Diefenbacher (1994), (c) United Kingdom — Jackson and Marks (1994), (d) Austria — Stockhammer et al. (1995), (e) The Netherlands — Rosenberg and Oegema (1995), and (f) Sweden — Jackson and Stymne (1996).

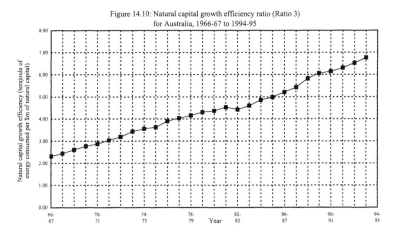

Figure 14.3 Per capita sustainable net benefits and per capita real GDP for Australia, 1966–1967 to 1994–1995.

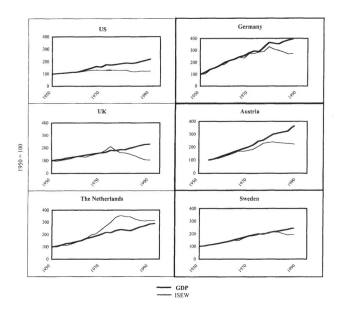

Figure 14.4 Comparison of GDP and ISEW for the United States, Germany, United Kingdom, Austria, The Netherlands, and Sweden. (Adapted from Jackson and Stymne, 1996. With permission.)

1970s for the United Kingdom, and 1980 for Sweden and the three European countries), the situation for each nation is essentially the same as the SNB index reveals for Australia. That is, despite GDP continuing to increase throughout the study period, sustainable economic welfare increases in line with GDP only for some time. Eventually the ISEW declines. A clear pattern is emerging that the macroeconomies of many of the world's developed nations, not just Australia, have long since exceeded their optimal scale. Indeed, given the length of time that the ISEW has been in decline for the United States and the United Kingdom, the macroeconomies of these two nations, like Australia, may well have exceeded their maximum sustainable scale. All told, Figures 14.3 and 14.4 provide alarming evidence of the problems now emerging for developed nations as a consequence of their failure to make the necessary transition from quantitative expansion (growth) to qualitative improvement (development).

14.4 Reform no.3: a measure of ecological economic efficiency

Two additional indicators of SD not revealed by the current system of national accounts are the efficiency with which natural capital, and the low entropy it provides, is transformed into human-made capital, and how well a nation is dealing with each of the subproblems that make up the ecological economic problem. The former can be ascertained, at the macro level, by calculating the ecological economic efficiency (EEE) ratio. To determine the latter, calculations are required of the service and maintenance efficiencies of human-made capital (Ratios 1 and 2) and the growth and exploitative efficiencies of natural capital (Ratios 3 and 4). Because the EEE ratio is the ratio of uncancelled benefits to uncancelled costs, its calculation requires no more than the information already used in the calculation of the SNB index. However, to calculate the four efficiency ratios, there is a need for three additional accounts — one for human-made capital, natural capital, and the throughput of matter–energy (resource input and waste output).

14.4.1 Account no.3 — the human-made capital account

In order to be consistent with Fisher's concept of income and capital, a human-made capital account is required to separate stocks from flows. To achieve this, four major varieties of human-made capital are included in the human-made capital account. They are consumer goods such as cars, furniture, and clothes; privately owned and publicly provided dwellings such as houses, museums, and libraries; producer goods such as plant, machinery, and equipment; and human labour.

Because it is impossible to add heterogeneous physical quantities, difficulties always arise when measuring the total stock of any form of capital. For instance, how does one add hammers, lathes, computers, carpenters, refrigerators, and buildings to obtain a single, well behaved physical index of human-made capital? One way to get around the problem is to use the *real* prices of human-made capital as a common denominator to conflate the entire stock into a single quantitative index expressed in real monetary values. However, in what has become known as the "Cambridge controversy," some observers question whether an index of this kind can adequately represent the physical quantity of capital (Scarth, 1988). There is, it seems, a problem with using monetary values to measure the physical quantity of capital when the quantity of capital depends on their individual prices; capital prices reflect their marginal product or marginal use values; and the marginal product or use values of capital are, in part, a function of the quantity of capital itself (Victor, 1991).

Another way of obtaining a single quantitative index of human-made capital is to sum the *available work* embodied in each item of human-made capital (measured in terms of an "electrical energy equivalent"). There is, unfortunately, a problem with using this method. While it is theoretically possible to estimate the available work embodied in newly produced goods, it is a near impossible task to estimate the available work embodied in already existing goods. So, too, is it impossible to estimate the available work lost as a consequence of the depreciation and consumption of human-made capital which, of course, needs to be subtracted from the cumulative total. Hence, until a suitable alternative to monetary values becomes available, the use of real prices to calculate a single quantitative index of human-made capital would appear justified. That having been said, because the aggregation of heterogeneous items into a well behaved index of human-made capital is only theoretically possible under very restrictive assumptions, there is still good reason to remain constantly wary of aggregate measures of capital (Burmeister, 1980).

The means by which the human-made capital account has been constructed for Australia can be found in Appendix 3: The Human-made Capital Account for Australia, 1966–1967 to 1994–1995. All items are measured in terms of their capitalised value based on 1989–1990 prices. The account is divided up into a number of distinct categories. Included in the private and public sector categories are dwellings and nondwelling construction and equipment (which includes machinery and equipment). Also included are business inventories, household residential dwellings, consumer durables, and a measure of all forms of human labour (paid, household nonpaid, volunteer, and unemployed labour).* Residential and commercial land is also included in the human-made capital account. Although not humanly produced, residential and commercial land yields service, not as a low entropy resource provider, but as an amenable location for living and as a convenient site for production and other commercial activities. The total

---

* Unemployed labour is included because the human-made capital stock account is designed to measure the total available stock, not simply the "working" stock.

stock as well as the average per capita ownership of human-made capital is included in Table 14.4.

Apart from a small decline in the year 1981–1982, Australia's stock of human-made capital increased in every financial year over the period of 1966–1967 to 1994–1995 ($2,084,139 million to $4,712,174 million). This indicates that the physical scale of the Australian macroeconomy expanded almost continuously throughout the study period. Despite this, the average per capita ownership of human-made capital declined slightly in the years 1981–1982, 1985–1986, and 1988–1989. Indeed, the overall rate of increase in per capita ownership was much smaller after 1976–1977 than it was in the 10-year period preceding it.

14.4.2 *Account no.4 — the throughput account*

The next recommended addition to the national system of accounting is the inclusion of a throughout account. Ideally, a throughput account would be divided into two subaccounts. The first would be an "input" account to record the annual flow of renewable and nonrenewable material and energy resources entering the macroeconomy. The second subaccount would be an "output" account to record the annual flow of high entropy wastes exiting the macroeconomy. The output account would also benefit by detailing the nature of the many wastes generated, the storage of some form of wastes, and the location and intensity of waste emissions.

Unfortunately, it has not been possible to establish a complete and comprehensive throughput account for Australia. There are three reasons for this. First, while there is adequate data regarding energy flows, data on the flow of material resources is substantially incomplete. Second, the data is expressed in totally incongruous physical units. Third, price information, which would otherwise allow a uniform measure of throughput to be calculated, is difficult to obtain. To get around this problem, Australia's total energy consumption was used as a proxy measure for throughput. Australia's total energy consumption and per capita consumption of energy is comprehensively revealed in Table 14.5. Although such a measure does not indicate anything about the quality of high entropy wastes, it does, by the very nature of the first law of thermodynamics, indicate something about the quantity of energy waste exiting the macroeconomy. Figure 14.5 is included to illustrate the growth in Australia's total energy consumption and its breakup into renewable and nonrenewable energy sources.

Table 14.5 shows that Australia's total energy consumption increased in every year from 1966–1967 to 1993–1994 except in 1982–1983 (column m). Perhaps, more importantly, per capita energy consumption increased virtually throughout the study period from 153.0 gigajoules per Australian in 1966–1967 to 234.0 gigajoules by 1993–1994 (column p). Clearly, Australia has failed to limit its growth in per capita energy consumption. Also, of major concern is Australia's heavy reliance upon the use of nonrenewable energy resources (crude oil, coal, and natural gas). No genuine transition

Table 14.4 Human-Made Capital Stock Account for Australia, 1966–1967 to 1994–1995

	Private Business Sector			Public Sector			Inventories			Statistical Discrepancy	Livestock Inventories
	Dwellings	Nondwelling Construction	Equipment	Dwellings	Nondwelling Construction	Equipment	Private Non-Farm	Farm	Public Authorities		
Year	A	B	C	D	E	F	G	H	I	J	K
1966–1967	116,038	57,987	63,980	6,910	127,707	24,555		31,319		-252	5,630
1967–1968	122,820	62,323	67,608	7,135	133,261	26,729		32,607		-877	6,046
1968–1969	130,643	67,350	72,064	7,371	139,407	28,275		35,736		-596	6,602
1969–1970	139,362	72,415	76,163	7,648	145,678	29,904		38,284		59	7,215
1970–1971	147,871	78,553	80,667	8,011	151,550	31,501		40,555		-1,011	8,064
1971–1972	157,172	83,823	84,649	8,215	158,352	33,428		40,182		-232	9,142
1972–1973	167,408	88,390	88,363	8,334	164,858	34,761		39,036		404	9,678
1973–1974	177,518	93,097	92,412	8,647	171,000	35,884		42,771		3,500	10,252
1974–1975	184,688	96,607	95,028	9,454	177,832	37,278		45,875		2,237	10,981
1975–1976	193,988	99,190	98,280	10,079	185,032	37,876		46,081		-1,283	11,283
1976–1977	204,776	101,600	101,563	10,503	192,829	37,272		48,599		-1,598	10,724
1977–1978	214,065	104,571	104,192	10,987	199,502	37,397		47,576		-445	9,951
1978–1979	223,435	108,155	108,823	11,321	206,123	37,340		50,220		-2,649	9,220
1979–1980	233,993	111,647	111,923	11,626	211,960	37,561		52,212		-2,622	8,926
1980–1981	246,201	116,815	116,818	12,012	216,592	38,228		53,106		-881	8,585
1981–1982	257,734	122,570	123,557	12,387	222,207	39,067		55,800		-3,449	8,367
1982–1983	265,022	126,824	126,736	12,986	227,416	40,166	49,750	687	1,704	-1,774	7,664
1983–1984	273,152	129,999	130,082	13,918	231,757	41,647	49,197	1,530	3,610	112	7,514
1984–1985	282,739	133,847	135,996	14,968	236,785	42,387	50,954	1,356	3,787	1,180	7,757
1985–1986	291,925	139,621	140,701	16,111	241,980	44,618	52,680	1,096	3,329	-326	7,507
1986–1987	299,145	145,796	145,288	17,174	246,960	46,595	51,123	1,189	2,918	31	7,592
1987–1988	307,494	154,506	151,130	17,890	250,383	47,458	52,505	1,254	4,166	531	7,603
1988–1989	318,984	163,925	159,560	18,439	253,137	48,514	55,835	1,524	2,083	5,075	7,817
1989–1990	329,035	174,231	165,634	19,342	256,288	51,148	57,600	1,300	5,600	4,173	8,140
1990–1991	336,790	181,556	167,574	20,039	259,100	52,545	54,723	1,186	6,816	1,533	8,384
1991–1992	344,054	185,928	167,755	20,671	261,318	53,970	53,103	1,309	6,410	-4,438	8,424
1992–1993	353,171	189,147	169,839	21,500	262,934	53,969	53,085	1,181	6,633	-4,277	8,277
1993–1994	364,199	192,518	173,449	21,723	263,818	53,945	53,968	1,055	6,530	-1,341	8,193
1994–1995	380,243	196,748	181,117	22,211	267,264	55,291	58,138	1,344	5,339	651	8,100

Table 14.4 Human-Made Capital Stock Account for Australia, 1966-1967 to 1994-1995 (Continued)

Year	Land Residential L	Land Commercial M	Labour Paid Labour Component of Labour Force N	Labour Unemployed Component of Labour Fource O	Labour Household Labour P	Labour Volunteer Labour Q	Household Sector Household Dwellings R	Household Sector Consumer Durables S	Human-Made Capital Stock ($m at 1989-1990 Prices) A1	Australian Population (Thousands) A2	Average Per Capita Human-Made Capital ($/person) A3
1966-1967	257,183	67,338	461,216	3,536	595,555	5,985	214,439	45,023	2,084,139	11,799	176,637
1967-1968	261,198	69,018	482,969	3,370	667,520	6,765	230,102	46,758	2,225,352	12,009	185,307
1968-1969	266,282	70,855	509,520	3,266	685,750	7,015	255,113	49,048	2,333,701	12,263	190,304
1969-1970	271,093	72,682	532,283	3,431	691,500	7,150	289,040	50,762	2,434,669	12,507	194,665
1970-1971	283,408	75,784	562,191	4,251	751,485	7,855	317,166	53,399	2,601,300	13,067	199,074
1971-1972	288,074	77,565	592,775	6,975	772,055	8,190	332,204	55,351	2,707,920	13,304	203,542
1972-1973	291,793	79,280	618,122	5,314	817,115	8,795	343,351	56,273	2,821,275	13,505	208,906
1973-1974	295,882	81,091	637,697	7,438	978,250	10,695	373,341	59,395	3,078,870	13,723	224,358
1974-1975	298,782	82,751	689,016	17,036	924,395	10,275	338,376	63,087	3,083,698	13,893	221,961
1975-1976	300,809	84,324	688,839	19,220	967,425	11,035	335,400	65,991	3,153,569	14,033	224,725
1976-1977	304,344	85,275	723,704	24,744	985,275	11,515	335,412	68,416	3,244,953	14,192	228,647
1977-1978	308,004	86,211	742,867	27,640	1,007,720	12,070	337,705	69,688	3,319,701	14,359	231,193
1978-1979	311,369	87,156	742,696	25,710	997,540	12,260	368,147	69,345	3,366,211	14,516	231,897
1979-1980	315,310	88,144	750,228	26,086	996,905	12,570	380,038	69,501	3,416,008	14,695	232,461
1980-1981	320,419	89,325	783,009	25,607	1,027,305	13,305	399,475	70,025	3,535,946	14,923	236,946
1981-1982	326,215	90,724	748,997	29,821	1,036,390	13,795	376,557	71,998	3,532,737	15,184	232,662
1982-1983	330,836	91,885	753,888	48,013	1,044,985	14,590	374,422	72,613	3,588,413	15,394	233,105
1983-1984	334,796	93,003	783,421	41,333	1,059,145	15,525	400,088	72,407	3,682,236	15,579	236,359
1984-1985	339,437	94,122	824,660	39,361	1,074,285	16,505	407,692	74,953	3,781,371	15,788	239,509
1985-1986	344,160	95,684	848,380	40,254	1,074,295	17,320	385,376	81,054	3,825,765	16,018	238,842
1986-1987	349,421	97,174	868,486	40,241	1,148,010	19,415	374,394	85,117	3,946,069	16,264	242,626
1987-1988	355,146	98,804	884,963	34,825	1,142,110	20,220	455,556	85,175	4,071,719	16,532	246,293
1988-1989	361,220	100,475	913,888	29,325	1,136,425	21,070	448,990	84,368	4,130,654	16,814	245,668
1989-1990	366,700	101,900	928,153	38,253	1,144,045	22,225	464,848	85,828	4,224,443	17,065	247,550
1990-1991	371,444	103,175	943,768	55,176	1,163,360	23,855	487,651	86,913	4,325,588	17,284	250,265
1991-1992	375,935	104,325	945,577	61,148	1,194,355	25,520	491,242	87,867	4,384,473	17,489	250,699
1992-1993	379,585	105,270	938,816	63,446	1,216,035	27,265	502,944	90,961	4,439,781	17,656	251,460
1993-1994	383,593	106,273	993,809	55,750	1,241,785	29,235	543,088	94,113	4,585,703	17,838	257,075
1994-1995	388,415	107,409	1,055,734	49,746	1,248,600	30,870	557,693	97,261	4,712,174	18,054	261,004

Table 14.5 Australian Energy Consumption, 1966–1967 to 1993–1994

Year	Crude oil and Other Refinery Products a	Natural Gas b	Black Coal c	Brown Coal d	Non-Renewable Energy (a + b + c + d) e	Wood Fuel f	Bagasse g	Hydro-Energy h	Solar Energy j	Renewable Energy (f + g + h + j) k	Total Energy Consumption (e + k) m	Australian Population (Thousands) n	Per Capita Energy Consumption (Gigajoules) (m/n) p
1966–1967	896.3	0.2	554.3	208.6	1613.8	115.2	49.4	27.4	0.0	192.0	1,805.8	11,799	153.0
1967–1968	961.0	0.2	572.0	214.9	1711.2	111.0	49.4	27.3	0.0	187.7	1,898.9	12,009	158.1
1968–1969	1,007.5	2.1	592.5	216.8	1836.1	106.6	53.3	29.9	0.0	189.8	2,025.9	12,263	165.2
1969–1970	1,077.5	29.5	612.5	229.0	1956.7	102.6	45.7	32.6	0.0	180.9	2,137.6	12,507	170.9
1970–1971	1,091.0	74.4	607.2	219.5	2018.9	98.9	50.2	42.3	0.0	191.4	2,210.3	13,067	169.2
1971–1972	1,136.8	102.2	609.2	238.6	2137.0	95.7	56.1	42.4	0.0	194.2	2,331.2	13,304	175.2
1972–1973	1,151.9	144.1	660.5	240.8	2257.1	92.3	55.5	42.8	0.1	190.7	2,447.8	13,505	181.3
1973–1974	1,230.1	172.5	662.9	262.6	2417.9	92.2	56.4	48.5	0.1	197.2	2,615.1	13,723	190.6
1974–1975	1,231.9	189.2	717.6	270.8	2490.6	89.6	59.7	54.4	0.2	203.9	2,694.5	13,893	193.9
1975–1976	1,233.7	211.3	702.1	286.7	2524.4	87.1	63.0	55.8	0.3	206.2	2,730.6	14,033	194.6
1976–1977	1,289.8	256.2	749.7	304.0	2706.1	81.7	68.2	49.2	0.4	199.5	2,905.6	14,192	204.7
1977–1978	1,331.0	283.1	776.8	297.4	2783.3	78.5	68.3	52.0	0.6	199.4	2,982.7	14,359	207.7
1978–1979	1,320.4	314.9	792.7	312.9	2852.4	80.1	59.9	57.7	0.8	198.5	3,050.9	14,516	210.2
1979–1980	1,312.8	362.6	846.5	321.0	2936.7	82.5	60.4	49.6	1.0	193.5	3,130.2	14,695	213.0
1980–1981	1,261.7	416.0	887.6	312.0	2938.9	84.0	68.5	53.4	1.3	207.2	3,146.1	14,923	210.8
1981–1982	1,276.1	461.9	876.7	357.7	3025.2	83.1	74.1	52.5	1.6	211.3	3,236.5	15,184	213.2
1982–1983	1,235.8	466.2	869.8	329.4	2918.5	85.1	71.9	45.6	1.8	204.4	3,122.9	15,394	202.9
1983–1984	1,261.5	490.0	913.7	316.5	3016.9	85.4	69.0	47.0	2.1	203.5	3,220.4	15,579	206.7
1984–1985	1,268.7	523.3	968.3	369.2	3152.6	86.1	72.9	55.5	2.5	217.0	3,369.6	15,788	213.4
1985–1986	1,247.7	570.7	981.4	350.1	3187.9	87.6	71.4	53.5	2.6	215.1	3,403.0	16,018	212.4
1986–1987	1,233.7	588.4	1,008.6	405.0	3297.1	88.7	73.4	53.2	2.4	217.7	3,514.8	16,264	216.1
1987–1988	1,321.4	610.6	1,035.9	424.9	3401.5	90.6	73.9	53.5	2.8	220.8	3,622.3	16,532	219.1
1988–1989	1,345.8	627.8	1,107.5	474.8	3600.5	93.9	80.7	54.6	2.4	231.6	3,832.1	16,814	227.9
1989–1990	1,386.7	688.0	1,132.1	450.7	3711.3	97.8	81.5	52.2	2.4	233.9	3,945.2	17,065	231.2
1990–1991	1,421.8	655.7	1,141.3	484.1	3707.2	100.1	78.2	58.7	2.4	239.4	3,946.6	17,284	228.3
1991–1992	1,440.4	678.7	1,168.2	497.3	3778.0	101.6	63.5	57.7	2.4	225.2	4,003.2	17,489	228.9
1992–1993	1,495.4	706.5	1,195.6	466.8	3832.5	104.7	78.0	61.6	2.4	246.7	4,079.2	17,656	231.0
1993–1994	1,504.9	733.4	1,197.4	486.8	3922.4	107.1	84.5	60.2	2.4	254.2	4,176.6	17,838	234.0
1994–1995	n.a.	n.a.	n.a.	n.a.	n.a.	n.a.	n.a.	n.a.	n.a.	n.a.	n.a.	18,054	n.a.

n.a. denotes not available.

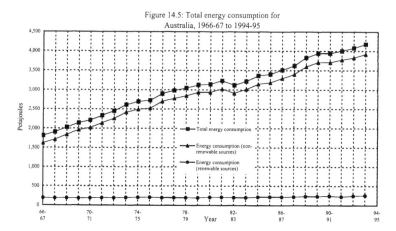

Figure 14.5   Total energy consumption for Australia, 1966–1967 to 1994–1995.

from nonrenewable to renewable energy resources appears to have been made by Australia over the study period. This raises serious doubts as to whether, in the long run, Australia could sustain even the current rate of throughput let alone the higher rates of throughput that will be required if energy demands are not subdued.

14.4.3 Account no.5 — the natural capital stock account

The final recommended account measures the stock of natural capital. It is principally designed to record the two major kinds of low entropy providing natural capital, namely, nonrenewable natural capital (agricultural land and subsoil assets), and renewable natural capital (forests and timber plantations, fish stocks, fixed capital livestock assets, and stored water). An inventory of natural capital is compiled in Appendix 4: The Natural Capital Account for Australia, 1966–1967 to 1994–1995.

The greatest difficulty associated with the construction of a natural capital account is determining the means by which its various elements should be measured. Not unlike human-made capital, one is again confronted with potential aggregation problems. Does one use physical or monetary values? As it turns out, both are required because neither physical estimates nor monetary values of natural capital bear any relationship to the capacity of natural capital to sustain its source, sink, and life-support functions. For example, and as the Lauderdale Paradox suggests, it is possible for the aggregate monetary value of a diminishing stock of natural capital to remain unchanged if, as it gets scarcer, its exchange value appreciates. In addition, many elements of the natural capital stock go unpriced by conventional markets. As for the elements that are priced, their exchange value is often distorted by externalities and a range of market imperfections.

There is, of course, a problem with quantitative measures of natural capital. Because various elements of the natural capital stock are entirely incongruous, it is impossible to add a tonne of timber and a tonne of iron ore. Moreover, while a one hectare timber plantation may provide a greater source of timber than the hectare of native forest it replaced, it is likely to provide fewer life-support services. Exotic systems are also likely to be less effective at assimilating wastes.

In compiling the natural capital account for Australia, market prices were utilised in the same way as they were for human-made capital. But we did so with the following in mind. First, an aggregate monetary measure of natural capital should only be used to calculate the growth and exploitative efficiencies of natural capital (Ratios 3 and 4). It should never be used to determine critical natural capital stock levels. This, as indicated at the beginning of the chapter, requires biophysical rather than economic assessments of natural capital and ecosystem services. Second, to overcome the potential measurement problems arising from the Lauderdale Paradox, real rather than nominal values must be used. This ensures the monetary values reflect any changes in the physical quantities of natural capital. While this does not indicate anything about the quality of natural capital, this is eventually captured by Ratios 3 and 4. For example, an increase in the growth efficiency of natural capital (Ratio 3) indicates that a given per unit quantity of natural capital can yield a greater rate of resource throughput. On the other hand, an increase in the exploitative efficiency of natural capital (Ratio 4) suggests that a given per unit quantity of natural capital can be exploited at the expense of fewer sacrificed natural capital services.

The total real monetary value of Australia's natural capital over the period 1966–1967 to 1994–1995 is indicated in Table 14.6 (column 16). Figure 14.6 shows Australia's natural capital stock and the breakup of the total stock into renewable and nonrenewable natural capital. Overall, Australia's natural capital stock continuously declined over the study period ($780,449 million in 1966–1967 to $608,913 million in 1994–1995). While there was a small increase in renewable natural capital ($154,963 million to $167,537 million), the stock of nonrenewable resources declined significantly ($625,486 million to $441,376 million). This decline is largely attributable to the diminution of subsoil assets. The average per capita ownership of natural capital fell even more alarmingly — from $66,145 per Australian in 1966–1967 to $33,727 per Australian in 1994-95.

Both Table 14.6 and Figure 14.6 show that Australia is not adhering to sustainability precept No. 1b. That is, Australia is failing to cultivate sufficient quantities of renewable resources to offset the depletion of nonrenewable resource stocks. What does this mean? In simple terms, it means an excessive portion of the proceeds from nonrenewable resource depletion is being used by Australians to finance current consumption expenditure. Too little is being set aside to invest in renewable resource substitutes in order to keep the stock of natural capital intact.

Table 14.6   Natural Capital Stock Account for Australia, 1966–1967 to 1994–1995

Year	Agricultural Land 1	Sub-Soil Assets				Timber Stocks			
		Metallic Mineral Stocks 2	Coal Stocks 3	Oil and Gas Stocks 4	Nonmetallic Mineral Stocks 5	Native Forest Timber 6	Woodland Timber 7	Plantation Timber – B/Leaved 8	Plantation Timber – Coniferous 9
1966–1967	67,693	200,680	187,073	154,723	15,317	15,770	27,755	38	2,729
1967–1968	68,054	199,719	186,951	152,868	15,149	15,379	27,559	41	2,813
1968–1969	68,193	198,678	186,813	150,844	14,973	14,988	27,364	45	2,896
1969–1970	68,763	197,407	186,660	148,651	14,767	14,597	27,168	48	2,980
1970–1971	69,180	196,101	186,496	146,125	14,540	15,270	26,973	52	3,063
1971–1972	69,431	194,828	186,315	142,900	14,306	15,942	26,777	55	3,147
1972–1973	69,472	193,466	186,111	139,382	14,106	16,615	26,582	58	3,230
1973–1974	69,570	192,025	185,907	135,435	13,893	16,684	26,386	62	3,314
1974–1975	69,444	190,556	185,686	131,489	13,688	16,754	26,191	65	3,398
1975–1976	69,597	189,020	185,445	127,423	13,470	16,824	25,996	69	3,481
1976–1977	68,319	187,519	185,178	123,216	13,227	16,848	25,800	72	3,565
1977–1978	68,027	185,952	184,906	118,865	13,012	17,134	25,605	76	3,648
1978–1979	68,555	184,465	184,622	114,582	12,820	16,416	25,409	78	3,796
1979–1980	68,888	182,815	184,343	110,494	12,608	15,982	25,409	82	4,001
1980–1981	68,861	181,198	184,020	106,510	12,371	15,975	25,409	86	4,128
1981–1982	68,221	179,610	183,683	102,640	12,176	15,972	25,409	72	4,329
1982–1983	67,248	177,800	183,314	98,987	11,977	15,965	25,409	74	4,314
1983–1984	67,915	175,945	182,929	94,549	11,819	16,003	25,409	72	4,415
1984–1985	67,999	174,006	182,491	89,470	11,638	16,148	25,024	67	4,473
1985–1986	65,094	171,906	182,002	84,284	11,431	16,138	25,021	70	4,656
1986–1987	65,469	169,906	181,458	79,150	11,236	15,965	25,017	80	4,779
1987–1988	65,608	167,735	180,959	74,129	11,030	15,960	25,013	89	4,957
1988–1989	64,899	165,596	180,404	69,609	10,822	16,016	24,809	103	4,091
1989–1990	64,538	163,318	179,817	64,532	10,602	15,957	24,806	164	5,158
1990–1991	64,329	160,782	179,206	59,476	10,400	15,956	24,802	181	5,236
1991–1992	64,774	158,124	178,557	54,549	10,259	16,050	24,798	200	5,325
1992–1993	63,954	155,492	177,898	49,734	10,101	15,917	24,794	224	5,330
1993–1994	65,205	152,888	177,240	45,219	9,927	15,917	24,794	250	5,341
1994–1995	64,399	150,314	176,539	40,413	9,711	15,914	24,790	265	5,369

Table 14.6 Natural Capital Stock Account for Australia, 1966–1967 to 1994–1995 (Continued)

Year	Wetlands, Mangroves, and Saltmarshes 10	Fish Stocks 11	Livestock (Fixed Natural Capital Assets) 12	Stored Surface Water Resources 13	Nonrenewable Natural Capital (1 to 5) ($m at 1989–1990 Prices) 14	Renewable Natural Capital (6 to 13) ($m at 1989–1990 Prices) 15	Natural Capital Stock (14+15) ($m at 1989–1990 Prices) 16	Australian Population (Thousands) 17	Average Per Capita Natural Capital (16/17) ($/person) 18
1966–1967	63,268	18,323	12,255	14,825	625,486	154,963	780,449	11,799	66,145
1967–1968	62,883	18,152	12,833	15,550	622,741	155,210	777,951	12,009	64,781
1968–1969	62,499	17,968	13,521	16,275	619,501	155,556	775,057	12,263	63,203
1969–1970	62,114	17,770	14,364	17,000	616,248	156,041	772,289	12,507	61,749
1970–1971	61,730	17,555	15,504	18,725	612,442	158,872	771,314	13,067	59,028
1971–1972	61,345	17,323	16,845	20,450	607,780	161,884	769,664	13,304	57,852
1972–1973	60,961	17,076	17,315	22,175	602,537	164,012	766,549	13,505	56,760
1973–1974	60,576	16,834	17,635	23,900	596,830	165,391	762,221	13,723	55,543
1974–1975	60,192	16,629	18,668	25,625	590,863	167,522	758,385	13,893	54,588
1975–1976	59,807	16,430	19,089	27,350	584,955	169,046	754,001	14,033	53,731
1976–1977	59,423	16,215	18,145	29,075	577,459	169,143	746,602	14,192	52,607
1977–1978	59,038	15,992	16,814	30,800	570,762	169,107	739,869	14,359	51,526
1978–1979	58,654	15,756	15,682	32,525	565,044	168,316	733,360	14,516	50,521
1979–1980	58,269	15,505	15,314	34,250	559,148	168,812	727,960	14,695	49,538
1980–1981	57,885	15,222	14,877	34,750	552,960	168,332	721,292	14,923	48,334
1981–1982	57,500	14,921	14,557	35,250	546,330	168,010	714,340	15,184	47,046
1982–1983	57,270	14,589	13,665	35,750	539,326	167,036	706,362	15,394	45,886
1983–1984	57,039	14,273	13,420	36,250	533,157	166,881	700,038	15,579	44,935
1984–1985	56,809	14,001	13,816	36,750	525,604	167,088	692,692	15,788	43,875
1985–1986	56,578	13,753	13,584	37,250	514,717	167,050	681,767	16,018	42,563
1986–1987	56,348	13,510	13,727	37,750	507,219	167,176	674,395	16,264	41,466
1987–1988	56,118	13,259	13,729	38,250	499,461	167,375	666,836	16,532	40,336
1988–1989	55,887	13,000	13,986	38,750	491,330	167,642	658,972	16,814	39,192
1989–1990	55,657	12,749	14,646	39,250	482,807	168,387	651,194	17,065	38,160
1990–1991	55,426	12,444	14,958	39,667	474,193	168,670	642,863	17,284	37,194
1991–1992	55,296	12,102	14,907	40,083	466,263	168,761	635,024	17,489	36,310
1992–1993	54,966	11,769	14,498	40,500	457,179	167,998	625,177	17,656	35,409
1993–1994	54,735	11,481	14,345	40,917	450,479	167,780	618,259	17,838	34,660
1994–1995	54,505	11,208	14,153	41,333	441,376	167,537	608,913	18,054	33,727

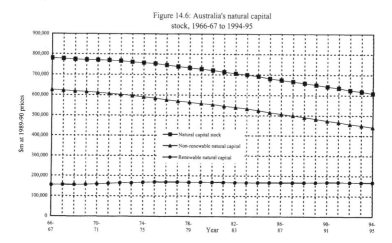

Figure 14.6   Australia's natural capital stock, 1966–1967 to 1994–1995.

14.4.4 Calculating the ecological economic efficiency (EEE) ratio

The EEE ratio is the ratio of uncancelled benefits to uncancelled costs as indicated by Eq. (6.2). An increase in the EEE ratio suggests an improvement in the efficiency with which natural capital is transformed into service-yielding human-made capital. The EEE ratio for Australia over the period 1966–1967 to 1994–1995 is indicated in Table 14.7 and by Figure 14.7. Both show the EEE ratio increasing from 2.41 in 1966–1967 to a peak in 1973–1974 of 2.85. The EEE ratio then declines to 1.86 in 1992–1993 before rising slightly to 1.94 by 1994–1995. By the end of the study period, the EEE ratio is much lower than its initial value (1.94 compared to 2.41). Interestingly, the trend movement of the EEE ratio closely follows that of the SNB index. This would indicate that the general decline in the SNB index after 1973–1974 is owing as much to the inefficient allocation of resources as to the depletion of natural capital and the inequitable distribution of income. In what ways inefficiencies have contributed to the decline in the SNB index is not altogether clear from the EEE ratio. This information is better revealed by the four efficiency ratios that make up the larger ecological economic problem. Hence, it is toward these efficiency ratios that we now direct our attention.

14.4.5 Ratio No.1 — the service efficiency of human-made capital

The service efficiency ratio is the ratio of uncancelled benefits to human-made capital. As explained previously, the ratio indicates how well the total stock of human-made capital contributes to the net psychic income of a nation. The service efficiency ratio for Australia is revealed in both Table 14.8 and by Figure 14.8. According to both, the service efficiency of Australia's human-made capital was 0.126 in 1966–1967 (an imputed service rate of 12.6 percent) and increased to a peak of 0.133 (13.3 percent) by 1972–1973.

Table 14.7 Ecological Economic Efficiency (EEE) Ratio for Australia, 1966–1967 to 1994–1995

Year	Uncancelled Benefits ($m at 1989–1990 Prices) a	Uncancelled Costs ($m at 1989–1990 Prices) b	Ecological Economic Efficiency (EEE) (a/b) c
1966–1967	262,604	108,941	2.41
1967–1968	282,955	112,783	2.51
1968–1969	294,982	117,325	2.51
1969–1970	306,626	122,234	2.51
1970–1971	332,253	126,379	2.63
1971–1972	347,796	130,889	2.66
1972–1973	375,721	135,406	2.77
1973–1974	401,247	140,830	2.85
1974–1975	399,145	145,994	2.73
1975–1976	405,147	150,465	2.69
1976–1977	394,748	155,458	2.54
1977–1978	391,491	159,879	2.45
1978–1979	386,822	166,098	2.33
1979–1980	367,176	171,878	2.14
1980–1981	402,596	176,772	2.28
1981–1982	412,641	182,179	2.27
1982–1983	392,506	186,465	2.10
1983–1984	406,435	190,905	2.13
1984–1985	406,845	197,158	2.06
1985–1986	415,546	202,278	2.05
1986–1987	431,871	206,712	2.09
1987–1988	457,198	211,576	2.16
1988–1989	454,235	216,213	2.10
1989–1990	447,106	224,165	1.99
1990–1991	435,977	228,793	1.91
1991–1992	448,429	233,163	1.92
1992–1993	445,187	239,767	1.86
1993–1994	455,638	243,227	1.87
1994–1995	479,482	247,456	1.94

Apart from a small rise between 1979–1980 and 1981–1982, Australia's service efficiency ratio effectively declined thereafter. By the end of the study period, the service efficiency ratio had fallen to 0.102 (10.2 percent).

Given that technological progress has undoubtedly increased the ability of human-made capital to directly yield service (e.g., televisions now provide colour images, microwave ovens cook food in a fraction of the time of conventional ovens, and cars are less noisy and considerably more comfortable than those gone by), why would the service efficiency of human-made capital decline over much of the study period? Closer examination of Australia's uncancelled benefit account shows that while there has been a marked increase in the capacity of human-made capital to provide a flow of psychic income, it has come at the expense of considerable growth in psychic outgo-related items. This suggests two things. First, since the early 1970s, a greater proportion of Australians are experiencing an unhealthy existential

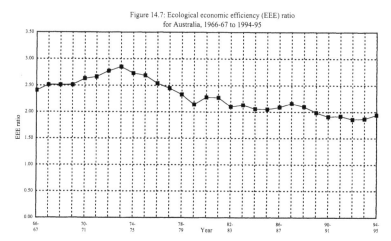

Figure 14.7 Ecological economic efficiency (EEE) ratio for Australia, 1966–1967 to 1994–1995.

imbalance. Second, Australia's uncancelled benefit (UB) curve has probably been shifting down rather than up over the last 20 or so years. Why? Not because there has been a lack of any technical design improvements in the stock of human-made capital, but because the increase in the quantity and quality of human-made capital has been accompanied by sharp increases in unemployment, commuting costs, noise pollution, crime, and a widening gap between the rich and poor.

14.4.6 Ratio no. 2 — the maintenance efficiency of human-made capital

The maintenance efficiency ratio is the ratio of human-made capital to the throughput of matter–energy required to maintain the stock of human-made capital intact. Because of an inability to compile a comprehensive throughput account, the maintenance efficiency ratio revealed in Table 14.9 and Figure 14.9 is a ratio of human-made capital to Australia's consumption of energy resources. Both indicate the human-made capital maintained for each peta-joule of energy consumed. They show that, in 1966–1967, a petajoule of energy was consumed to maintain $1,154.2 million of human-made capital. This increased to a maximum of $1,176.9 million by 1970–1971. The value then declined very gradually to a low of $1,070.8 million in 1989–1990. By 1993–1994, it had marginally recovered to $1,098.0 million — still lower than the initial 1966–1967 figure. This evidence, plus the large overall increase in total energy consumption between 1966–1967 to 1993–1994, indicates that most technological innovation in Australia over the last 30 years has probably been of the throughput-increasing variety. Little progress seems to have been made in terms of maintenance efficiency-increasing innovation.

Table 14.8 Service Efficiency Ratio for Australia, 1966–1967 to 1994–1995

Year	Uncancelled Benefits ($m at 1989–1990 Prices) a	Human-Made Capital Stock ($m at 1989–1990 Prices) b	Service Efficiency (Ratio 1) (a/b) c
1966–1967	262,604	2,084,218	0.126
1967–1968	282,955	2,225,336	0.127
1968–1969	294,982	2,333,664	0.126
1969–1970	306,626	2,434,694	0.126
1970–1971	332,253	2,601,308	0.128
1971–1972	347,796	2,707,881	0.128
1972–1973	375,721	2,821,402	0.133
1973–1974	401,247	3,078,897	0.130
1974–1975	399,145	3,083,765	0.129
1975–1976	405,147	3,153,736	0.128
1976–1977	394,748	3,244,834	0.122
1977–1978	391,491	3,319,686	0.118
1978–1979	386,822	3,366,186	0.115
1979–1980	367,176	3,415,954	0.107
1980–1981	402,596	3,535,902	0.114
1981–1982	412,641	3,532,698	0.117
1982–1983	392,506	3,588,392	0.109
1983–1984	406,435	3,682,337	0.110
1984–1985	406,845	3,781,412	0.108
1985–1986	415,546	3,825,621	0.109
1986–1987	431,871	3,946,174	0.109
1987–1988	457,198	4,071,769	0.112
1988–1989	454,235	4,130,459	0.110
1989–1990	447,106	4,224,443	0.106
1990–1991	435,977	4,325,569	0.101
1991–1992	448,429	4,384,513	0.102
1992–1993	445,187	4,439,826	0.100
1993–1994	455,638	4,585,737	0.099
1994–1995	479,482	4,712,150	0.102

14.4.7 Ratio no. 3 — the growth efficiency of natural capital

The growth efficiency ratio is the ratio of the input or consumption of energy resources to the stock of natural capital from which it is exacted. The ratio indicates how well natural capital is able to provide the flow of energy resources required to produce and maintain the stock of human-made capital. The growth efficiency ratio revealed in Table 14.10 and Figure 14.10 represents the terajoules of energy entering Australia's macroeconomy for every dollar of available natural capital (one petajoule equals 1,000 terajoules). Both Table 14.10 and Figure 14.10 reveal that, apart from the financial year 1982–1983, the growth efficiency of natural capital increased in every year over the study period. The ratio began at a value of 2.31 in 1966–1967 and increased to a value of 6.76 by 1993–1994.

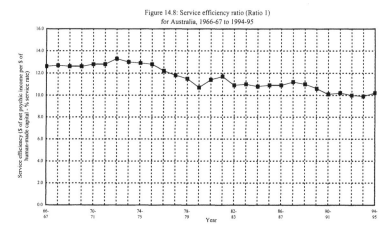

Figure 14.8   Service efficiency ratio (Ratio 1) for Australia, 1966–1967 to 1994–1995.

The increase in the growth efficiency ratio suggests Australia has become more efficient in obtaining an energy offtake from natural capital as input into the macroeconomy. Such a suggestion is misleading for a number of reasons. First, much of Australia's energy input comes from nonrenewable sources which, as indicated by Table 14.5, has increased over the study period. Second, the actual stock of nonrenewable natural capital has steadily declined. Thus, the growth efficiency ratio has risen largely as a consequence of Australia having increased the depletion rate of its non-renewable resource stock.

Given that a nation is ultimately dependent on renewable energy, a more cogent growth efficiency ratio is a *renewable* natural capital growth efficiency ratio. This can be calculated by excluding the nonrenewable resource com-ponent of the natural capital stock and including only the consumption of renewable energy. Australia's renewable natural capital growth efficiency ratio for the period 1966–1967 to 1993–1994 is revealed in both Table 14.11 and by Figure 14.11. They show that the renewable natural capital growth efficiency ratio changed very little between the period 1966–1967 and 1983–1984 (1.24 in 1966–1967 and 1.22 in 1983–1984). However, by 1993–1994, the ratio had increased to a value of 1.52. This increase probably reflects the impact of stricter pollution standards as well as a lagged response to the oil price shocks of 1973 and 1979. While this increase is encouraging, it must be seen in the context of Australia's continuing reliance on and increase in the consumption of nonrenewable energy. One would be hard pressed to conclude that the capacity of Australia's natural capital to provide a sustain-able flow of energy had in any way significantly increased.

Table 14.9 Maintenance Efficiency Ratio for Australia, 1966–1967 to 1994–1995

Year	Human-Made Capital Stock ($m at 1989–1990 Prices) a	Total Energy Consumption (Throughput) (Petajoules) b	Maintenance Efficiency (Ratio 2) (a/b) c
1966–1967	2,084,218	1,805.8	1,154.2
1967–1968	2,225,336	1,898.9	1,171.9
1968–1969	2,333,664	2,025.9	1,151.9
1969–1970	2,434,694	2,137.6	1,139.0
1970–1971	2,601,308	2,210.3	1,176.9
1971–1972	2,707,881	2,331.2	1,161.6
1972–1973	2,821,402	2,447.8	1,152.6
1973–1974	3,078,897	2,615.1	1,177.4
1974–1975	3,083,765	2,694.5	1,144.5
1975–1976	3,153,736	2,730.6	1,155.0
1976–1977	3,244,834	2,905.6	1,116.8
1977–1978	3,319,686	2,982.7	1,113.0
1978–1979	3,366,186	3,050.9	1,103.3
1979–1980	3,415,954	3,130.2	1,091.3
1980–1981	3,535,902	3,146.1	1,123.9
1981–1982	3,532,698	3,236.5	1,091.5
1982–1983	3,588,392	3,122.9	1,149.1
1983–1984	3,682,337	3,220.4	1,143.4
1984–1985	3,781,412	3,369.6	1,122.2
1985–1986	3,825,621	3,403.0	1,124.2
1986–1987	3,946,174	3,514.8	1,122.7
1987–1988	4,071,769	3,622.3	1,124.1
1988–1989	4,130,459	3,832.1	1,077.9
1989–1990	4,224,443	3,945.2	1,070.8
1990–1991	4,325,569	3,946.6	1,096.0
1991–1992	4,384,513	4,003.2	1,095.3
1992–1993	4,439,826	4,079.2	1,088.4
1993–1994	4,585,737	4,176.6	1,098.0
1994–1995	4,712,150	n.a..	n.a.

n.a. denotes not available.

14.4.8 Ratio no. 4 — the exploitative efficiency of natural capital

The final ratio is the natural capital exploitative efficiency ratio. This ratio is a measure of the opportunity cost of natural capital services foregone relative to the stock of natural capital available for exploitation. The exploitative efficiency ratio is calculated by dividing the estimated monetary value of natural capital by the uncancelled cost of economic activity. The larger/smaller is the ratio, the smaller/larger is the opportunity cost of natural capital services sacrificed per dollar of available natural capital (valued at 1989–1990 prices). Australia's exploitative efficiency ratio is indicated in Table 14.12 and by Figure 14.12. Both reveal that the exploitative efficiency ratio declined in every year between 1966-67 and 1994–1995. This result

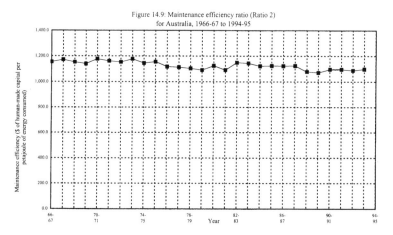

Figure 14.9 Maintenance efficiency ratio (Ratio 2) for Australia, 1966–1967 to 1994–1995.

suggests that the opportunity cost of exploiting natural capital for the throughput of matter–energy increased continuously over the study period. The decline in the exploitative efficiency ratio was considerable. The ratio began at a value of 7.2 in 1966–1967 and declined to a value of 2.4 by 1994–1995. While this result probably overstates the true increase in the opportunity cost of natural capital exploitation, it does reflect Australia's heavy reliance on nonrenewable resources and the lack of reinvestment into renewable resource substitutes. Both have greatly reduced the source function of natural capital, the sacrifice of which has made the largest contribution to the loss of natural capital services (see Figure 14.2). Clearly, a policy designed to cultivate renewable resource substitutes instead of using the proceeds of nonrenewable resource depletion to finance current consumption would go a long way toward reversing or at least decelerating the decline in Australia's exploitative efficiency ratio.

14.5 Reform no. 4: supplementary sustainable development indicators

As useful as economic indicators are in measuring a nation's SD performance, there are a number of aspects that, when taken together, suggest the need for supplementary SD indicators. To begin with, the list of items used to calculate the SNB index is not entirely comprehensive. Owing to measurement difficulties and a lack of appropriate data, many welfare-related factors were overlooked (e.g., the disutility of work and the existence values of natural capital). As for some of the items that were included, ascertaining monetary estimates required a great deal of subjective judgement. In addition, it was necessary at times to employ extrapolation and interpolation

Table 14.10 Natural Capital Growth Efficiency Ratio for Australia, 1966–1967 to 1994–1995

Year	Total Energy Consumption (Throughput) (Terajoules) a	Natural Capital Stock ($m at 1989–1990 Prices) b	Natural Capital Growth Efficiency (Ratio 3) (a/b) c
1966–1967	1,805,800	780,449	2.31
1967–1968	1,898,900	777,951	2.44
1968–1969	2,025,900	775,057	2.61
1969–1970	2,137,600	772,289	2.77
1970–1971	2,210,300	771,314	2.87
1971–1972	2,331,200	769,664	3.03
1972–1973	2,447,800	766,549	3.19
1973–1974	2,615,100	762,221	3.43
1974–1975	2,694,500	758,385	3.55
1975–1976	2,730,600	754,001	3.62
1976–1977	2,905,600	746,602	3.89
1977–1978	2,982,700	739,869	4.03
1978–1979	3,050,900	733,360	4.16
1979–1980	3,130,200	727,960	4.30
1980–1981	3,146,100	721,292	4.36
1981–1982	3,236,500	714,340	4.53
1982–1983	3,122,900	706,362	4.42
1983–1984	3,220,400	700,038	4.60
1984–1985	3,369,600	692,692	4.86
1985–1986	3,403,000	681,767	4.99
1986–1987	3,514,800	674,395	5.21
1987–1988	3,622,300	666,836	5.43
1988–1989	3,832,100	658,972	5.82
1989–1990	3,945,200	651,194	6.06
1990–1991	3,946,600	642,863	6.14
1991–1992	4,003,200	635,024	6.30
1992–1993	4,079,200	625,177	6.52
1993–1994	4,176,600	618,259	6.76
1994–1995	n.a.	608,913	n.a.

n.a. denotes not available.

techniques. A question mark, therefore, hangs over the accuracy of some of the values.

Second, the co-evolutionary paradigm informs us that the qualitative state of a nation's macroeconomy depends very much on the nature of its underlying parameters. Unfortunately, while the SNB index and the ecological economic efficiency ratios convey a lot of information about the current manifestations and immediate effects of past and present human activities, they say very little about the likely future impact of current activities. This greatly reduces their policy-guiding value.

Third, from a development perspective, the means by which net psychic income was measured for Australia cannot be directly related to any notion of objective value. Throughout the book, development has been perceived

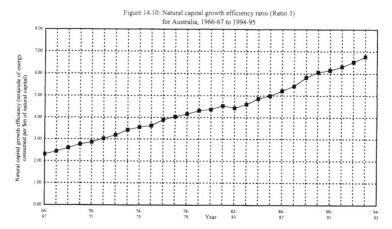

Figure 14.10   Natural capital growth efficiency ratio (Ratio 3) for Australia, 1966–1967 to 1994–1995.

as a qualitative condition to be understood in the context of a set of objective value-based standards and principles. Yet much of what supposedly contributes to the net psychic income of a nation is unlikely to be objectively desirable, indeed, it is likely to be illth. For instance, personal consumption expenditure (Item A) includes the consumption of junk food, tobacco products, and alcohol, which is often of little benefit to the consumer.

Fourth, a per capita measure of sustainable net benefits does not say a great deal about the distribution of net benefits. Who has ownership and access to human-made capital is, of course, a very important SD consideration. While distributional concerns were dealt with to some degree by weighting personal consumption expenditure in line with changes in the distribution of income, the final result cannot provide a transparent indication of the spread of net benefits enjoyed by Australian citizens.

Finally, the SNB index does not indicate whether the macroeconomy has exceeded the maximum sustainable scale. Being an economic indicator of SD, the SNB index merely counts the cost of sacrificed natural capital services. It does not indicate whether the total loss of natural capital services has exceeded a sustainability threshold. Consequently, the SNB index needs to be accompanied by a satellite account based, not on monetary measures of natural capital such as the natural capital account discussed earlier in the chapter, but on biophysical assessments. This would allow one to better ascertain whether the net benefits from economic activity can be ecologically sustained.

In order to supplement the SD indicators put forward so far in this chapter, a number of additional indicators are recommended. In what follows, a brief description of some of these indicators is given along with the rationale behind their inclusion. To make the task easier, the supplementary indicators are divided into two categories, one dealing with development

Table 14.11 Renewable Natural Capital Growth Efficiency Ratio for Australia, 1966–1967 to 1994–1995

Year	Total Renewable Energy Consumption (Throughput) (Terajoules) a	Renewable Natural Capital Stock ($m at 1989–1990 Prices) b	Renewable Natural Capital Growth Efficiency (a/b × 1,000) c
1966–1967	192.0	154,963	1.24
1967–1968	187.7	155,210	1.21
1968–1969	189.8	155,556	1.22
1969–1970	180.9	156,041	1.16
1970–1971	191.4	158,872	1.20
1971–1972	194.2	161,884	1.20
1972–1973	190.7	164,012	1.16
1973–1974	197.2	165,391	1.19
1974–1975	203.9	167,522	1.22
1975–1976	206.2	169,046	1.22
1976–1977	199.5	169,143	1.18
1977–1978	199.4	169,107	1.18
1978–1979	198.5	168,316	1.18
1979–1980	193.5	168,812	1.15
1980–1981	207.2	168,332	1.23
1981–1982	211.3	168,010	1.26
1982–1983	204.4	167,036	1.22
1983–1984	203.5	166,881	1.22
1984–1985	217.0	167,088	1.30
1985–1986	215.1	167,050	1.29
1986–1987	217.7	167,176	1.30
1987–1988	220.8	167,375	1.32
1988–1989	231.6	167,642	1.38
1989–1990	233.9	168,387	1.39
1990–1991	239.4	168,670	1.42
1991–1992	225.2	168,761	1.33
1992–1993	246.7	167,998	1.47
1993–1994	254.2	167,780	1.52
1994–1995	n.a.	167,537	n.a.

n.a. denotes not available.

and the other with ecological sustainability. The chapter concludes with a means by which the supplementary indicators can be used to assess and rank the SD performance of different countries.

14.5.1 Supplementary development indicators

It was established in Chapter 2 that development involves the qualitative improvement in the human condition. Thus, in order to establish a suitable list of supplementary development indicators, it is necessary to identify the factors and relationships that reflect changes in an individual's quality of life. To do this, use was again made of Ekin's (1990) classification of

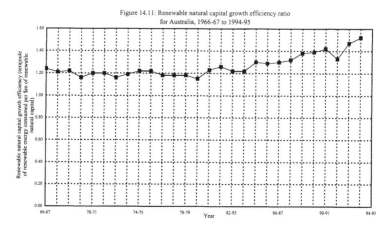

Figure 14.11   Renewable natural capital growth efficiency ratio for Australia, 1966–1967 to 1994–1995.

services/disservices in terms of the "having," "being," "doing," and "relating" modes of human experience. Many of the development indicators listed next are taken from the *Human Development Report* (1994) published annually by the United Nations Development Programme (UNDP). The list of recommended development indicators includes the following.

- *Real per capita Hicksian income (having).*   While the use of any income measure brings to bear some of the problems just discussed, it hardly seems possible to avoid income in some way as a development indicator. However, unlike the UNDP, which relies on real GDP as a development indicator, a measure of real Hicksian income should instead be used (calculated as per reform measure No. 1).
- *Percentage of population living above the absolute and relative poverty lines (having).*   The use of poverty lines based on the distribution of income can be used to ascertain: (a) the percentage of a nation's population living in *absolute* poverty (i.e., earning an income below what is required to live a comfortable and decent existence), and (b) the percentage of a nation's population living in *relative* poverty (i.e., earning, for example, one-half the nation's average income). These indicators are required for the following reasons. First, a reduction in the percentage of the population living under the absolute poverty line indicates an improvement in the "standard of living" among the poor that is not always accurately reflected by an increase in per capita income. Second, even when the incidence of absolute poverty is in decline, if the gap between rich and poor is widening, it is possible for the percentage of a nation's population living in relative poverty to be increasing. While reductions in absolute poverty are a more concrete reflection of human development than reductions in relative

Table 14.12 Natural Capital Exploitative Efficiency Ratio for Australia, 1966–1967 to 1994–1995

Year	Natural Capital Stock ($m at 1989–1990 Prices) a	Uncancelled Costs ($m at 1989–1990 Prices) b	Natural Capital Exploitative Efficiency (Ratio 4) (a/b) c
1966–1967	780,449	108,941	7.2
1967–1968	777,951	112,783	6.9
1968–1969	775,057	117,325	6.6
1969–1970	772,289	122,234	6.3
1970–1971	771,314	126,379	6.1
1971–1972	769,664	130,889	5.9
1972–1973	766,549	135,406	5.7
1973–1974	762,221	140,830	5.4
1974–1975	758,385	145,994	5.2
1975–1976	754,001	150,465	5.0
1976–1977	746,602	155,458	4.8
1977–1978	739,869	159,879	4.6
1978–1979	733,360	166,098	4.4
1979–1980	727,960	171,878	4.2
1980–1981	721,292	176,772	4.1
1981–1982	714,340	182,179	3.9
1982–1983	706,362	186,465	3.8
1983–1984	700,038	190,905	3.7
1984–1985	692,692	197,158	3.5
1985–1986	681,767	202,278	3.4
1986–1987	674,395	206,712	3.3
1987–1988	666,836	211,576	3.2
1988–1989	658,972	216,213	3.0
1989–1990	651,194	224,165	2.9
1990–1991	642,863	228,793	2.8
1991–1992	635,024	233,163	2.7
1992–1993	625,177	239,767	2.6
1993–1994	618,259	243,227	2.5
1994–1995	608,913	247,456	2.4

poverty, it has been shown that happiness can be as much a function of relative income as absolute income (Easterlin, 1974). Hence, the welfare increase resulting from reductions in absolute poverty must be weighed up against the possible welfare decrease caused by large increases in relative poverty.

- *Home ownership rate (having)*. The national home ownership rate not only serves as a good indicator of the distribution of wealth, it says a great deal about the affordability of the most vital of all service-yielding assets — the family home.
- *National self-sufficiency (having)*. Although international trade can, depending on the circumstances, be of great national benefit, a nation's ability to provide for itself is a distinct advantage over a nation

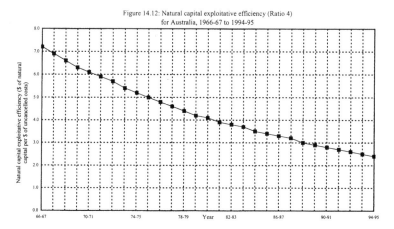

Figure 14.12   Natural capital exploitative efficiency (Ratio 4) for Australia, 1966–1967 to 1994–1995.

that relies heavily on international trade. Not only does national self-sufficiency protect a country from the vagaries of international market prices, it reflects a multiskilled workforce and the ability to efficiently produce a wide variety of goods. National self-sufficiency can be determined by calculating import expenditure as a percentage of GDP. The lower the percentage, the higher is a nation's self-sufficiency.

- *Average life expectancy at birth (being).*
- *Infant mortality rate (being).*
- *Adult literacy levels (being).*   Literacy levels are an important indicator of a nation's human capital as well as the capacity of its citizens to satisfy their self-actualisation needs.
- *Population density (being).*
- *Suicide rate (being).*   Suicide is a growing problem in many countries that reflects a decline in the self-worth and the sense of belongingness of individual citizens.
- *Average number of leisure hours enjoyed per person (doing).*   The number of leisure hours indicates the potential of a nation's citizens to pursue benefit-yielding pursuits not related to either work or consumption.
- *Average weekly hours worked (doing).*   This and the previous item are important because it is possible for income levels to be much the same in two countries but for the citizens in one nation to be working longer and harder owing to a lower hourly wage.
- *Assault rate (relating).*
- *Burglary rate (relating).*   An increase in such crimes as assault and burglary indicates a more violent and insecure place to live and a breakdown in social capital and other social support structures.

While a change in any particular indicator reveals something specific about the development of a nation's citizens, a more general development perspective can be gained by combining them into a single development index. The efficacy of this approach has been recognised by the UNDP with the construction of a Human Development Index (HDI). The HDI involves the aggregation of three development indicators — adult literacy, life expectancy, and income — and is constructed in the following manner (UNDP, 1994). First, for any given country, a "deprivation" indicator is calculated for each of the three variables. This is achieved by identifying a minimum and maximum value for each variable and then placing a country in the 0–1 range defined by the difference between the maximum and the minimum values. Second, an average deprivation indicator is calculated by taking a simple average of the three deprivation indicators (note that the UNDP assumes that adult literacy, life expectancy, and income make an equal contribution to human development). The HDI is finally calculated by subtracting the average deprivation indicator (a value somewhere between 0 and 1) from 1. As a guide, the HDI and the average deprivation indicator for the highest ranked country in 1991, Japan, was 0.993 and 0.007, respectively.

A similar approach can be adopted by aggregating all the supplementary development indicators listed above.* The only major difficulty with using this approach is having to assign weights for each indicator based on their perceived contribution to the human development process. While there is little doubt that average life expectancy is of greater importance than national self-sufficiency, conjecture will always surround the relative contribution of average life expectancy and the infant mortality rate. Indeed, even where there is little argument over which factors make the greatest development contribution, a great deal of subjectivity will always be involved when determining their relative weights. For example, how much more of a contribution does average life expectancy make compared to national self-sufficiency? To some extent, this problem can be dealt with by calculating a number of

---

* There is, however, a need to alter the way in which the deprivation indicator is calculated for the infant mortality rate, population density, average weekly hours worked, and the suicide, assault, and burglary rates. Because the remaining indicators reflect positive aspects of the development process, the deprivation indicator for each indicator can be calculated as per the UNDP approach, that is, if I_{ij} is the deprivation indicator for the jth country with respect to the ith variable, then (UNDP, 1991, page 88):

$$I_{ij} = \frac{\left(\max X_{ij} \pm X_{ij}\right)}{\left(\max X_{ij} \pm \min X_{ij}\right)}$$

As a consequence of the infant mortality rate, population density, average weekly hours worked, and the suicide, assault, and burglary rates being negative indicators, the deprivation indicator must be calculated on the basis of how much each variable differs from the minimum rather than maximum value. Thus, if I_{nj} is the deprivation indicator for the jth country with respect to the nth negative variable, then,

$$I_{nj} = \frac{\left(X_{nj} \pm \min X_{nj}\right)}{\left(\max X_{nj} \pm \min X_{nj}\right)}$$

aggregate development indexes with different weights assigned to each of the development indicators. It then could be left to the observer to decide which of the aggregate development indexes is the best indicator of human development.

In order to assign different weights to each of the development indicators, they can be separated into different categories according to their likely contribution to the human development process. Should three categories be chosen, the indicators in the first category would be assigned the highest weight, the second category of indicators a less weight, and the third category the lowest weight. As an instructive example, the development indicators could be categorised as follows.

- *Category 1* — real per capita Hicksian income; the percentage of population living above the absolute poverty line; average life expectancy, infant mortality, and adult literacy.
- *Category 2* — the percentage of population living above the relative poverty line, population density, the average number of leisure hours enjoyed per person, and average weekly hours worked.
- *Category 3* — national self-sufficiency, the home ownership rate; the suicide rate; the assault rate, and the burglary rate.

Finally, unlike the HDI, which has a value between 0 and 1, the aggregate development index being proposed here would have a value between 0 and 100. This can easily be achieved by multiplying the final index figure by 100.

14.5.2 *Supplementary sustainability indicators*

As has been pointed out a number times throughout the book, ecological sustainability is dependent on the state of natural capital stocks and the comparison of resource and waste flows relative to the capacity of the ecosphere to regenerate low entropy matter–energy (its source function) and to assimilate high entropy matter–energy (its sink function). We have, therefore, chosen physical indicators that reveal something about the consumption of various material and energy resources, the generation of various forms of waste and pollution, and the availability of various renewable and nonrenewable natural capital stocks. However, as useful as these indicators are, particularly when considered at a per capita level, they fail to convey sufficient information about the integrity of both natural capital stocks and ecological systems (i.e., the life-support function of natural capital). They also reveal little about the magnitude of human activity relative to the supporting ecosphere. For this reason, two well-publicised sustainability concepts are included in the list of sustainability indicators. They are the percentage human appropriation of a nation's terrestrial net primary product (NPP) and a measure of a country's "ecological footprint." The list of supplementary sustainability indicators is, therefore, as follows.

- *Total and per capita energy consumption (source function).* All economic activity requires the consumption of energy. There is, however, a limit on how much energy any nation can continue to consume. A measure of the total and per capita consumption of energy indicates the stress that the nation as a whole and the individual citizen is having on the ecosphere.
- *Total and per capita arable land (source function).* It has been said that a measure of how well a society manages its resources can be measured by the state of its soil or land resources. While vegetation clearance can increase the availability of arable land, so too can the area of arable land decline owing to soil erosion, acidification, and dryland salinity, all of which are the result of poor land management.
- *Total and per capita area of timber stocks (source function).*
- *Total and per capita stock of renewable water resources (source function).* Water is probably the most precious of all resources. It not only assists in the provision of food, it is used to generate electricity and meet the demands of manufacturing industries;.
- *Carbon dioxide, nitrogen oxide, and sulfur oxide emissions (sink function).* Air pollution emission levels indicate a nation's contribution to such problems as acid rain and global warming;
- *Airborne particulate levels (sink function).*
- *Water eutrophication levels (sink function).* A measure of water eutrophication levels is necessary given that a measure of the total stock of renewable water resources can overstate the quantity of water available for immediate use.
- *Per capita solid waste to landfill (sink function).* Countries which reuse and recycle a high percentage of unwanted materials have a lower per capita rate of solid waste going to landfill sites.
- *Area of remnant vegetation as a percentage of the total land mass of a nation (life-support function).* The life-support function of a nation's natural capital depends on its biodiversity and thus the health of its ecosystems. While it is possible to ascertain the health of individual ecosystems (Costanza, 1992), it is a nigh impossible task to determine the overall ecosystem health of an entire nation. One guide to a nation's ecosystem health is the percentage of a nation's total land mass remaining as remnant vegetation.
- *Percentage human appropriation of the nation's terrestrial net primary product (life-support function).* Vitousek et al. (1986) have estimated that, in the early 1980s, the total human appropriation of the planet's net terrestrial primary product (NPP) was approximately 40 percent. This is around one-half of the 80 to 90 percent level that, if reached, is likely to destroy the functional integrity of the ecosphere (Rees and Wackernagel, 1994). National estimates of the percentage appropriation of the NPP serve two purposes: first, as an indicator of the life-support services and integrity of a country's natural capital stock; and second,

as a means of estimating the size of the macroeconomy relative to the supporting ecosphere;

- *Ecological footprint (source, sink, and life-support function).* A country's ecological footprint is a measure of the area of land required to sustain its consumption of food, forestry products, and energy resources in terms of both required resource inputs and inevitable waste outputs (Rees and Wackernagel, 1994; Rees, 1996; Wackernagel and Rees, 1996; Wackernagel et al., 1999). Because countries differ in terms of population and land area, an international comparison of ecological footprints requires that the estimated area of required land for each country be divided by its population to obtain a per capita ecological footprint, and an estimation of the available area of land per person or the country's biocapacity (both measured in hectares per person).* The former is important because it allows one to make international comparisons of per capita consumption pressures on the ecosphere. For example, in 1997, the United States had a per capita ecological footprint of 10.3 hectares per person compared to 0.5 hectares per person for Bangladesh (Wackernagel et al., 1999). This suggests the average American has nearly 21 times the impact on the ecosphere than the average Bangladeshi. The latter is important because, should it exceed the former, the nation in question cannot sustain its per capita consumption levels and must eventually rely on the importation of material and energy resources. In other words, it has an ecological "deficit." In 1997, the Unied Sates had an available bio-capacity of 6.7 hectares per person and so had an ecological deficit of –3.6 hectares per person. Bangladesh, on the other hand, had an available biocapacity of 0.3 hectares per person and an ecological deficit of –0.2 hectares per person. One of the advantages of the ecological footprint concept is that it highlights the importance of population and population density. For example, Australia, Canada, and Finland have relatively high per capita footprints (9.0, 7.7, and 6.0 hectares per person) but, because of their low population density, have ecological surpluses (5.0, 1.9, and 2.6 hectares per person).

While there are some methodological problems associated with the ecological footprint concept (see van den Bergh and Verbruggen, 1999), the concept is widely accepted as one of the most effective and ecologically meaningful ways of appraising the sustainability performance of different nations. For this reason, the ecological surplus/deficit of a country can serve as a useful means for judging its sustainability status. Judging countries in this way avoids the need to obtain an aggregate sustainability indicator, as was done with the supplementary development indicators.

---

* The country's biocapacity is calculated on the basis that at least 12 percent of the total land area must be set aside exclusively for biodiversity preservation (WCED, 1987; Noss and Cooperrider, 1994).

There is, however, one adjustment that needs to be made before assessing the sustainability performance of different nations. Because very poor countries, such as Bangladesh, India, and Nigeria, have very low per capita ecological footprints, it follows that they do not have large ecological deficits (–0.2, –0.3, and –0.9 hectares per person, respectively). From a sustainability perspective, each of these countries ranks above much wealthier nations with slightly higher ecological deficits. This is misleading because one has to ask themselves what it is that is "almost" being sustained. For the majority of the people living in these countries, not much that contributes toward their development. Clearly, an urgent increase in consumption growth is needed in these countries. However, this would have the immediate effect of increasing their respective ecological deficits.

To ascertain a more appropriate measure of the ecological surplus/deficit of poor countries, the 1997 world average ecological footprint of 2.8 hectares per person can be subtracted from the national biocapacity on the assumption that an ecological footprint of this magnitude is required to enjoy a sufficient per capita consumption level. An ecological footprint of 2.8 hectares per person also closely corresponds to the ecological footprint of countries classed by the UNDP as having a "medium" human development status. Making this adjustment increases the ecological deficits of Bangladesh, India, and Nigeria to –2.5, –2.3, and –2.2 hectares per person.

To further assist in the assessment of a nation's sustainability status, countries can be classified into six different sustainability categories. The gradation of each category is based on the 1997 estimate of the world's biocapacity of 2.0 hectares per person. The six sustainability categories and some of the countries included in them are:*

- *Very strongly sustainable* — countries with an ecological surplus of 4.1 or more hectares per person (Iceland 14.3, New Zealand 12.8, Australia 5.0, and Peru 4.9).
- *Strongly sustainable* — countries with an ecological surplus of 2.1 to 4.0 hectares per person (Brazil 3.6 and Finland 2.6).
- *Weakly sustainable* — countries with an ecological surplus of 0.0 to 2.0 hectares per person (includes such countries as Canada 1.9, Colombia 1.3, Sweden 1.1, Argentina 0.7, Ireland 0.6, Chile 0.4, Malaysia 0.4, and France 0.1);.
- *Weakly unsustainable* — countries with an ecological deficit of –0.1 to –2.0 hectares per person (includes such countries as Indonesia –0.2, Costa Rica –0.3, Austria –1.0, Hungary –1.0, Mexico –1.4, Turkey –1.5, Spain –1.6, and Thailand –1.6).
- *Strongly unsustainable* — countries with an ecological deficit of –2.1 to –4.0 hectares per person (includes such countries as Nigeria –2.2, India –2.3, the Russian Federation –2.3, Bangladesh –2.5, Greece –2.6,

---

* The ecological surplus/deficits of the countries indicated are from Wackernagel et al. (1999, Table 4). An adjustment has been made to the ecological surplus/deficits of Peru, Colombia, Chile, Indonesia, Costa Rica, Mexico, Thailand, Nigeria, India, and Bangladesh.

Switzerland –3.2, Germany –3.4, Japan –3.4, United Kingdom –3.5, United States –3.6, and Belgium –3.8,).
- *Very strongly unsustainable* — countries with an ecological deficit of –4.1 or more hectares per person (Hong Kong –5.1 and Singapore –6.8).

The sustainability status and relative positions of the countries revealed previously are unlikely to remain as indicated. Because of the likelihood of significant increases in both the consumption levels and population of many developing countries over the next few decades, one can expect large rises in their per capita ecological footprint as well as large decreases in their per capita biocapacity. This translates to a probable deterioration in the ecological surplus/deficit position of most developing countries and their gradual relegation to a lower sustainability category. Conversely, there is likely to be little if any deterioration in the ecological surplus/deficit position of most wealthy countries. Indeed, given the relatively stable population of some of these countries (e.g., Sweden, Finland, Germany, and the United Kingdom) and their lessened need for growth, an improvement in their ecological surplus/deficit position may eventually be observed.

14.5.3 Ranking the sustainable development performance of different countries

Assessing and ranking the SD performance of different countries becomes a simple process of combining the ecological footprint concept and the proposed aggregate development index. Because ecological sustainability is a prerequisite for any nation to claim genuine SD status, countries should first be categorised according to their sustainability performance. This means categorising them in terms of the six sustainability categories outlined previously. Within each category, countries can then be ranked in accordance with their aggregate development index. In terms of an overall SD performance, a nation belonging to the weakly sustainable category would rank above a nation belonging to the weakly unsustainable category even if it had a lower aggregate development index value.

Interestingly, when the SD ranking of the different countries is considered closely, the rationale for adjusting the ecological surplus/deficits of poor nations seems well justified. Following the adjustment, the United States outranks Bangladesh, India, and Nigeria on the basis of its far superior development status, though not on sustainability grounds. Prior to the adjustment, Bangladesh, India, and Nigeria outrank the United States — an unrealistic result despite the United States' massive ecological footprint and correspondingly large ecological deficit. Nevertheless, the United States still ranks behind such countries as Brazil, Colombia, Costa Rica, and Thailand, indicating that, despite its much higher development status, achieving SD requires more than just a high standard of living. It requires a standard of living that can be ecologically sustained.

chapter fifteen

Toward sustainable development: policy conclusions and prescriptions

15.1 Structure of the chapter

Chapter 15 outlines the types of policies required to achieve SD. In keeping with the policy framework outlined in Chapter 13, the chapter begins with a range of policy initiatives to directly resolve the intermediate macro policy goals of allocative efficiency, distributional equity, and ecological sustainability. By applying a separate policy instrument in each case, these initial policies go a considerable way toward promoting outcomes of the ecological economic efficiency kind. However, because an ecological economic efficiency context is defined in relation to a society's dogmatic belief in objective value, two additional policies are prescribed to establish and maintain the market's moral capital foundations — the very foundation upon which the market's individualistic ethos has the propensity to deplete. The first involves the introduction of functional property rights to ensure that one's right to the use, employment, and consumption of property is conditional upon the performance of duties and obligations. The second involves the introduction of a corporate charter as a means of imposing conditions on the range and nature of a corporation's activities.

As briefly alluded to in Chapter 6, international trade provides the opportunity for a nation to beneficially shift the uncancelled benefit (UB) and uncancelled cost (UC) curves and increase sustainable net benefits. Through specialisation in commodity production and waste disposal, international trade can lead to a level of sustainable net benefits greater than what a nation can generate through domestic production alone. However, if trade is conducted on the basis that an unfettered international market is indisputably desirable, it can reduce a nation's sustainable net benefits, if not render net benefits unsustainable. A number of policy initiatives are prescribed to promote an international trading environment more conducive to achieving SD.

In the next part of the chapter, policies are prescribed to deal with a number of policy-related aspects discussed throughout the book. These include

- The facilitation of efficiency-increasing technological progress which, unlike the throughput-increasing variety, beneficially shifts the UB and UC curves and guarantees a rise in sustainable net benefits.
- The restoration of some degree of central bank control over the nation's money supply in view of the falsely conceived nexus between money (debt) and human-made capital (wealth).
- The severing of the GDP-employment and GDP-poverty links to overcome two of the more critical economic growth imperatives.
- The promotion of an international consensus on SD issues;
- A consideration of the contentious issue of discounting and its implications for sustainable resource use.

Finally, as indicated in Chapter 12, the implementation of SD policies is made difficult by the tendency of the global system, of which the macro-economy is a part, to react to any change in a conservative, identity-preserving fashion. In order to take this co-evolutionary factor into consideration, a policy timetable is provided at the end of the chapter. The timetable tentatively outlines a period over which the policies prescribed in this chapter can be introduced gradually.

15.2 Policy prescriptions to resolve the macro policy goal of allocative efficiency

15.2.1 Establishing a set of properly defined property rights

Property rights are a legal set of nonprice rules that govern the manner in which property owners may utilise, exploit, and exchange the natural or human-made capital in their possession. A clearly defined and enforceable set of property rights is essential to the efficient and effective operation of markets. In most market economies, property rights are well defined, at least in terms of three critical aspects pertaining to them — namely, the universality, exclusivity, and the transferability of property (DeSerpa, 1988). Consequently, there is little to prescribe in terms of the legislative mechanisms required to define, institutionalise, and enforce property rights. Nevertheless, problems do exist in terms of the actual nature of and extent to which property rights are assigned. It is in relation to this aspect of property rights that policy changes are required.

15.2.1.1 Exclusivity

Consider the issue of *exclusivity*. Property rights exist to provide the owner of property exclusive rights to its utilisation, exploitation, and transferability. They have importantly served to establish and maintain a reward-based incentive structure necessary to facilitate the efficient allocation of resources (DeSerpa, 1988). This aside, does it not matter for what and for how long one should have exclusive rights to property? In most market economies, the perception is that it does not, because property rights are assigned to virtually anything capable of ownership and, in the majority of instances, for an indefinite period. However, this can lead to a number of problems, not the least being the inefficient allocation, unjust distribution, and unsustainable use of scarce resources.

For example, it is widely believed that allocative efficiency can be increased by giving legal private goods status to what has public goods or open access resource characteristics. Sometimes referred to as a Coasian redefinition of property rights, it is argued that by endowing private ownership status to resources previously devoid of exchange value, one can improve the allocation of abused resources and overcome such undesirable outcomes as the "Tragedy of the Commons." While, under certain conditions, a Coasian redefinition of property rights can genuinely improve the efficiency of resource use, it is wrong to believe that a legal redefinition of property rights can alter or extinguish an object's inherent public goods characteristics. Indeed, Coasian solutions all too often result in both the unsustainable use of resources and distributional inequities. Unsustainability invariably results because it is erroneously assumed that ecological systems are unaffected by their dissection into smaller parcels of privately owned land. For instance, the private ownership of a pristine forest may guarantee a sustainable flow of commercially viable timber but not the ecosystem services it provides. Inequity can often result because the private ownership of natural and human-made capital with public goods characteristics deprives many people access to a range of services they previously enjoyed. Despite these shortcomings, Coasian solutions give the impression of having addressed the problems they are designed to resolve because, on the whole, an improvement in allocative efficiency is observed. Yet there are three good reasons for doubting whether such an improvement always follows.

First, the exclusivity conferred by a Coasian redefinition of property rights can lead to an underprovision of human-made public goods and, as explained previously, the destruction of important ecosystem services provided by naturally occurring public goods. Thus, giving private ownership to objects with public goods characteristics can lead to inefficiencies in itself. Second, it can also lead to the deprivation of use value-increasing knowledge and other important forms of information (e.g., genetic information). Knowledge, existing as it does in a variety of forms, exhibits public good characteristics. Its availability enhances the potential emergence of efficiency-increasing technological progress. However, because knowledge and other important forms of information are subject to private ownership under existing patent and copyright laws, this potential can be greatly reduced. In the third instance, exclusivity can reduce the contestability status of markets, which is necessary to keep firms' economic profits at the normal level and facilitate an efficient use of resources. This is particularly so in the case where private property rights result in the exclusive ownership of producer goods in industries where *sunk costs* are highly prevalent. Sunk costs, as Demsetz (1968) has shown, constitute a more potent market entry barrier than economies of scale. Consequently, considerable market power can be conferred to firms with exclusive ownership rights to sunk cost producer goods.

All told, in prescribing policies to deal with the issue of exclusivity, it is necessary to keep in mind the fundamental reasons for conferring exclusive ownership to property in the first place. These are

a. The need to establish a suitable incentive structure by rewarding individuals for their service-providing and/or efficiency-increasing efforts.
b. To permit the legal owner of property to engage in mutually beneficial exchange.
c. To enable individuals to enjoy the fruits of their efforts.

This having been recognised, it is then necessary to ask the following. What is required to ensure the preceding conditions are met without denying access to the public goods characteristics of knowledge and of some forms of natural and human-made capital, without reducing the potential emergence of efficiency-increasing technological progress, and without diminishing the contestability status of markets? In the past, it has been assumed that to meet conditions (a), (b), and (c) then, for all but pure or near pure public goods, where provision is best left to government instrumentalities, the preferred option is to confer exclusive private ownership and private provision of property in all its various forms. Of course, in the case of private goods, such an option is a desideratum. But, because impure public goods are similarly impure private goods, property rights must be assigned so as to adequately reward the creator of an object's increased use value while also ensuring its public goods characteristics remain available to all.

Consider knowledge again. The facilitation of use value-increasing knowledge is best dealt with, not by intellectual property rights, such as patents and copyrights, but by intellectual royalty rights. Intellectual property rights confer exclusive access and use of patented knowledge. Intellectual royalty rights, if introduced, would not. Intellectual royalty rights would confer exclusive rights to the contributors of new knowledge insofar as they would seek to reward them for their use value-increasing endeavours. This would ensure that knowledge capable of furthering the public good is not exclusively confined to those who might otherwise use it for private gain alone. Of course, a royalty-based system of rights will not always be the most practical way of striking an appropriate balance between private ownership of, and public access to, impure public or private goods. Where this is so, alternative reward mechanisms such as prizes, grants, or bonuses should instead be considered (Daly, 1996).

15.2.1.2 Transferability

With regards to the *transferability* of property rights, it needs to be recalled that many rights exist as a means of institutionalising a society's objective value-based standards and principles. They, as such, constitute the objective standards — the bedrock — upon which an allocatively efficient outcome can be properly judged and assessed. For this reason, there must be at least some limitation on the nature and extent to which property can be transferred from one individual to another. Exactly how much depends on a particular society's objective value-based standards and principles. Because policy setting in relation to this aspect of property rights is better addressed in terms of the establishment and maintenance of a market's moral capital foundations, it will be dealt with in a later section of the chapter.

15.2.1.3 Universality

The *universality* of property rights is essential for two reasons. First, it ensures the rights to ownership and exchange are applied equally to all citizens. Second, it guarantees that what is exchangeable is property and not the rights to property itself. Of the three critical aspects pertaining to the assignment of property rights, universality is, in most market economies, adequately institutionalised. Therefore, no policy prescriptions are necessary in this regard.

15.2.2 Minimising transaction costs

A transaction cost is the value of the time and/or limited resources used to gain the information required to make a beneficial form of exchange and to engage in the bargaining process itself. As explained in Chapter 9, where transaction costs exceed the potential gains from exchange, exchange does not take place. A reduction in transaction costs can increase the efficiency of the exchange process by increasing the number of potential exchanges where the net gains are positive. Clearly, there is a need for policies to minimise transaction costs where it is both feasible and practicable.

15.2.2.1 The generation and dissemination of information and knowledge

The most important of all transaction costs requiring policy attention are those pertaining to the provision and availability of something already covered in this chapter — knowledge. Knowledge is an important transaction cost factor because its widespread and accessible provision reduces the need for prospective market participants to seek, sort, and interpret the information required to make informed market decisions. Because of the public goods characteristics of knowledge, governments are required to play a significant role in both its generation and dissemination. This requires five main things on the part of governments.

First, government funded research levels must be maintained. Publicly funded research ensures the generation of potential use value-increasing knowledge considered uneconomical for the private sector to finance. Second, the information generated by government statistical agencies (such as the Australian Bureau of Statistics in Australia) must be made accessible and affordable to all. Third, governments must actively participate in all prominent forms of information disseminating media — e.g., television, radio, and so on. Without government-owned broadcasting agencies, "uneconomical" forms of information, that is, information unable to attract large audiences, is unlikely to be broadcast in sufficient quantities. Fourth, there is a need for laws that guarantee the diversity of private media ownership. A diversity of ownership ensures the dissemination of a wide range of opinions and the availability of a diverse range of information. Finally, adequate legislation is required to ensure important knowledge is not unduly withheld at the public expense. While a system of intellectual royalty rights outlined in this chapter would play an important role in this regard, there is also a need for adequate labelling of goods to allow consumers to make informed choices regarding product quality and the means and whereabouts of their manufacture.

15.2.2.2 Removing unnecessary nonprice rules and regulations

Nonprice rules and regulations often complicate the bargaining process and increase transaction costs. Many, therefore, need to be removed. Having said this, policy makers should not set out to extinguish all existing nonprice rules. Like property rights, many nonprice rules exist as means of institutionalising a society's objective value-based standards and principles and, thus, play an indispensable role in the evolution of an appropriate market domain. Clearly, nonprice rules of this nature need to be maintained and, whenever possible, be updated and improved.

Ironically, most contemporary societies preside over market economies with too few objective value-based nonprice rules but too many regulatory controls with little if any objective value basis. The irony exists because in opposing moral and biophysical restraints on the market economy, free marketeers precipitate the very collective intervention they wish to avoid — a reflection, again, of the falsely perceived conflict between freedom and authority. Unfortunately, the emerging plethora of rules and regulations are largely *ex post facto* responses to the operational legacy of a market machine that continues to erode its functional prerequisites. Very few of the emerging rules and regulations appear to be contributing to a more appropriate market domain.

Clearly, there is a need to simplify the bargaining process and reduce the number of unnecessary nonprice rules and regulations. This can be achieved by establishing a periodic review process to identify legislative nonprice rules that impede the efficient and effective operation of markets. Once identified, such nonprice rules must either be modified appropriately or be extinguished altogether. Over time, one would expect the review process and any subsequent modification of legislative nonprice rules to be greatly simplified by a concerted attempt to regenerate and maintain the market's functional prerequisites.

15.2.2.3 *Reducing the cost of contract enforcement*

For contractual arrangements to serve positively in the facilitation of mutually beneficial exchange, it must be possible for affected parties to enforce contractual conditions and to seek damages where such conditions are transgressed. Better still, it is important that the cost of contractual enforcement and redress be kept to a minimum. Only then can one be confident that the potential gains from exchange will exceed the total value of resources expended over the entire length of the terms of exchange. For this to occur, two things are necessary. First, because both property rights and the terms of any contractual arrangements are legal concerns, legal services must be readily accessible and affordable to all. Because the poor will always have difficulty affording legal services, they must be provided with subsidised legal aid in the same way as basic health care. Second, there is a need for appropriate legislation and independent authorities to protect parties from damages caused through negligence (e.g., product liability and other consumer protection laws).

In most developed countries, laws to protect individuals from negligent behaviour are adequately provided. Nonetheless, because of the current rate of technological change and the potential for surprising events to emerge in the future, there is always the need to introduce new laws to deal with new and novel circumstances. The periodic review process described previously is an ideal opportunity to examine existing laws and to identify areas requiring new legislative measures.

15.2.3 *Establishing and maintaining the contestability status of markets*

Whereas economies of scale largely determine the efficient number of firms in an industry, it is the prevalence of sunk costs that confers market power. Markets characterised by the relative absence of sunk costs are deemed "contestable" because their absence permits the easy entry and exit of potential market participants.* Potential competition, like actual competition, is able to discipline market incumbents by allowing a transient profit opportunity to be readily captured by a potential market entrant. It is the relative absence of sunk costs that ensures contestable market outcomes have the allocative efficiency qualities of more traditional competitive market

---

* A sunk cost is a cost that cannot be recovered. For a detailed and rigorous explanation of contestable markets and the importance of sunk costs, see Baumol et. al., (1982).

structures — that is, market prices approximate the marginal cost of production and firms earn normal economic profits.

Provided policies can be designed to minimise the exclusive ownership of, and access to, sunk cost producer goods, there is no need to focus a national competition policy exclusively on the actual number of firms in any given industry, as has been the traditional approach in most countries. Indeed, because economies of scale have a significant bearing on the efficient number of firms in an industry, attempts to maintain a large number of actual market participants can lead to suboptimal size firms and a lessening of efficiency. This aside, there will always be a need to outlaw anticontestable behaviour and to prohibit mergers where sunk costs unavoidably remain a significant barrier to market entry.

15.2.3.1 Dealing with sunk costs

To establish and maintain the contestability status of markets, sunk costs need to be detached from the serving or utilising firm. There are two ways to achieve this. First, the sunk costs can be borne by a government instrumentality as is normally the case with highways, railways, and airports (Bailey, 1981). The sunk cost producer goods can then be leased to firms on a short-term transferable licence basis on the presumption that the highest bidders are the most efficient producers. The transferability of the licences ensures the cost of the lease is a fixed rather than a sunk cost. Second, should this approach be neither possible or practicable, the government can mandate that sunk costs be borne by a consortium. Consortiums already share in the provision of international broadcasting satellites and, in some countries, the provision of telecommunications networks. The sunk cost producer goods can again be leased to utilising firms in the manner described previously.

There is, unfortunately, one major problem with the preceding policy. It has the clear potential to reduce the incentive for utilising firms to qualitatively improve sunk cost producer goods, in particular, improvements that would otherwise increase the service and maintenance efficiency of human-made capital. Why make advances in sunk cost producer goods if you do not have exclusive ownership of them? To overcome this problem, royalty rather than property rights should be conferred to those who develop improved sunk cost producer goods. By granting royalty rather than property rights, the superior producer goods remain accessible to those who wish to access and utilise them, thereby ensuring the contestability status of markets, while the creator of the superior facilities receives an ongoing but limited pecuniary payment as a reward for their efficiency-increasing efforts.

15.2.3.2 Dealing with instances where sunk costs cannot be detached from the utilising firm

In the case of many capital-intensive industries, the exclusive access to sunk cost producer goods cannot be detached from the serving or utilising firm. It is often too expensive or too risky for either governments or a private consortium to bear the cost of providing them. Because it is under these circumstances where economies of scale pose the greatest potential for exploitation of market power, antitrust legislation is required to prohibit any anticompetitive behaviour by market incumbents, and to limit mergers where they are likely to substantially lessen competition. To ensure the success of antitrust legislation, a well funded and independent statutory authority is required to scrutinise market participants and assess impending mergers. Because prevention is better than a cure, penalties for breaches of antitrust legislation must be large enough to serve as a deterrent to anticompetitive behaviour.

15.2.3.3 *Market deregulation and the privatisation of government instrumentalities*

The recent global trend towards market deregulation and the corporatisation and privatisation of government instrumentalities reflects the strong desire of policy makers to facilitate the efficiency-increasing benefits of potential competition and private ownership (Brown, 1996).* While market deregulation and privatisation appear to be a positive step toward increasing the efficient operation of markets, there are a couple of aspects concerning both that warrant close attention.

First, most of the reform measures introduced by policy makers to increase the level of potential competition have focussed entirely upon the freeing up of highly regulated markets. While such measures have successfully exposed previously legalised monopolies and duopolies to market forces, many have failed to adequately deal with sunk cost considerations. Consequently, deregulatory policies have left many markets less contestable than they could otherwise be and, as a consequence, more vulnerable to the abuse of potential market power.

Second, the quest for increased efficiency has led to a dramatic rise in the privatisation of government instrumentalities in the belief that privately owned business enterprises operate more efficiently than their publicly owned counterparts. Despite apparent evidence to support such a belief (Hutchinson, 1991; Vining and Boardman, 1992; Clare and Johnston, 1992; Megginson et al., 1994), sufficient doubt exists to conclude that privatisation itself cannot guarantee efficiency gains (Bishop and Thompson, 1992). This doubt exists largely because efficiency is often empirically evaluated on the basis of profitability and/or sales volume per employee or per asset. As Brown (1996) points out, private enterprises can be expected to outperform government instrumentalities in these two areas because the public sector, by its very nature, has multiple objectives of which profit maximisation is rarely one of them. Hence, efficiency comparisons between private and public enterprises are invariably weighted in favour of private enterprises. Not

---

* Privatisation involves the sale and subsequent transfer of public assets to the private sector. Corporatisation, on the other hand, involves the introduction of private enterprise-based objectives and practices to the operations of government instrumentalities.

surprisingly then, in circumstances where private enterprise-based practices have been incorporated into the operational objectives of government instrumentalities (of which profitability and sales volume usually become important considerations), operational efficiency has dramatically improved (EPAC, 1995). What does this indicate? It indicates that a properly conceived and executed form of market deregulation (i.e., one where sunk costs are also appropriately dealt with) is likely to have far greater efficiency-increasing benefits than the mere transference of assets from public to private ownership (Bishop and Thompson, 1992).

Finally, assume that the privatisation and corporatisation of government instrumentalities leads to improvements in their operational efficiency. Before any such initiatives can be implemented, it must first be demonstrated that the instrumentality in question is not a natural monopoly or the provider of public goods. Should either or both be the case, the private ownership or the introduction of private enterprise-based objectives and practices will lead to a socially undesirable quantity and an inflated market price for the good in the case of the enterprise being a natural monopoly, or underprovision in the case of it being the supplier of a public good. Only government ownership of natural monopolies and government provision of public goods can prevent the potential for market failure arising from private sector involvement. Thus, in order to improve the operational efficiency of publicly owned natural monopolies and government instrumentalities providing public goods, an alternative means to privatisation is required. Introducing incentives for public sector employees such as rewards and bonuses for improvements in both productivity and efficiency are likely to be of considerable assistance in this regard.

15.2.4 Labour market and corporate reform

Labour is a resource in the sense that it constitutes an agent in the transformation of natural to human-made capital. How effectively labour is used in this transformation process is measured by the change in the real unit cost of labour, which is calculated by dividing the cost of employing labour by the productivity of labour. Provided the percentage increase in the latter exceeds that of the former, the real unit cost of labour declines and, *ceteris paribus*, so does the marginal cost of production. In view of the obvious importance of lowering production costs, the key to a successful labour market reform process is the implementation of policies that boost the productivity of labour. Increases in labour productivity are obtained by improving the knowledge and skills possessed by human labour. They are also obtained by improving the logistical employment of labour and/or by establishing workplace arrangements that increase the commitment, motivation, and responsibility of workers. How the preceding can best be facilitated is open to considerable debate. While specific labour market policies will not be prescribed here, a couple of things are worth mentioning.

First, there is little doubt that flexible labour markets do much to increase labour productivity. They make it easier for employers to utilise labour when they most require it and for employees to acquire productivity-enhancing skills. The flexibility of labour markets can be greatly increased by moving toward enterprise-based arrangements between employers and employees and away from the traditional setting of award conditions based on trade or professional classifications. This is because enterprise-based arrangements allow award conditions to better reflect the idiosyncratic needs of individual firms. Having said this, enterprise-based arrangements must always be subject to minimum award conditions to ensure efficiency is not increased at the expense of objective value-based standards and principles pertaining to conditions of work (e.g., occupational health and safety standards).

Second, there is a need to focus policy attention on the legal and formalised nature of the modern corporation. The legal nature of modern corporations has served to perpetuate the existence of three conflicting parties — namely, corporate owners (stockholders), corporate managers, and employees. In doing so, it has the led to the evolution of industrial relations systems that, in a strict Hegelian sense, are dialectical in nature. As will be explained in the next chapter, dialectical systems are systems with a contradiction implicit in their origin and subsequent development. It is because of this contradiction that dialectical systems hinder rather than advance the knowledge-building process (Boulding, 1970a). From a labour productivity perspective, a dialectical industrial relations system impedes efficiency-increasing technological progress and the capacity of labour to contribute to the more efficient transformation of natural to human-made capital.

To assist in the transition to a nondialectical industrial relations system, the legal and formalised nature of a corporation needs to be altered to ensure all parties share common rather than disparate corporate interests. One way to achieve this is to change the nature of corporate ownership, particularly in relation to stock holding. Long ago, Tawney explained the rationale behind private ownership and why so many have sought to protect one's right to own property. To wit (Tawney, 1921, page 64):

> "Property reposed, in short, not merely upon convenience, or on the appetite for gain, but on a moral principle. It was protected not only for the sake of those who owned, but for the sake of those who worked and of those whom their work provided. It was protected because, without security for property, wealth could not be produced or the business of society carried on."

However, in most contemporary societies, property no longer exists solely as an aid to creative work. It invariably serves as an instrument for the acquisition of pecuniary gain or the exercise of power where, in many instances, there is "no guarantee that pecuniary gain bears any relation to service, or power to responsibility" (Tawney, 1921, page 66). Tawney, there-fore, believed that, as much a society should preserve the private ownership and control of the property accruing from the toil of one's efforts, so it should abolish the gains from "passive" ownership accruing from the toil of some-one else's efforts.

In order to put Tawney's view of property into practice, the legal nature of a corporation should be altered in the following manner. First, stockhold-ers would receive an annual interest payment equivalent to the opportunity cost of financial capital as is presently the right of a debenture or bondholder. As a passive part-owner of a corporation's assets, a stockholder would not receive a share of the corporation's profits unless, of course, they are also a manager or employee of the firm. Second, the stockholder would be stripped of the right to control the management of the corporation's assets. Third, management responsibility would be assigned to managers and employees who are actively involved in the creative utilisation of the corporation's assets. Finally, managers and employees would be the sole beneficiaries of the profits generated by the toil of their efforts (less the sum of the interest payment to stockholders).

Should the legal nature of the corporation to be altered in the preceding fashion, managers and employees would be better placed to establish more productive workplace arrangements. Of course, the question still remains as to what kinds of workplace arrangements increase the operational efficiency of the firm. Empirical evidence suggests the most effective workplace arrangements are those with incentive-based means of remuneration, for example, profit-sharing arrangements (Weitzman, 1984; Estrin, 1986; Blandy and Brummitt, 1990). Such workplaces are typically characterised by a high level of workplace democracy, harmony, and trust, all of which facilitates a long-term commitment on the part of everyone concerned to the future success of the organisation. For this reason, profit-sharing and other incen-tive-based workplace arrangements should be more actively encouraged.

15.2.5 *Ameliorating externalities*

Under ideal conditions, markets have the capacity to generate Pareto efficient outcomes. Because ideal conditions do not exist, markets will sometimes allocate resources inefficiently. That is, markets will from time to time "fail." As indicated in Chapter 9, market failure arises because of the "public goods" characteristics of certain types of human-made and natural capital, the exist-ence of natural monopolies and other imperfectly contestable markets, and the presence of positive and negative externalities.

To a large degree, the policies required to deal with the first two sources of market failure have already been outlined and discussed. My focus of attention is, therefore, on the amelioration of externalities. Two categories of externalities need to be addressed. The first is a category of externalities where the nature and likely extent of any spillover benefits or costs can be specified *a priori*. The second category of externalities involves spillover effects resulting from a co-evolutionary change in the underlying parameters of social, ecological, and macroeconomic systems. In these circumstances, the nature and extent of any spillover benefits or costs are, in principle, unpredictable. As explained in Chapter 12, both categories of externalities demand an entirely different policy response. Whereas Pigouvian taxes and subsidies, tradeable permits, and Coasian solutions are able to effectively deal with the first category of externalities, they are ineffective at ameliorating externalities with a co-evolutionary origin.

In terms of the first category of externalities, the means by which they can best be ameliorated has long been researched and debated by economists. Because the various solutions put forward by economists have their own advantages and disadvantages, the choice of solution ultimately depends on the circumstances contributing to the externality in question. The following solutions are prescribed to deal with the below described circumstances.

- *Property rights (Coasian) solutions.* This is most applicable where the externality in question involves private rather than public goods (see Section 15.2.1.); where only a small number of easily identifiable parties are affected, where the spillover effects are nonpervasive, and where the transaction costs of a potential Coasian bargaining process are negligible.
- *Pigouvian taxes and subsidies.* This is the preferred solution in circumstances where the spillover effects are pervasive, where the number of affected parties is large and diverse, where the identification of affected parties is difficult, and where the transaction costs of a potential Coasian bargaining process are prohibitively large. Pigouvian remedies are also preferred in circumstances where Coasian solutions fail to adequately compensate injured parties and/or facilitate environmental rehabilitation. Taxes, for instance, allow governments to compensate injured parties and, if necessary, rehabilitate the natural environment. However, the choice between the two must be considered in light of the fact that Coasian solutions are much simpler and cheaper to implement. Pigouvian solutions are considerably more complex and costly because, in order to implement them, it is necessary to employ bureaucrats to administer the process, to calculate appropriate taxes and subsidies, and to identify and measure the source of all spillover benefits and costs.
- *Tradeable permits.* Tradeable permits are referred in circumstances where the spillover effects of certain activities threaten the long-term sustainability of the ecosphere's source, sink, and life-support services. Tradeable permits ameliorate externalities in much the same way as a Pigouvian tax. For example, the premium paid for a permit during an initial auctioning process induces resource depleters and polluters to reduce the spillover effects of their actions. Thus, the spillover costs that had previously escaped market valuation are captured and internalised without a Pigouvian tax having to be calculated. The

advantage of employing a tradeable permit scheme in circumstances where the ecosphere's instrumental services are at stake is that, by placing a limit on the quantity of permits initially auctioned, the rate of resource throughput can be restricted to one that is ecologically sustainable. This allows the macro policy goal of allocative efficiency to be resolved without compromising the macro policy goal of ecological sustainability — a consequence of a tradeable permit scheme incorporating the separate policy instruments of relative prices and quantitative throughput restrictions. A proposed system of tradeable resource exaction permits is explained in greater detail later in the chapter.

15.2.5.1 Internalising the user cost of nonrenewable resource depletion
Sustainability precept No.1b demands the setting aside of some of the proceeds from nonrenewable resource depletion to cultivate renewable resource substitutes. To determine the amount to be set aside, it is necessary for a nonrenewable resource earmarked for depletion to be converted into a perpetual income stream. This, according to El Serafy (1989), requires a finite series of earnings from the sale of the resource to be converted into an infinite series of true income such that the capitalised values of the two series are equal. To achieve this, an income and capital component of the finite series of earnings must be identified. If correctly estimated, the income component constitutes an amount that can be consumed without any fear of undermining the capacity to sustain the same level of consumption over time. The capital component, on the other hand, becomes the amount that needs to be set aside each year to ensure a perpetual income stream of constant value, both during the life of the resource as well as following its exhaustion. It is this capital component that constitutes the true "user cost" of resource depletion.*

At present, there is no requirement on the part of nonrenewable resource depleters (e.g., miners and drillers) to set aside some of their profits to cultivate renewable resource substitutes. As a consequence, the user cost of nonrenewable depletion is not being fully reflected in nonrenewable resource prices. This constitutes a negative externality. In order to internalise the externality, nonrenewable resource depleters should be legislatively required to deposit the necessary portion of their depletion proceeds into a capital account. The capital account could be held by a government authority (hereafter referred to as the "Environmental Agency") that would oversee both the depositing of the necessary portion of depletion proceeds and the cultivation of renewable resource substitutes. (Note that the same authority can also be assigned the responsibility to set and police environmental guidelines; to set, reimburse, and confiscate environmental assurance bonds; and to administer a tradeable resource exaction permit scheme, all of which are discussed later in the chapter).

---

* For more on the user cost of nonrenewable resource depletion and how the income and capital components should be calculated, see Appendix 2, Item 1.

15.2.5.2 *Environmental assurance bonds*

As explained in Part III of the book, uncertainty and human ignorance are part and parcel of a co-evolutionary world. Complexity and genotypic change forbid human beings from accurately predicting the full costs associated with the manipulation of natural capital. Where a substantial portion of the costs of human activities are unpredictable in principle, that is, where the worst case scenario cannot be estimated *a priori*, a precautionary or risk-averse approach is to prohibit them. However, should the environmental costs be predictable in broad terms, prohibition is only necessary if the worst case ecological damage is intolerably large (e.g., it exceeds environmental threshold levels). If it is not, but uncertainty still abounds, environmental assurance bonds should be introduced.

A system of assurance bonds can be administered in the following manner. In addition to charging economic agents for known ecological damage (e.g., Pigouvian taxes on pollution), a bond can be levied at an amount equal to the current best estimate of the largest potential future loss of natural capital services (Costanza and Perrings, 1990). The bond is held in trust by a government authority (the Environmental Agency) for as long as the activity under scrutiny is likely to impact on the ecosphere. Over this period, the economic agent receives a payment equal to the opportunity cost of financial capital. How much of the bond is finally returned to the economic agent depends on the eventual ecological damage. Should no additional damage ensue, the bond is reimbursed in full. Where there is additional ecological damage, only part of the bond is returned. The bond is completely confiscated if the largest potential loss of natural capital services eventuates. The retained monies can be used to finance the rehabilitation of natural capital and, where necessary, to compensate people directly affected by excessive ecological damage.

The great advantage of environmental assurance bonds over Pigouvian taxes in circumstances where externalities have a co-evolutionary origin is their ability to shift potential future costs to the present (Costanza, 1994). This greatly increases the incentive of economic agents to commit resources to minimise the cost of future ecological damage which reduces the off-loading of costs onto other parties. Provided markets are sufficiently contestable, environmental assurance bonds force firms to minimise their ecological impact in order to compete with rival firms (because the confiscated portion of the bond constitutes an additional cost of production). This encourages the more efficient allocation of resources by facilitating increases in both natural capital growth and exploitative efficiencies (Ratios 3 and 4). Of course, environmental assurance bonds can only improve the overall level of allocative efficiency. As a market-based policy instrument, environmental assurance bonds cannot be relied on to achieve ecological sustainability.

Assurance bonds are most likely to be levied by the Environmental Agency on the following economic agents.

- Resource exactors (e.g., miners, well drillers, and lumberers) and ag-riculturalists (e.g., pastoralists, horticulturalists, and irrigators), be-cause the impact on ecological life-support services depends on how well resources are exacted and how well land and water resources are utilised for agricultural purposes.
- Heavy industrial polluters, in particular, those involved in activities generating toxic and highly durable substances.
- Tourism operators who conduct activities in ecologically sensitive areas (e.g., areas such as the Great Barrier Reef and the Kakadu wet-lands in Australia).
- Transport companies where the movement of certain goods can prove hazardous if they are negligent (e.g., the Exxon Valdez oil spill in Alaska).

15.3 Policy prescriptions to resolve the macro policy goal of distributional equity

15.3.1 Minimum and maximum levels of income and wealth

Distributional equity requires a fair and just distribution of both the incom-ing resource flow as embodied in human-made capital — *income*, and the accumulated stock of human-made capital — *wealth*. To achieve distribu-tional equity, it is necessary to guarantee each citizen a minimum level of income. At the very least, a minimum level of income must ensure each citizen has access to the basic necessities of life. Preferably, a minimum income would afford each citizen a number of "luxury" goods. This would not only help to satisfy some of the higher-order needs of individual citizens, it would allow them to accumulate a stock of wealth should they choose to limit their consumption of goods to the basic necessities.

For two reasons, the actual lower limit on income would depend on the income and wealth of the nation in question. First, a nation's income and wealth determines its capacity to deliver any predetermined minimum level of income. The United States, for example, has the capacity to provide each of its citizens a much higher minimum income than India. Second, as a nation becomes progressively wealthier, the minimum income level must be increased to pre-vent the disparity between rich and poor becoming ethically undesirable.

A minimum income is best provided by way of a desirable mix of a money payment and in kind benefits. The money payment, as the largest component of a minimum income, would be a standard minimum payment designed to replace existing welfare payments such as the old age and disability pensions, unemployment benefits, and higher education allowances. The remaining component of the minimum income would come in the form of nontransfer-able vouchers.* These would afford individuals a limited range of necessities such as health and dental care, basic accommodation, limited free use of public

---

* The vouchers would be nontransferable to prevent people from using them irresponsibly.

transport (e.g., five free journeys per week), and basic legal representation. Other vouchers issued would contribute to the cost of important services such as tertiary education and vocational-based training.

Because an equitable distribution of income and wealth implies a limit on the disparity between a nation's richest and poorest citizens, distributional equity also requires maximum limits on both income and wealth. However, unlike the notion of a minimum income, calls to impose limits on the maximum permissible levels of income and wealth have received very little support (Daly, 1991b; and Pizzigati, 1992). This lack of support stems from two widespread beliefs. The first is the belief in the infinite growth potential of macroeconomic systems. If macroeconomies can grow indefinitely, what need is there for a maximum limit on personal income and wealth? Of course, it is because macroeconomic systems cannot continue to expand, that the *real* income and wealth of a nation cannot continue to grow. Consequently, any continual increase in the *real* income and wealth of the rich must eventually come at the expense of the real income and wealth of the remaining population. Clearly, maximum limits on income and wealth need to be set simply to prevent a growing disparity between the rich and poor.

The second reason is the belief that a maximum limit on income blunts the incentive to achieve. Why contribute to the accumulation of the nation's wealth if, beyond a certain personal income level, you are forbidden to share in it? As reasonable as this objection seems, there are a couple of things worth considering. To begin with, only a small percentage of a nation's population is likely to be affected by a maximum limit on income. Hence, should a maximum income lessen the incentive to achieve, the cost is certain to be far outweighed by the benefits of a more equitable distribution of income. Next, economists have long pointed out the probable existence of a backward-bending labour supply curve. This curve, and the theory behind it, suggests a higher wage induces the rich to eventually opt for more leisure and less work — the consequence of the income effect of high wages exceeding the substitution effect (Samuelson et al., 1992). Because this points to higher wages acting as a "natural" disincentive for the rich to increase effort levels, a maximum limit on income would serve its equity function at little or no cost at all.

15.3.1.1 Determining minimum and maximum limits on income

As previously indicated, a minimum level of income must afford each citizen no less than the basic necessities of life. Because it is also necessary to maintain an ethically desirable gap between rich and poor, a minimum income must be set at an acceptable percentage of per capita national income (say 50 percent of either per capita Hicksian income or the per capita SNB index). Setting the minimum in this way ensures the lower limit increases with every rise in per capita national income.

The maximum limit on income is best ascertained by capping personal income at an amount equal to an order-of-magnitude difference between a nation's richest and poorest citizens. Amongst those calling for a maximum limit on income, Pizzigatti (1992) believes a factor of 10 range is the most appropriate. Daly (1996) agrees by pointing out that a factor of ten is comparable to income ranges between the lowest and highest paid employees in the civil service, the military, and universities. As for any implications for worker incentives, a factor of 10 range approximates the much lower order-of-magnitude difference in wages of executives and assembly line workers found in the Japanese automotive industry where it is quite evident that efficiency and international competitiveness have not been compromised (Daly, 1991b).

Of course, in setting a maximum limit on income, it is the order-of-magnitude difference between the rich and the poorest welfare recipient that matters, not the gap between the rich and lowly paid workers. For this reason, a factor of 20 range may be a more appropriate distributional benchmark. It is also more likely to be socially accepted. Assume, therefore, that a factor of 20 difference is chosen. With a minimum income set at 50 percent of per capita national income, the maximum would be set at an amount equal to 10 times the per capita national income. This would mean that once an individual's income reached 10 times the per capita national income, any additional income would be taxed at the marginal rate of 100 percent.

It should be pointed out that setting a maximum limit on income in this fashion does not rule out small increases over time in the *nominal* incomes of the rich. To recall, increases in the service-yielding qualities (use value) of human-made capital can lead to increases in their relative price and, therefore, an increase in a nation's *nominal* income. For a given population, this translates to a higher per capita nominal income and, if set at 50 percent of the per capita nominal income, a higher minimum income for the nation's poorest citizens. With a factor of 20 difference in the income of a nation's richest and poorest citizens imposed, a higher minimum income allows for a higher maximum income. Again, however, it must be stressed that only nominal and not real income can continue to increase in the long run and, even then, only while it is possible to continually increase the use value of human-made capital.

15.3.1.2 Determining maximum limits on wealth

Determining a desirable maximum limit on wealth is a more difficult exercise. Because it is possible for an individual to possess virtually no physical wealth, an order of magnitude difference between the richest and poorest citizens cannot be employed to impose an explicit maximum limit on personal wealth. However, with appropriate policy settings in place, it is at least possible to prevent the wealth gap between rich and poor widening to ethically undesirable levels.

Consider what is likely to occur if

a. A maximum income is set at 10 times per capita national income (20 times the minimum income).

b. Quantitative restrictions are imposed on the incoming resource flow to ensure ecological sustainability.

For starters, a maximum income would limit one's ability to acquire additional human-made capital. Next, for any given levels of maintenance, growth, and natural capital exploitative efficiency (Ratios 2, 3, and 4), throughput restrictions would set a limit on the maximum physical scale of the macroeconomy. This would limit, at any given moment, the accumulated stock of national wealth. Because personal wealth can only increase if one's ability to acquire human-made capital exceeds their consumption/depreciation of human-made capital, it is clear that both (a) and (b) would restrict the ability of the richest citizens to accumulate additional wealth. But, given that a 100 percent marginal income tax rate would not apply to anyone with an income less than the maximum, there would be no similar restriction on the less wealthy, at least while there is room for further expansion of the macroeconomy.

This having been said, the wealth gap between rich and poor can still increase if the difference between the acquisition of and consumption/depreciation of human-made capital is greater for the rich than the poor — which it can be if the rate of consumption by both classes is the same. To assist in maintaining, if not bridging the wealth gap, quantitative throughput restrictions and a maximum limit on income must be supplemented by wealth taxes (e.g., inheritance, land, and capital gains taxes). In passing, wealth taxes need to be carefully designed to ensure the rich are not penalised for the poor opting to recklessly consume its wealth — something the poor should have no need to do if a minimum income can afford them something more than the mere basic necessities of life.

15.3.2 Financing a minimum income

How and in what way redistribution mechanisms can be used to finance a minimum income is limited because redistribution mechanisms such as contestable markets and profit sharing arrangements do not directly redistribute income and wealth. Only taxes, transfers, and subsidies can be relied upon to achieve concrete redistribution targets. With this in mind, a minimum income should be financed out of general government revenue raised essentially from the following taxes and revenue raising measures.

- A progressive negative/positive income tax on incomes up to the maximum income limit.
- A 100 percent marginal tax rate on all personal income above the maximum limit.
- Wealth taxes such as inheritance, land, and capital gains taxes.
- Assurance bonds confiscated from operators inflicting ecological damage in excess of permissible limits.

- Revenue raised from the government auction of tradeable resource exaction permits.
- Revenues raised from a procreation levy imposed on violators of a transferable birth licensing scheme.
- Tariffs imposed on imported goods where market prices do not incorporate the social and environmental costs arising from the method of their production.

Most of the above measures have already been explained or are outlined later in the chapter. For now, attention will be focused on a negative/positive income tax arrangement. It would probably best function in the following manner. First, each household would receive a minimum income in the form of a standard money payment and a limited number of nontransferable vouchers to access a range of previously described services. Second, all income (e.g., wages, salaries, royalties, dividends, and interest) below the maximum income limit would be subject to a flat marginal tax rate. If preferred, the tax can be graduated in order to maintain the existing progessivity of the overall tax structure.

At a particular income level, the amount of income tax paid by each household exactly offsets the value of the minimum income it receives. This so-called "break-even" income level is equal to the inverse of the tax rate multiplied by the minimum income. Thus, if the minimum guaranteed income is $10,000 and the marginal tax rate is 40 percent, the break-even income is $1/0.4 \times \$10,000$ which is $25,000. Between the break-even and the minimum income, a household receives a net payment — a negative tax — from the government. For example, if a household's income is $20,000, its tax bill is $8,000 which, when the $10,000 minimum income is included, means the household receives a negative tax of $2,000. Conversely, beyond the break-even income level, a household pays more in tax than it receives from the government. It is, therefore, required to pay a positive tax. For example, if a household's income is $40,000, its tax bill is $16,000 which, when the $10,000 minimum income it receives is subtracted, means the household pays a positive tax of $6,000. The size of the negative and positive taxes paid depends, of course, on the income level earned. For instance, the closer is a household's income to the minimum income level, the larger is the negative tax it receives from the government. Similarly, the greater is the difference between a rich household's income and the break-even point, the larger is the positive tax it is required to pay. In all, any income earned by the members of a household increases the household's total income except where the household's income exceeds the maximum income limit.

Despite strong support for a negative/positive income tax arrangement, it is subject to two main objections. First, it is argued that current marginal tax rates on income would have to be greatly increased to afford each and every household a minimum income (Fischer et al., 1988). This, it is suggested, would dilute worker incentive and discourage efficiency-increasing efforts. There is little doubt that, under existing arrangements, undesirable income tax rises would be required. However, many of the revenue raising measures listed previously do not currently exist and, if introduced, could conceivably raise enough additional revenue to eliminate the need for large income tax rises. Indeed, there is every likelihood that governments would be in a position to reduce the marginal tax rate on all incomes below the maximum permissible level. If so, this would serve to facilitate rather than discourage efficiency-increasing efforts.

In the second instance, detractors point out that a minimum income encourages idleness and, in doing so, leaves many of the poor entrenched in a so-called "poverty trap" (Murray, 1984). They, as a consequence, argue for a reduction in the existing level of welfare payments and would clearly oppose a minimum income of the magnitude we have just proposed. If these detractors are right, as evidence suggests they are, why do so many people idly accept the bare necessities of life when paid employment provides a means to wealth accumulation and financial independence? The answer to this question is twofold. To begin with, the lack of equivalent educational opportunities has made it difficult for the poor to compete on equal terms in the labour market. Second, the alternative to a minimum income is usually a meaningless and undignified form of employment, even if it yields them a higher income. In all, because the poor have difficulty in securing anything but low paid and undesirable occupations, idle acceptance of welfare payments has, in effect, become a rational form of optimising behaviour. Clearly, a fair and just solution to idleness is not a stick-wielding dose of lower welfare payments but a policy designed to provide the poor with ample opportunity to secure meaningful and dignified forms of employment. Of course, no matter what is done to increase the opportunities for the poor, a small number of malingerers will always opt to exist on the bare necessities of life. However, this should never be seen as a reason for denying the genuinely underprivileged a minimum income.

15.3.3 Other important redistribution mechanisms

15.3.3.1 Contestable markets
The means by which contestable markets assist in the redistribution process has been explained previously many times. To reiterate, contestable markets keep economic profits at the normal level and, in doing so, extend profit-making opportunities across the entire macroeconomy. This prevents economic profits from falling in the hands of a select few. The policies to establish and maintain the contestability status of markets were discussed in Section 15.2.3.

15.3.3.2 Profit-sharing arrangements
As indicated earlier in the chapter, property no longer exists as a means to creative work. It often serves as a passive means to pecuniary gain. The growth in ownership of passive property has effectively led to two social classes — one with a primary interest in the pecuniary gains from passive

property ownership, the other reliant upon active work as a means to income (Tawney, 1921). The previously prescribed policy initiative of limiting the pecuniary gain of a stock-holder to an annual payment equivalent to the opportunity cost of financial capital and having the profits shared amongst those directly involved in the active and functional use of property (e.g., managers and workers) can go a long way toward widely distributing corporate profits amongst a nation's citizens. In doing so, it can greatly reduce the growing gap between rich and poor.

15.3.3.3 The free and universal entitlement to basic rights and privileges

Without doubt, the most fundamental of all universal entitlements is one's right to the basic necessities of life. Unfortunately, providing a minimum income to afford each citizen the bare essentials does not guarantee fairness and justice — the defining condition for all outcomes to be truly equitable. Without a comprehensive set of free and universal rights and privileges, individuals are still open to unjust forms of exploitation and abuse (e.g., inadequate occupational health and safety conditions). To avoid this, a country needs to have in place something equivalent to a Bill of Rights. Exactly what a Bill of Rights should include depends entirely on a nation's objective value-based standards and principles. While some countries already have a Bill of Rights, it is debatable whether many are sufficiently comprehensive or appropriately drafted to protect individual citizens from unjust practices and to ensure individual dignity. For example, prolonged high levels of unemployment in most countries reflects the extent to which many people have been denied the right to a meaningful and dignified form of employment. True, the inevitable existence of unemployment owing to cyclical economic factors does render one's right to employment unenforceable in a strict legal sense. However, recognition of such a right would at least force policy makers into making full employment a high priority objective, something increasingly overlooked in recent decades.

15.4 Policy prescriptions to resolve the macro policy goal of ecological sustainability

15.4.1 Tradeable resource exaction permits

In order to resolve the macro policy goal of ecological sustainability, policies must adhere to the sustainability precepts established in Chapter 4 concerning the maintenance of the ecosphere's source, sink, and life-support functions. To impose quantitative throughput restrictions without subverting the macro policy goals of allocative efficiency and distributional equity, tradeable resource exaction permits are recommended. The permit scheme can be administered in the following manner. First, in order to satisfy sustainability precept No. 1, the previously mentioned Environmental Agency would need to determine the desirable rate of nonrenewable resource depletion (h_{NR}), and the maximum sustainable rate of renewable resource exaction (h_R). Should the Environmental Agency be instructed to adopt a precautionary approach to sustainable resource use, a sustainability buffer can be incorporated into the scheme by setting exaction limits at 80 percent of the maximum sustainable rate. As an indirect consequence of setting exaction limits, the Environmental Agency would also be determining, as per the law of conservation of matter–energy, the maximum permissible quantity of high entropy waste (W). This would allow it to partially satisfy sustainability precept No. 2.*

The rights to low entropy resources would be inscribed in a limited number of transferable exaction permits. Given the potential for novel ecological feedbacks to alter the maximum sustainable rate of throughput, each permit would have a limited life of one year. This would provide the issuing authority with the necessary flexibility to vary the maximum permissible rate in line with the changing sustainable rate. At the beginning of each year, the permits would be auctioned off to the highest bidders. To ensure resource markets remain suitably contestable, buyers would be limited to a certain number of resource exaction permits as well as the number of available permits for a given resource type.

To purchase low entropy resources, possessors of resource exaction permits would convene with low entropy resource suppliers (e.g., well drillers, miners, and lumberers) in a normal market environment. Under the proposed conditions of the scheme, permit holders would forfeit all or part of their permits depending on whether or not they purchased their fully allotted resource quota. Prior to their expiration, permits can be resold at any time and to anyone except, of course, those already in possession of the maximum number of permits. A scheme of this type allows environmentalists to exercise their preference for a rate of resource throughput less than the maximum permissible rate. For example, they could elect to purchase exaction permits and abstain from their use, thereby facilitating a macroeconomic adjustment to a biocentric rather than egocentric optimum.

Provided resource markets are sufficiently contestable, low entropy resource prices should closely approximate the marginal cost of resource exaction. As such, all resource suppliers should earn, at best, normal economic profits. Indeed, because resource suppliers would be forced to charge the lowest possible resource prices, only the most efficient operators would be able to survive in the long run. Moreover, because resource suppliers would be required to lodge an assurance bond with the Environmental Agency, they would be compelled to minimise the ecological impact of their operations because any assurance bond subsequently confiscated would constitute a cost that could render the resource seller uncompetitive. Thus,

---

* It would not fully satisfy sustainability precept No. 2 because exaction limits have no direct control over the qualitative nature of high entropy waste.

by keeping resource markets contestable, it is possible to induce increases in the exploitative efficiency of natural capital (Ratio 4).

The fact that resource suppliers are forced to charge the lowest possible resource prices gives the impression that resource prices would be lowered under the proposed scheme. This is not so. As explained in Chapter 8, if the incoming resource flow exceeds the maximum sustainable rate prior to throughput restrictions being enforced, the lower overall rate causes resource prices to reflect the genuine absolute scarcity of the total resource flow as well as the relative scarcity of each low entropy resource type. This leads to a rise in low entropy resource prices. Contestable resource markets would simply ensure the sustainable incoming resource flow is made available to resource buyers at the lowest and most efficient price.

The absolute scarcity factor forcing up the price of low entropy resources is sometimes referred to as an absolute scarcity rent. Because the monetary value of this rent would be effectively equal to the premium paid for a resource exaction permit by resource buyers, absolute scarcity rents would be captured, not by resource suppliers, but by the Environmental Agency during the auctioning process. Indirectly, the absolute scarcity rents would serve as a throughput tax and induce increases in the service and maintenance efficiencies of human-made capital (Ratios 1 and 2). As previously mentioned, the revenue raised via the auctioning of resource exaction permits could be used to partially finance a minimum income. It could also be used to finance such things as

- Public investments in natural capital, in particular, ecosystem rehabilitation and conservation and the cultivation of additional renewable resource stocks.
- Public research into augmenting the ecosphere's negentropic potential (i.e., ways to increase the regeneration rate and waste assimilative capacities of natural capital).
- Public research into improved resource exaction and pollution abatement techniques to increase the growth and exploitative efficiencies of natural capital (Ratios 3 and 4).
- Public research into recycling, durability, and waste minimisation technologies to increase the maintenance efficiency of human-made capital (Ratio 2).
- The subsidisation of private research initiatives.
- Tax incentives to encourage private investment in renewable resource stocks.
- Subsidies and tax incentives to encourage agriculturalists to adopt sustainable land use practices.
- Compensation for land owners prohibited from clearing native vegetation.
- Grants to land owners to control noxious and introduced fauna and flora.

- The purchase of private land deemed to have sufficient ecological significance to warrant National Park preservation.
- A reduction in the marginal tax rate on incomes below the maximum limit.

15.4.2 Setting and policing of environmental guidelines

Although tradeable resource exaction permits allow for quantitative restrictions on the entropic rate of resource throughput, they have no way of ensuring the ecological disruption involved in resource exaction activities is kept within sustainable limits. Nor are they able to control the qualitative nature of high entropy wastes subsequently generated by production and consumption activities. Hence, tradeable resource exaction permits are unable to ensure sustainability precepts No. 2 and 3 are fully satisfied. To help overcome this problem, environmental guidelines governing the conduct of resource exaction operations as well as the generation and insertion of high entropy wastes need to be established. Of course, environmental guidelines cannot prevent at least some ecological or natural capital damage because, as explained in Chapter 4, resource exactions and waste insertions have unavoidable and largely indeterminate impacts on the ecosphere. However, environmental guidelines can be used to prohibit all practices and substances where the range of possible ecological impacts are unknown or where the known effects are significant enough to endanger the sustainability goal. They can also be used as a benchmark for setting and charging pollution taxes to control the *qualitative* nature of all high entropy wastes; assessing the conduct of resource suppliers, agriculturalists, and producers of human-made capital in order to determine whether environmental assurance bonds should be reimbursed or confiscated; and determining what portion of the absolute scarcity rents raised by the sale of resource exaction permits should be used to rehabilitate damaged ecosystems and cultivate additional renewable resource stocks. The responsibility of establishing environmental standards and policing the operational conduct of producers and resource suppliers would be assigned to the previously mentioned Environmental Agency.

15.4.3 Sustainable use of agricultural land

Agricultural land is not directly harvested but exploited for its propagating qualities. Thus, unlike a flow of timber that is sustained by ensuring the harvest rate does not exceed its capacity to regenerate, the sustainable use of agricultural land cannot be achieved via direct controls on the incoming resource flow. Short of having to directly regulate all agricultural activities, the sustainable use of agricultural land is best achieved by encouraging agriculturalists to adopt sustainable land use practices. This is best facilitated by a policy mix that places the practical responsibility of sustainable land use on the land owner and the financial responsibility largely on the government. Having the financial responsibility rest predominantly with the government is perfectly reasonable and legitimate given that the condition of sustainability is a public good and, like all public goods, requires government intervention.

The first major component of the policy mix would be the use of subsidies and tax incentives to assist farmers in the financing of sustainable land use practices. The second, as previously mentioned, would be the levying of assurance bonds on agriculturalists. This places the practical responsibility of sustainable land use on the land owner insofar as any land degradation caused by agricultural activities leads to the full or part confiscation of the bond and, thus, a penalty for the failure of the land owner to fulfil his or her stewardship responsibility. The final component of the policy mix would involve strict controls over land clearance. Because native vegetation plays a critical role in the maintenance of soil productivity as well as important ecosystems, future land clearance needs to be strictly controlled if not entirely prohibited. This is already the case in the Australian state of South Australia where the Native Vegetation Clearance Act of 1990 has resulted in the cessation of wholesale land clearance. Under this particular Act, land owners require permission to clear native vegetation — which is often denied — however, the land owner is suitably compensated for the loss of agricultural production. Land owners are also provided with funds to fence off and manage native vegetation. Legislation of this type needs to be introduced nationally, particularly given the rate at which native vegetation continues to be cleared in the states of Queensland and New South Wales (Biodiversity Unit, 1995).

15.4.4 Preservation and restoration of critical ecosystems

To maintain the level of biodiversity necessary to sustain the ecosphere's life-support services, critical ecosystems require preservation. Ideally, somewhere in the vicinity of 20 percent of a nation's land area should, at the very least, be preserved as a habitat for wildlife. This target can be reached through the establishment of additional National Parks and the introduction of legislation to control native vegetation clearance on private land, such as the previously described Native Vegetation Clearance Act of 1990.

Unfortunately, ecosystem preservation is not enough to guarantee biodiversity maintenance. Indeed, one of the more significant threats to native fauna and flora is the impact of introduced species. In Australia, for instance, introduced foxes and cats continue to decimate and endanger native marsupial species. Rabbits also have had an extraordinary impact on semi-arid ecosystems, whilst the European carp has greatly affected the major rivers of the Murray-Darling Basin. There are also a number of introduced plant species that continue to have a dramatic ecological impact on an island continent already stressed by a growing human presence. Clearly, noxious fauna and flora that threaten or endanger native species must be controlled. National Park authorities or some similar government authority therefore should be assigned the role of exotic species management in all publicly owned and preserved areas. In addition, tax incentives should be provided to enable land owners to play their part in exotic species management and ecosystem rehabilitation.

15.4.5 Transferable birth licenses

As previously explained, ecological sustainability demands an ecologically sustainable human population, that is, a population that can be sustainably provided with a sufficient per capita stock of human-made capital. A sustainable population target can be achieved via the introduction of transferable birth licenses. Like resource exaction permits, transferable birth licenses are a means of imposing quantitative restrictions — this time on the human birth rate — without unduly subverting the macro policy goals of allocative efficiency and distributional equity.

A transferable birth licensing scheme was first proposed by Boulding (1964). It has since been revisited by Heer (1975) and Daly (1991b). The transferable birth licensing scheme proposed here would operate as follows. First, each person would be freely allotted a license to have one child. This means a couple would be freely entitled to two children.* Should someone die before using their license, the license would be willed to a person of their choice. In the event that a child dies, the child's license would be passed onto the next of kin who would be free to have another child. In both instances, the inherited license would lose its transferable status. The change in status would avoid any problems that could arise from having an explicit monetary value on the life of permit-holding individuals.

A critical element of a transferable birth licensing scheme is deciding how to penalise nonlicensed procreators. For obvious reasons, the penalty must be severe enough to minimise the number of violations. On the other hand, if the penalty is considered unacceptably excessive, it is unlikely that the scheme would ever gain widespread support. A penalty that is likely to be least resisted is a procreation levy. The levy would be set at a given amount (X) above the going market price for a transferable birth license. The levy would be adjusted either quarterly or half yearly in line with the changing price of birth licenses over time. Setting a procreation levy in this manner would limit the number of violations by ensuring the cost of a license is always cheaper than the cost of violating the scheme. Why have a child when not in possession of a birth license if the penalty exceeds the amount required to purchase a license to do so? Because not all offenders would be able to pay the procreation levy in full, they would instead be required to pay the levy over time through the tax system (e.g., by increasing the marginal tax rate on an offender's income).

How much is a birth license likely to cost? Because the intranational supply of licenses is effectively fixed, the market price of a birth license would depend largely on the fluctuating intranational demand for children.

---

* In heavily populated countries with high population growth rates, it may be necessary to limit each couple to one child.

Thus, in many high population growth countries, the market price of a transferable birth license would be high. In many wealthy countries, where fertility rates are very low, birth licenses would be relatively inexpensive. Indeed, where the fertility rate in some countries is already below replacement levels, birth licenses would have virtually no value at all.

On the whole, there has been considerable resistance to a transferable birth licensing scheme. Many simply regard the legal transferability of birth licenses as morally repugnant. Moreover, they claim the scheme is advantageous to the rich and, as such, is unjust. Such opposition overlooks three things. First and foremost, while the scheme permits the legal transfer of birth licenses, the rights to procreate are freely and equally provided to both rich and poor. Second, while the scheme is advantageous to the rich insofar as they are better able to afford a would be expensive license, the rich always have an advantage, the extent of which depends on the gap between rich and poor, not on the scheme itself (Daly, 1991b). In fact, because the poor are not obliged to sell their licenses, a transferable birth licensing scheme reduces any advantage the rich may have. Last, should it be the rich who have more children, the high cost of rearing a child will serve to bridge the gap between rich and poor.

15.4.6 *Immigration*

One cannot address the population problem without dealing with the contentious issue of immigration. To achieve a desired population target, immigration numbers must be restricted. Exactly how much depends on the desired population target, the rate of population replacement, and emigration numbers. Normally, the rate of population replacement would vary in line with the fluctuating intranational demand for children. However, should a transferable birth licensing scheme be in place, the replacement rate is unable to fluctuate as freely. Indeed, it must be less than 100 percent because some licences would go unused given the loss of transferability status on inheritance.

If, for a moment, it is assumed that a nation's desired population is the population it has at the present and that it would remain steady with a 100 percent replacement rate, the permissible level of net immigration would be determined by the amount the population replacement rate fell short of the 100 percent maximum rate. Because this gap is unlikely to be significant in wealthy countries then, unless the emigration rate is very high, immigration numbers would be have to be kept very low. For wealthier nations wishing to meet their humanitarian requirements, there is likely to be little room for immigration beyond the admission of political and economic refugees. In impoverished nations, where infant mortality is high and life expectancy low, the gap between the population replacement rate and a 100 percent maximum rate is likely to be much greater. In the event that emigration numbers are high, this leaves considerable room for immigration although, of course, the desire for people to migrate to a poor nation is usually very low.

15.5 Policies to establish and maintain the market's moral capital foundations

Designing policies to establish and maintain the market's moral capital foundations is best achieved by examining the nature in which the market mechanism first gained prominence and the process that has since depleted its moral capital foundations. As previously explained, one of the features of modern property ownership is the lack of any requirement on the part of property owners to perform various duties and obligations. This was not always the case. Indeed, as Tawney pointed out,

> "The idea that the institution of private property involves the right of the owner to use it, or refrain from using it, in such a way as he may please, and that its principal significance is to supply him with an income irrespective of the duties which he may discharge, would not have been understood by most public men, and if understood, would have [.....] hesitated neither to maintain those kinds of property which met these obligations nor to repress those uses of which it appeared likely to conflict with them" (Tawney, 1921, page 63).

The critical change in the nature of property ownership occurred following the eighteenth and nineteenth century enactment of property rights laws that diluted the moral limitations on the pursuit of economic self-interest (Wilber, 1974). This had the effect of severing the previously existing nexus between private property ownership and public welfare. True, the market was in no way left entirely devoid of its moral capital foundations. However, as a mere legacy of a pre-industrial past, moral capital was left vulnerable to the "success" of the market's individualistic ethos which, over time, has systematically dismantled the market's invisible hand.

15.5.1 Minimising the depletion of moral capital through functional property rights and a corporate charter

To reestablish and maintain the market's moral capital foundations, functional property rights legislation must be introduced. The move toward functional property rights would involve two major changes to existing property rights. First, as previously explained, the emphasis of property ownership needs to be changed from the passive to the active mode. Second, property rights must be redefined to ensure one's rights to the use of property and any subsequent pecuniary gain from exercising these rights is conditional upon the performance of duties and the provision of service (i.e., having served a collectively determined set of objective value-based standards and principles).

One way to introduce functional property rights is to establish a corporate charter. A corporate charter would regulate business activity by detailing the permissible range and nature of a corporation's activities. To some extent, a corporate charter implicitly exists in most countries. For instance, numerous laws exist to ensure corporations and smaller business enterprises adhere to environmental, social, and workplace standards and conditions. However, these laws pertain to the process of commodity production, not to the nature of the commodities subsequently produced. Indeed, apart from product liability legislation, few if any laws exist in relation to the carrying out of commodity production in accordance with collectively determined standards and principles. There is, however, no reason why collectively determined standards and principles cannot be applied to business activities. In many countries, they are already applied to the public sector in the form of community service obligations. Community service obligations are standards and conditions that are imposed on government business enterprises as a means of achieving socio-political goals. In addition, professional standards are currently required of medical practitioners, lawyers, engineers, accountants, and so on. For a corporate charter to re-establish the nexus between private property ownership and public welfare, similar standards of conduct and performance need only be enforced in a more general sense on all forms of property ownership and on all forms of economic activity. A corporate charter thus would be used to abolish all forms of private property for which no function is performed in the same way one's professional license is repealed for having failed to carry out a minimum standard of service.

15.5.2 The need for functional property rights to incorporate both rigidity and flexibility

To maintain the market's moral capital foundations, functional property rights must be sufficiently rigid to limit the erosion of the standards by which the production of physical commodities should serve. At the same time, functional property rights must be designed to deal with changes in a society's belief system. As belief systems are systematically revised, so is a society's concept of the ultimate end and, thus, its objective value-based standards and principles. Consequently, what are construed to be permissible and functional business activities will change over time. To ensure the constant evolution of a corporate charter, functional property rights must be sufficiently malleable to allow for their appropriate redefinition. Provided a suitable balance between rigidity and flexibility is struck, the evolution of a more appropriate market domain should follow.

15.5.3 Re-embedding the macroeconomy into the social and ecological spheres

One of the more useful roles of a system of functional property rights is its ability to deal with externalities arising from a co-evolutionary change in the underlying parameters of social, ecological, and macroeconomic systems. In Chapter 12, it was pointed out that solutions to co-evolutionary-induced spillover effects require the institutionalisation of solidarity amongst members of society. This, it was argued, would ensure the incorporation of the "interest of others" into the decision-making domain of individuals, thereby reducing the incidence of cost shifting onto other parties. Albeit incompletely, functional property rights are able to institutionalise a sense of community solidarity by re-embedding the economic or market sphere of influence back into the social and ecological spheres. In doing so, functional property rights have the capacity to partially reverse the Great Transformation — a process that led to the embedding of the social and ecological spheres into the somewhat unruly domain of the market. If properly defined, functional property rights ensure an appropriate balance between price-determining decisions and price-determined outcomes and, in doing so, further facilitate the evolution of a more appropriate market domain.

15.6 *International trade policy*

Given the rapid transition toward a globalised economy, there is probably no other policy-related area of greater importance to SD than international trade. As indicated in Chapter 6, international trade offers the potential for countries to enjoy a higher level of sustainable net benefits than that enjoyed from domestic production alone. However, for a number of reasons, ecological economists believe that international trade is contributing to a decline in sustainable net benefits, if not rendering the net benefits of most countries unsustainable. It should be pointed out that ecological economists are in no way suggesting that countries should abandon international trade. Nonetheless, to understand the position of ecological economists and to ensure that international trade is ultimately beneficial to all, the significance of these factors needs spelling out. Only then can the rationale behind the policy measures about to be prescribed be fully understood.

15.6.1 *The arguments for and against free international trade*

Economists have long promoted the "free" international trade dogma on the following premises.

- Free trade maximises the competitive pressure on producers, irrespective of their location, thereby facilitating a more efficient allocation of scarce global resources.
- Free trade lowers the cost of production and leads to the availability of cheaper commodities for all.
- Free trade facilitates a high growth rate of macroeconomic systems, thereby making everybody better off.

- Should the growth promoted by free trade result in ecological degradation, an eventually richer world is better placed to rehabilitate the environment and introduce measures to prevent ecological degradation occurring in the future. Advocates of free trade now claim empirical support for their position following the discovery of an inverted U-shaped relationship between a nation's GDP and its concentrations of certain air and water pollutants. The so-called "Kuznets curve" depicting this relationship indicates a decline in pollution once GDP reaches a particular threshold level (World Bank, 1992; Selden and Song, 1994; Grossman and Krueger, 1995).
- Because international trade is governed by the principle of *comparative advantage*, free trade is mutually beneficial. Even nations with an absolute disadvantage in the majority of commodities they produce can benefit from international trade by specialising in those with which they have a comparative advantage.
- Should unbalanced trade occur, that is, should some nations accumulate large foreign debts, it is the result of a large number of bad private decisions made by domestic debtors who, in due course, are unable to service their accumulated debts; and/or bad public policy. In all, a nation's foreign debt is not the direct fault of international trade (Pitchford, 1990).

Ecological economists are in no disagreement with the notion that specialisation and trade, when governed by the principle of comparative advantage, can be mutually beneficial.* What ecological economists wish to highlight is that international trade is now dictated by the principle of *absolute advantage*. As a consequence, the benefits of trade do not necessarily accrue to all participating nations. To understand the ecological economists' position, one must go back to a basic premise underlying the rationale for international trade. In 1817, Ricardo pointed out that the comparative advantage argument for free trade rests entirely on the immobility of capital between nations. For instance, the principle of comparative advantage can never operate within the confines of a national economy because capital is always free to move to locations offering the most profitable investment opportunities. Hence, at the intranational level, investment and the allocation of resources is always governed by absolute profitability. Of course, Ricardo promoted free trade because, in 1817, the mobility of capital was severely limited. But improvements in transport and communication technology, the subsidisation of transport costs owing to underpriced fuels, and fewer institutional restrictions on the international flow of capital have all contributed to the massive liberalisation of international capital flows. As a result, international trade is now governed by the law of absolute advantage

---

* There are some exceptions. For example, excessive specialisation can reduce the range of available occupations as well as increase a nation's vulnerability to rapid variations in international market prices.

in all but a few exceptional instances (e.g., tourism and some forms of agriculture).*

Does it matter that international trade is now governed by a different principle? After all, no national economy has been brought to ruin because intranational trade is governed by the principle of absolute advantage. For a number of reasons, yes, it does matter. First, free trade governed by the principle of absolute advantage can increase the world's net benefits but decrease the net benefits accruing to some nations. Only when the principle of comparative advantage governs specialisation and the location of capital investment can international trade be virtually guaranteed to be mutually beneficial.

Second, intranationally, all production and exchange activities are subject to basically the same nonprice rules, including any efficient national policy of cost internalisation. Consequently, no one producer can gain an unfair advantage over another by paying an equivalent form of work a lower wage, by polluting when another producer cannot, or by paying a much lower rate of tax. To gain a competitive advantage, a producer must be genuinely more efficient than its nearest competitors. The same, however, cannot be said of the international market. This is because the international market is not a formally instituted market. To recall from Chapter 9, markets are a set of social and cultural institutions within which a large number of commodity exchanges between buyer and seller take place. Because social and cultural institutions rarely exist beyond national boundaries, commodity exchanges between the international buyer and seller take place in a domain largely free of institutional constraints. Consequently, the price-determining parameters of the global market come to rest at the lowest common denominator (Daly and Cobb, 1989). In doing so, the price-determining parameters of national markets within which domestic production takes place are, for many countries, grossly incommensurate with those of the global market. To some extent, this is not a bad thing. On the positive side, it is desirable for relative price signals to reflect the difference in price-determining market parameters brought about by variations in economic efficiency. If a domestic producer is inefficient because a foreign producer is better at producing a similar commodity, the variation in relative prices should ensure the survival of the latter and the demise of the former. On the negative side, it is undesirable for a domestic producer to close down simply because it cannot compete with a foreign operator subject to much weaker social and environmental standards. Yet this is precisely what an unfettered globalised market promotes because the free mobility of capital allows nationally instituted nonprice rules and a policy of cost internalisation to be avoided by transnational corporations.** Furthermore, the competitive pressure to

---

* Tourism and agriculture differ because, unlike factory production, natural scenery and desirable growing conditions cannot be relocated.

lower the cost of production leads, not always to increased efficiency, but often to the erosion of objective value-based standards and principles at the national level. The recent move in Australia to reduce the minimum wage, to allow mineral exploration in National Parks, to lower tax rates in line with taxation regimes of its nearest Asian neighbours, and to slash public funding to hospitals, schools, and universities is symptomatic of the degenerative impact of a global free trade environment governed by the law of absolute advantage.

Third, because highly mobile capital will generally flow to locations with an absolute advantage in production, the potential for large trade imbalances is significantly high. The same does not occur when capital is effectively immobile because the level of international lending and borrowing required to run up unserviceably large foreign debts is precluded. Of course, there are a number of countries with very large foreign debts that, on the surface at least, appear quite serviceable. Indeed, according to Pitchford (1990), because many accumulated debts are largely the result of numerous rational arrangements between domestic borrowers and foreign lenders, most foreign debts are of no great concern. While Pitchford is probably correct in his assessment that foreign debt is more or less the aggregation of many rational micro decisions, is it true to say that this translates to macro rationality? Probably not. As explained elsewhere in the book, micro rationality can often lead to macro irrationality. With regard to foreign debt, macro irrationality will often ensue because transactions between two individuals across international borders have no reason to be commensurate with the collec-

---

** A number of studies have been undertaken to verify or repudiate the theory that capital moves to locations with weaker community and environmental standards. The majority of these studies support the position that differences in labour costs account for at least some industrial flight (Leonard, 1988; Hodge, 1995; Garrod, 1998; Ratnayake and Wydeveld, 1998). However, almost all the studies lead to the conclusion that environmental stringency has virtually no impact on the choice of production location (Dean, 1992; Pearce and Warford, 1993; Jaffe et al., 1995; Garrod, 1998). The reason for this, it seems, is that the cost of adjusting to environmental standards is small for all but a few highly pollutive industries and avoiding such costs through relocation is almost always absorbed by the cost of relocation itself (Leonard, 1988; Stevens, 1993). For some observers, however, the lack of conclusive statistical evidence means the verdict is still out on whether variations in environmental standards cause industrial flight (Hodge, 1995; Field, 1998; Ratnayake and Wydeveld, 1998).

We belong to the latter school of thought largely because studies of the so-called "pollution haven hypothesis" have only focussed on the relocation of firms from the North to the South. They have not considered the following three potential manifestations of industrial flight. First, how many new industries have established themselves in the South where, if there was a lack of variation between Northern and Southern environmental standards most would have emerged in the North? To what extent is the low adjustment cost to environmental standards in the North owing to standards falling short of what is required to meet sustainability requirements (in which case, if standards were suitably tightened, the cost differential would be significant enough to induce the relocation of capital to the South)? How much has the threat of offshore relocation served to prevent the introduction of more exacting environmental standards in the North, or indeed, led to the watering down of existing standards? Until these questions have been suitably answered, the apparent lack of any mass relocation of industries from North to South cannot be used to disclaim the pollution haven hypothesis.

tively determined standards and principles of the countries in which they reside, in particular, standards and principles relating to the macro policy goals of distributional equity and ecological sustainability. The last goal is of particular interest because the serviceability of a large foreign debt usually requires a nation to pursue a policy of export-led growth which often leads to the unsustainable liquidation of its natural capital stock.

Finally, as indicated by Figures 14.3 and 14.4, the growth facilitated by international trade does not always make a nation richer. Nor does continued growth mean an inevitable improvement in environmental quality as implied by an inverted U-shaped GDP-pollution curve. The fact that many countries with a relatively high per capita GDP are experiencing a decline in the level of some pollutants could just as easily reflect a shift in dirty production from the rich North to the poor South as transnational corporations avoid stricter Northern environmental standards. In view of the ease with which dirty production can be shifted from one location to another, the true test of a proposed link between growth and environmental quality is a curve depicting the relationship between Gross World Product (GWP) and global concentrations of various air and water pollutants. Because of the lack of available data on the world volume of most major pollutants, this relationship has yet to be confirmed (Chapman, 1999). What is known is that the global emission of four major pollutants, namely, carbon dioxide, sulfur dioxide, methane, and nuclear fuel waste are increasing at an exponential rate (Chapman et al., 1995). Given that the GWP is rising steadily, there is no evidence at this stage to suggest that growth eventually leads to an improvement in environmental quality.

15.6.2 Policies to gain the sustainable net benefits of international trade

Exactly what can be done to increase the sustainable net benefits from international trade? Whatever the solutions, it is because the international market is a very large and highly complex subsystem of the global system that the major stakeholders are likely to react to proposed changes in a conservative, identity-preserving way. When coupled with the binding constraints associated with current commitments to international agreements such as the General Agreement on Tariffs and Trade (GATT) and the now established World Trade Organisation (WTO), it is clear that the ability of organisations, communities, and even governments to autonomously act in ways consistent with achieving SD will be difficult. Indeed, any prescribed policy is unlikely to be successful without a prior international consensus on trade issues relating to SD.

15.6.2.1 *Restoring comparative advantage by restricting the mobility of financial capital*

To increase the sustainable net benefits from international trade, it is necessary to restore comparative advantage as the principle governing a nation's trade with the rest of the world. Because, it is unwise to limit the movement of human-made capital, comparative advantage can be restored by reducing the incentive to shift producer goods where absolute profitability is highest internationally (as opposed to where relative profitability is highest, which is both desirable and in need of encouragement). This can be achieved by forcibly restricting the international mobility of financial capital. A restriction on the mobility of financial capital reduces industrial flight insofar as the mobility of financial capital is necessary to gain financially from the international movement of producer goods. If one cannot easily move the profits generated elsewhere back to their home country, there is little incentive to have producer goods located overseas. A restriction in the international mobility of financial capital therefore shifts the concern of capitalists to where producer goods can be best located domestically — a choice that is still dictated by the principle of absolute advantage at home but by the principle of comparative advantage internationally. For example, one would now invest in the production of goods where his or her home country has a comparative advantage (despite it perhaps having an absolute disadvantage) but choose an efficient domestic location where the cost of production is minimised.

One way of restricting the mobility of financial capital is to introduce an IMPEX system of foreign exchange management (Iggulden, 1996). To operate the IMPEX system, an IMPEX facility would be formally established by a national government. Ideally, the IMPEX facility would come under the supervision of a central bank. Unlike the present system, the IMPEX system would operate under five critical rules.

1. Every international transaction must pass through the IMPEX facility.
2. All foreign currency must be exchanged for IMPEX dollars ($IMP).
3. The purchase of foreign currency requires the possession of IMPEX dollars.
4. No spending on imports is permitted unless there is sufficient "earned" foreign exchange available on the day (held in the form of IMPEX dollars) and only if importers are willing to pay the price being asked by the possessors of IMPEX dollars.
5. The buying and selling of the IMPEX dollars of any particular country is only open to the citizens of that country.

The exchange process would operate as follows. When foreign currency enters a country, the earner receives IMPEX dollars based on the exchange rate between the domestic currency and the foreign currency earned. For instance, if an Australian exporter earns $US100 and the going exchange rate between an Australian and American dollar is $US1.00 = $Aus1.20, the Australian receives $IMP120 from the IMPEX facility. The owner of the $IMP120 is free to either purchase another foreign currency to import foreign produced goods or to sell the $IMP120 to an Australian who does. The Australian could not sell the IMPEX dollars to a foreign national.

In the day-to-day buying and selling of IMPEX dollars, an IMPEX rate would establish itself relative to the domestic currency. For example, assume the demand for IMPEX dollars exceeds its supply such that $Aus140 is required to purchase $IMP100. The going IMPEX rate would be 1.40. Should the amount of IMPEX dollars demanded by Australians increase relative to the IMPEX dollars earned, the IMPEX rate would appreciate (i.e., go higher than 1.40). On the other hand, if the demand for IMPEX dollars fell relative to its supply, the IMPEX rate would depreciate (i.e., go lower than 1.40). What, then, would an Australian have to do if they required $US100 to import American goods? Unless they were already in possession of IMPEX dollars, they first would be required to purchase $IMP120 as per the going exchange rate between the Australian and American dollar of $US1.00 = $Aus1.20. Second, to purchase $IMP120, they would be required to part with $Aus168 as per the going IMPEX rate of 1.40 (i.e., $US100 × 1.20 × 1.40). Thus, in order for an Australian to import $US100 worth of American goods, it would cost him or her $Aus168, not $Aus120 as per usual.

Clearly, there are two currency markets in place. One is the traditional foreign exchange market. The other is a domestic IMPEX market. In the international foreign exchange market, exchange rates fluctuate as per normal because although the IMPEX system ensures a nation's overall trade is balanced, it still allows a country to have trade imbalances with individual countries. For example, Australians may choose to import more American goods and fewer Japanese goods, thereby leading to a deterioration in the terms of trade with the United States but an improvement with Japan. Because Australians would demand more $US and less Japanese Yen then, *ceteris paribus*, the $Aus should depreciate relative to the $US and appreciate relative to Japanese Yen.

One of the important aspects of the IMPEX system is that it establishes a natural incentive or disincentive to export and import. Consider the effect of an Australian IMPEX rate of 1.40 caused by Australia's demand for foreign exchange exceeding its foreign currency earnings. On top of any advantage/disadvantage that might exist owing to existing exchange rates, Australian imports will be subject to a price disadvantage of 40 percent as evidenced by the $Aus168 required to import $US100 worth of American goods instead of $Aus120. Australian exports, conversely, will have a 40 percent price advantage. The high IMPEX rate deters import spending while providing an incentive to boost exports and replace imports. However, in the absence of any exogenous factors affecting the demand and supply for different currencies, the incentive/deterrent factor generated by the IMPEX system gradually disappears. For instance, as the earning of foreign currency rises and spending on imports falls, the demand for foreign exchange progressively declines along with the IMPEX rate. Eventually, the IMPEX rate approaches something near parity (an IMPEX rate of 1.00).

One final point, an IMPEX system of foreign exchange does not have to be as rigid as suggested here. It is possible for a government to make available for purchase a small amount of "unearned" IMPEX dollars. This would allow for small trade imbalances and incorporate some degree of flexibility into the system. All the same, given the potential for foreign debt to become economically and ecologically unserviceable, there would need to be strict controls on the amount of unearned IMPEX dollars issued by the IMPEX facility. Ideally, unearned IMPEX dollars would cease to be issued once the accumulated foreign debt reached a small fraction of a nation's income, say 2 or 3 percent of its Hicksian income.

15.6.2.2 Facilitating efficiency-increasing rather than standards-lowering competition

Assume, for a moment, that the principle of comparative advantage has been successfully restored. Increasing the sustainable net benefits from international trade still requires the prices of all goods sold in domestic markets to reflect full social and environmental costs. Failure to do so will, if nothing else, lead to the specialisation in certain goods where countries do not have a genuine comparative advantage. To ensure domestic markets reflect full social and environmental costs, national governments must have the power to impose tariffs when the price of a foreign good does not incorporate the social and environmental costs of production.* This does not mean that tariffs should be used to protect inefficient domestic producers from foreign competition. Nor does it mean that access to a nation's markets should be denied to foreign producers who are not subject to the same social and environmental standards and conditions (unless the production of the foreign good involves slavery, child labour, etc.). It simply means that a nation should have the right to impose tariffs on foreign goods to allow domestic producers to accurately assess to which goods it has a comparative advantage, and protect domestic producers from any cost advantage that foreign producers enjoy as a consequence of operating under a weaker set of social and environmental standards.

---

* While ensuring market prices fully reflect all benefits and costs is widely accepted on efficiency grounds, the idea of the nation state determining what and by how much something constitutes a benefit or cost is not. Some observers argue that this allows the people of one country to impose its will over the people of another. This is not true. Applying a tariff to reflect costs that have not been included in the price of a foreign produced good does not prevent the foreign producer from operating in the manner it does. Nor does it prevent the foreign producer from importing its goods into the country applying the tariff. In any case, as Daly (1999) has pointed out, international trade implies trade between nations. This implies that the basic unit remains the nation. Thus, despite trade occurring between individuals from two different nations, it is the standards upon which trade takes place that ought to take precedent over individual tastes and preferences. If not, one effectively allows international trade to erase national boundaries and the standards collectively determined by the people living within them. This, unfortunately, seems to be the current trend resulting from the rapid movement towards globalisation.

At present, the WTO allows countries to impose tariffs on imported goods only if a similar duty is charged on domestically produced goods. This gives the appearance that a country can use tariffs to protect collectively determined standards and principles and any efficient policy of social and environmental cost internalisation. While it is true that tariffs can be used to protect the latter, they cannot be used to protect the former. Existing WTO arrangements protect an efficient policy of cost internalisation because internalisation essentially involves the imposition of a tax on domestic producers whose spillover costs are borne by their fellow citizens. If a foreign producer is not subject to a similar internalisation policy, the foreigner's cost advantage can be offset by a compensating tariff.

Nevertheless, existing WTO arrangements do not protect collectively determined standards and principles in circumstances where the policy in question increases the cost of domestic production but involves no direct imposition of a duty or tax. For example, while the use of tradeable exaction permits to limit the incoming resource flow increases resource prices and the cost of domestic production, it does not involve the direct imposition of a throughput tax. Similarly, the cost of having to comply with strict occupational health and safety standards increases the cost of production but involves no explicit imposition of a tax on employers. In both instances, there is a good reason for not imposing taxes. Ecological sustainability and distributional equity are not allocation problems and cannot be resolved through a policy of cost internalisation. Unfortunately, as a consequence of not having imposed taxes on domestic producers, governments are unable to impose tariffs on imported goods that are not subject to the same environmental and workplace standards and conditions. In response, governments often dilute the collectively determined standards that place domestic producers at a relative cost disadvantage. All told, the competitive pressure exerted by existing WTO arrangements is as much standards lowering as it is efficiency increasing.

One way to facilitate efficiency-increasing competition is to have all nations categorised by the WTO according to their domestic social and environmental standards. A compensating tariff structure then could be established to allow countries to impose tariffs on lower ranked countries but have tariffs imposed on its own products by higher ranked countries. The lower is the ranking of one country relative to another, the larger would be the permissible compensating tariff. Where two countries belonged to the same category, the imposition of tariffs on each other's goods would not be permitted. The compensating tariff structure and the categorisation of each country would be determined independently by the WTO, with each signatory nation having representatives in the organisation. The imposition of tariffs outside the tariff structure would be prohibited unless sanctioned by the WTO itself. This would prevent nations from imposing tariffs for undesirable protectionist reasons. It would also prevent countries from gaining a cost advantage by lowering its domestic social and environmental standards. Of course, countries would not be prevented from lowering their standards.

Standards setting is, after all, the prerogative of the nation state. However, any potential cost advantage would be immediately offset by the imposition of higher tariffs resulting from the nation's relegation to a lower ranked category.

15.7 Additional policy conclusions and prescriptions

15.7.1 Facilitating efficiency-increasing technological innovation

As explained a number of times, once a country's macroeconomy reaches its optimal scale, the application of throughput-increasing technological progress no longer increases sustainable net benefits. Further increases in sustainable net benefits can only be achieved through the use of efficiency-increasing technological progress. Given that many countries appear to have exceeded their optimal macroeconomic scale, it would appear that technological progress has been overwhelmingly weighted toward the throughput-increasing variety. There is, therefore, a need for policies to induce the development and application of efficiency-increasing technological progress. Virtually all the policy measures outlined in this chapter serve to facilitate this crucial transition. However, some of the more important policy measures include

1. *Reductions in income tax rates.*   The income received from wages and profits depends very much on the relative prices obtained from the sale of newly produced commodities. Because the relative price of a given commodity is largely a function of its service-yielding qualities, improving the quality of all newly produced commodities can increase money incomes. Unfortunately, as necessary as income taxes are, they somewhat penalise the generation of income. In doing so, they dampen the incentive to improve the quality of all newly produced commodities and, in turn, discourage increases in the service efficiency of human-made capital (Ratio 1). Hence, with the exception of a 100 percent tax on incomes above the maximum permissible level, income tax rates should, whenever possible, be reduced.

2. *Tradeable resource exaction permits.*   Resource exaction permits achieve two things. First, by imposing quantitative restrictions on the rate of resource throughput, they prevent the incoming resource flow from exceeding the maximum sustainable rate. This, by itself, discourages throughput-increasing technological progress. Second, they increase the price of low entropy resources. This forces producers, as low entropy resource users, to improve the technical efficiency of the production process. Because it also encourages improvements in the recyclability, durability, and operational efficiency of all newly produced commodities, resource exaction permits facilitate increases in the maintenance efficiency of human-made capital (Ratio 2).

3. *Environmental assurance bonds.*   Environmental assurance bonds force resource suppliers and polluters to commit resources to minimise the ecological impact of their operations. This facilitates increases in both the growth and exploitative efficiencies of natural capital (Ratios 3 and 4). In many ways, these first three measures constitute the centrepiece of what is commonly referred to as "ecological tax reform" — a means by which the emphasis on the source of tax revenue is transferred from income (something desirable) to throughput (something deplored).

4. *Profit-sharing arrangements.*   By having the profits shared amongst those directly involved in the active and functional use of property (e.g., managers and workers), the incentive to improve the service-yielding qualities of newly produced commodities is increased. In doing so, profit-sharing arrangements facilitate an increase in the service efficiency of human-made capital (Ratio 1).

5. *Contestable markets.* Unless markets are sufficiently contestable, firms are not compelled by the actions of rival firms to operate efficiently. Clearly, policies that increase the contestability status of markets, such as the detachment of sunk costs and the prohibition of anticompetitive mergers, can greatly facilitate the transition towards efficiency-increasing technological progress. Because competition from foreign producers further increases the contestability status of domestic markets, international trade can also play an important role in facilitating this transition. Exactly how much, of course, depends on the extent to which its potential standards-lowering impact is minimised.

15.7.2 Discounting

Individuals express their preferences not only in terms of their desire for different commodities, but also in terms of when they desire benefits and costs. Because of uncertainty, impatience, and human mortality, it is customary for benefits and costs to matter more if they emerge sooner rather than later. The tendency to weight the present over the future is commonly referred to as *discounting*. If it is known how much an individual prefers the present to the future, one can determine their personal discount rate. For example, consider someone presented with the following choice: receive $100 now or $110 in a year's time. If he or she is indifferent between the two, one can safely assume that his or her discount rate is 10 percent.* As for someone with a discount rate of 20 percent, he or she would prefer the $100 payment now because $120 would be required in a year's time to feel equally well off. Conversely, anyone with a discount rate of less than 10 percent would opt for the $110 payment in a year's time.

Because the tendency to discount future benefits and costs has potential implications for the natural environment, the role of discounting remains a

* This assumes the nonexistence of inflation that would otherwise reduce the real value of the $110 received in a year's time.

contentious issue in the SD debate. Environmentalists have long criticised the use of high discount rates when using benefit–cost analyses to assess the viability of human activities. The reasons for their criticism are as follows. First, because the ecologically destructive impact of human activities often takes many years to emerge (e.g., soil erosion, fresh water eutrophication, and global warming), so also do the measured environmental costs. Should these future environmental costs be heavily discounted then, because the benefits of most projects occur immediately or shortly after their completion, the use of discounting in benefit–cost analyses increases the approval rate of ecologically destructive projects. Second, the beneficial effects of environmental rehabilitation also have the propensity to take many years to eventuate. For example, the ecological benefits of reafforestation are only fully realised once a forest reaches maturity. Because the major portion of the total cost of reafforestation is incurred in the project's initial stages, the heavier discounting of benefits compared to costs reduces the incentive to rehabilitate damaged ecosystems. Third, environmentalists point to the widely held belief among many economists (e.g., Clark, 1976) that it is "economically rational" to harvest a renewable resource to extinction should its regeneration rate be exceeded by the discount rate. Hence, high discount rates effectively amount to a large number of renewable resources being rationally harvested to extinction. To ensure ecological sustainability, environmentalists believe that only low discount rates should be used to assess the viable exploitation of renewable resources.

While there is nothing fundamentally wrong with the environmentalists' criticism of discounting, it is wrong to conclude that low discount rates work in favour of keeping natural capital intact. For example, while low discount rates promote low resource-intensive activities, they also promote activities with long-term payoffs and substantial upfront costs. In doing so, low discount rates increase the number of economically viable projects. Should the number of projects undertaken subsequently increase, it is possible for the overall rate of resource throughput to rise — the very thing that needs to be minimised, not maximised, to avoid the unsustainable diminution of natural capital.

How, then, should the socially optimal discount rate be determined? As it turns out, determining the socially optimal discount rate is as futile as determining the socially optimal price of each and every resource. This is because discount rates are the means to achieving an intertemporally efficient allocation of resources which, as explained in Chapter 7, can readily co-exist with ecological unsustainability. Thus, selecting the discount rate is no more likely to achieve a sustainable use of natural resources than selecting the Pigouvian tax rate to ensure market prices reflect the full cost of resource depletion and waste generation. In sum, ecological sustainability is achieved by imposing quantitative restrictions on the incoming resource flow, not by fine tuning discount rates or resource prices and pollution charges. It is highly probable that quantitative throughput restrictions would lower discount rates just as they would probably increase low entropy resource prices.

However, should discount rates remain high, so be it. While high discount rates make viable more human activities, the restriction of the incoming resource flow to the maximum sustainable rate limits the number of activities ultimately undertaken. Unfortunately, the current debate over discounting continues to rage because many observers fail to recognise the difference between intertemporal efficiency and ecological sustainability. They also fail to realise that achieving intertemporal efficiency and ecological sustainability requires a separate policy instrument — a socially optimal discount rate in the former instance (which is better left to the market than a remote bureaucrat) and quantitative throughput restrictions in the latter. If this were better understood, the unwarranted attention that discounting receives would quickly cease.

15.7.3 Central bank control of a nation's money supply

So long as a nation's money supply continues to grow at the rate of interest, there is a perceived need to expand the macroeconomy in order to avoid hyperinflation and/or widespread debt repudiation. Because the long-run expansion of the macroeconomy is both unsustainable and undesirable, it is necessary to closely scrutinise and restrict the growth of a nation's money supply. How much the money supply should be restricted, or equivalently, to what extent it should be permitted to expand, depends on whether one is referring to the nominal or real money supply. The nominal money supply is the nominal or face value of the stock of money. The real money supply is a measure of the purchasing power of money and is calculated by dividing the nominal money supply by an aggregate price index, such as the Consumer Price Index (CPI). Ideally, the change in the nominal money supply should correspond with changes in the nominal exchange value of the entire stock of human-made capital. The real money supply should increase in line with changes in the real exchange value of human-made capital, that is, in keeping with changes in its physical magnitude.

Given that the maintenance of an optimal scale requires the macroeconomy to be largely nongrowing, there is a much greater need to restrict the real money supply than the nominal money supply. Indeed, with the incoming resource flow restricted to the maximum sustainable rate, a larger stock of human-made capital and, therefore, a larger real money supply is only permitted if there are increases in the maintenance, growth, and exploitative efficiency ratios (Ratios 2, 3, and 4). Because there are limits to the increase in these ratios, the growth of the real money supply must eventually cease altogether. It is a somewhat different story for the nominal money supply. This is because the nominal exchange value of the entire stock of human-made capital can also rise as a consequence of increases in the service efficiency of human-made capital (Ratio 1). Increases in other price-increasing influences, such as taxes, tradeable exaction permits, and assurance bonds can also inflate the nominal exchange value of human-made capital. It is because these price-increasing influences can take effect at any time and the fact there is a question mark over the limit to increases in Ratio 1 that, unlike the real money supply, an indefinite albeit controlled increase in the nominal money supply is permissible.

In theory, absolute control over the money supply requires two specific measures. The first is a 100 percent reserve requirement of all banks. This prevents banks from creating money *ex nihilo* (Soddy, 1926; Fisher, 1935). The second is a central bank to create and destroy money on behalf of the government. In reality, the sheer magnitude of the task renders absolute central bank control of the money supply inefficient at best, and ineffective at worst. As such, a 100 percent reserve requirement is an undesirable way to achieve money supply targets. Central bank control over a nation's money supply is best confined to the manipulation of the monetary base and the reserve-deposit ratio, with the latter set somewhere between 25 and 50 percent.* This would greatly reduce the money-creating flexibility that banks currently enjoy.** Nonetheless, it would still give banks considerable freedom to create and destroy money, indeed, enough freedom to frustrate a central bank's desire to meet its money supply targets. To prevent this, the central bank should pre-announce its monetary targets and be given the power to revoke banking licences or increase the reserve-deposit ratio of offending banks. This would put transgressors at a great disadvantage relative to well-behaved banks and, in doing so, serve as a potent disincentive for banks to act in an unruly fashion.

Because the achievement of money supply targets requires tight control of the monetary base, two additional factors must be closely considered. The first is an assessment of the most appropriate means of financing government expenditure. While the use of tax revenue and the borrowing of money from the public through the sale of government securities does not affect the monetary base, an increase in the amount of money borrowed from the central bank causes the monetary base to expand, thereby increasing the money supply. In view of the need to keep a tight reign on the money supply, the borrowing of money from the central bank to finance government expenditure will have to be used more sparingly.

The second additional factor with the potential to influence the monetary base is central bank intervention in foreign exchange markets. Often central banks buy and sell the domestic currency to manipulate exchange rates. In doing so, central banks increase or decrease the supply of the domestic currency which, in turn, can unduly increase or decrease the money supply. Should the previously described IMPEX system of foreign exchange

---

* The money supply (MS) is basically equal to the monetary base (MB) times the inverse of the reserve–deposit ratio (r). That is, $MS = MB \times 1/r$. If the monetary base is $100 million and the reserve–deposit ratio is 25 percent, the money supply equals $100 million $\times 1/0.25 = $400 million. A doubling of the monetary base results in a doubling of the money supply (i.e., $200 million $\times 1/0.25 = $800 million). A doubling of the reserve-deposit ratio results in a halving of the money supply (i.e., $100 million $\times 1/0.5 = $200 million).

** As of July 1993, Australia's reserve–deposit ratio was just 0.019. This constitutes a minuscule 1.9 percent fractional reserve requirement on the part of banks.

management be introduced, exchange rate fluctuations (particularly from currency speculation) would be significantly reduced. This would diminish the need for central bank intervention in the foreign exchange market, thereby making monetary targeting a much simpler task.

15.7.4 Severing the GDP-employment link and the GDP-poverty link

Under present institutional arrangements, the largely nongrowing economy required to maintain an optimal scale equates to unemployment and increased poverty. For example, it is generally considered that growth rates of 3 to 4 percent are required simply to maintain absolute poverty and unemployment rates at existing levels. Thus, despite growth rendering most people worse off, its continuation is justified on the grounds that the cost of little or no growth is intolerably painful. It is little wonder that ecological economists refer to the modern obsession with growth as a "growth addiction" when the withdrawal symptoms from its abstinence (increased unemployment and poverty) are considered worse than the deadly cost of growth itself. To help overcome the modern addiction for growth, it is necessary to make appropriate institutional changes to sever the GDP-employment and the GDP-poverty links. This will allow a nation to move toward a largely nongrowing economy without the majority of its citizens having to suffer from poverty and unemployment.

15.7.4.1 Severing the GDP-employment link

The most plausible way to sever the GDP-employment link is to minimise the need for paid forms of employment. This can be achieved by focussing economic activity on improving the quality of human-made capital and reducing the rate at which both the stock is consumed and new items are produced. As mentioned earlier, this increases money incomes, in particular hourly wages, because the higher use value of all newly produced goods means they obtain, *ceteris paribus*, higher relative prices. From an employment perspective, higher wages correspond to a lessened need for work and an increased potential for job-sharing.

There are, however, three main obstacles preventing most people from reducing their need for work. The first is the unequal distribution of wealth. The distribution of wealth is important because wealth provides for its owner a flow of income without the need for excessively laborious work. Because the majority of a nation's citizens possess very little wealth, they are compelled to work long hours to obtain a share of the annual flow of newly produced goods (received in the form of a money income that can then be used to claim some of the annual throughput flow). The second major obstacle is the degenerative influence that an unfettered global market has on wages and conditions of employment. So long as international trade continues to apply standards-lowering pressures it will be difficult to bring about the increase in hourly wages required to reduce the need for work and to share the workload across the entire labour force. The third major obstacle is the inflexibility of labour markets. Caused mainly by archaic industrial relations systems, labour market inflexibility forces a great number of people to work more hours than they would like at a time when unemployment in many countries is unacceptably high. With these obstacles in mind, the following policy measures are the most critical in severing the GDP-employment link.

1. *Quantitative throughput restrictions.* As previously explained, quantitative throughput restrictions induce a movement toward efficiency-increasing technologies. Unlike the throughput-increasing varieties, efficiency-increasing technologies aim to procure the highest level of service (net psychic income) per unit of resource throughput. This facilitates the establishment of a high service-yielding and durable stock of human-made capital that reduces the need for additional production and work and increases the potential for job sharing.
2. *Wealth taxes.* By redistributing wealth, wealth taxes reduce the need for many people to work long hours to obtain a share of the annual flow of newly produced goods.
3. *Reductions in payroll taxes and income tax rates.* By decreasing labour costs, a reduction in payroll taxes encourages employers to substitute toward labour and away from producer goods (i.e., payroll taxes alter the cost-minimising combination of production factors in favour of labour). Income tax rate reductions increase the incentive to improve the quality of all newly produced commodities. This leads to higher commodity prices and higher hourly wages which reduces the need for long hours of work. It also increases the potential for job sharing.
4. *Profit-sharing arrangements.* Profit-sharing arrangements have two major benefits. First, by increasing labour productivity, they facilitate higher rates of efficiency-increasing innovation which again leads to higher hourly wages. Second, profit-sharing arrangements assist in the redistribution of wealth. Both further reduce the need for long hours of work.
5. *Population growth control.* If the labour force grows, so does the number of jobs needed to prevent unemployment. While labour force growth has been partly fuelled by increases in the participation rate, especially by increases in the number of females seeking employment, its main source is the increase in human population. The transferable birth licensing scheme that was previously prescribed and strict immigration controls reduce the growth of a nation's labour force. This, in turn, reduces the need for continued economic growth to provide jobs to meet the growing demand for work.
6. *International trade policy — a compensating tariff structure.* Compensatory tariffs imposed on nations with lower social, workplace, and environmental standards can overcome the standards-lowering

pressure in international markets that otherwise undermine domestic attempts to sever the GDP-employment link.

15.7.4.2 *Severing the GDP-poverty link*

The three most critical factors in severing the GDP-poverty link are the distribution of income and wealth and the quality of the stock of human-made capital. For instance, if the distribution of wealth is equitable then, because the need for a money income is reduced, so too is the need to maintain high levels of GDP to alleviate poverty. If the distribution of income is equitable, that is, there is an ethically desirable order-of-magnitude difference between a nation's richest and poorest citizens, there is a reduced need for high levels of GDP to enable the poor to receive an income capable of providing the basic necessities of life. Finally, if the quality and the durability of the stock of human-made capital is improving over time, money incomes can be maintained, if not increased, without the need to hasten the rate of production and consumption. Given that the six policy measures outlined to sever the GDP-employment link have the effect of redistributing wealth and facilitating improvements in the stock of human-made capital, they would again be useful in severing the GDP-poverty link. However, to further weaken this link, they must be supplemented by another policy measure already prescribed, namely, a guaranteed minimum income.

15.7.5 *International consensus on sustainable development issues*

Sustainable development is a global phenomenon. National borders do not insulate countries from global environmental problems caused by resource depletion, waste pollution, and biodiversity erosion at the regional or national levels. The best efforts of an individual nation to move toward SD will undoubtedly come undone if the rest of the world opts for continued economic and population growth. Clearly, the success of policies at the national level is conditional upon similar policies being introduced in most, if not all, remaining countries. For this to be achieved, it is necessary to establish an international consensus on such issues as population control, international trade, global resource consumption, and Greenhouse gas emissions. Once a consensus is reached, it will then be possible to establish a range of workable, multilateral treaties and agreements, most of which will need to incorporate a North-to-South redistribution of wealth. Such agreements will also need to make allowances for the much needed growth in impoverished nations that are still in the immature stage of the development process. Any country serious about achieving SD can assist in establishing an international consensus on the previously mentioned issues by articulating the undesirability of continued growth as well as the existence of a desirable growth alternative, namely, an optimal macroeconomic scale.

15.7.6 A timetable to gradually introduce sustainable development policies

The tendency of the global system to react to change in a conservative, identity-preserving fashion, demands that SD policies be gradually rather than immediately introduced. For instance, throughput restrictions to limit the incoming resource flow to the maximum sustainable rate cannot be introduced overnight, even if there is a strong desire to do so. The following is a recommended timetable for introducing some of the policy measures prescribed in this chapter. It may be possible to implement some policies more rapidly than recommended here. Others may have to be introduced more gradually. All told, the rate of policy implementation must be one that best facilitates a smooth transition toward SD.

1. *Quantitative throughput restrictions.* Restrictions on the incoming resource flow should initially be set at their present rate. This rate should be maintained for approximately three to 5 years beyond which it should be reduced over a 10 to 15 year period until it equals the maximum sustainable rate. If a sustainability buffer is preferred, the transition period may have to be extended for a further 3 to 5 years.

2. *Environmental assurance bonds and internalisation of the user cost of non-renewable resource depletion.* Environmental assurance bonds and legislation to force resource depleters to set aside a portion of their profits to establish renewable resource substitutes can be introduced over a three to five year period. This would give the public a preparatory period prior to the introduction of both measures.

3. *Pigouvian taxes/subsidies and Coasian solutions to ameliorate externalities.* On the whole, measures of this type can be introduced immediately. However, where Pigouvian taxes are required to control the qualitative nature of high entropy waste, they are likely to require a gradual phasing in over a three to five year period.

4. *Transferable birth licensing scheme.* In wealthy nations, where fertility rates are at or below replacement levels, the movement toward a transferable birth licensing scheme should pose few if any structural problems. Opposition is only likely to come from those who have a moral objection toward the scheme. Provided the objection can be overcome, the scheme can be introduced immediately. In countries with high fertility rates, it may be necessary to allot additional licences to permit each adult an extra half child (thereby allowing three children per family instead of the preferred two). Over a 5 to 10 year period the scheme can be adjusted to ensure each adult is allotted just one licence.

5. *Minimum income.* A minimum income to ensure each citizen the basic necessities of life should be introduced immediately.

6. *Maximum permissible income.* The 100 percent tax on the maximum permissible income can be introduced immediately. However, the recommended factor of 20 difference between the minimum and maximum income needs to be phased in. For the first 5 years, it can exist at a factor of 50 difference (still a considerable limit on some present incomes). It can then be reduced to a factor of 40, 30, and 20 difference over a 10 to 15 year period.

7. *Income tax rate reductions.* The speed and extent to which income tax rates can be reduced for incomes below the maximum depends largely on two things. The first is the amount of revenue collectively raised by a 100 percent tax rate on incomes above the maximum limit, the auctioning of tradeable exaction permits, confiscated assurance bonds, procreation levies, and compensatory tariffs. Second, it depends on the amount of government revenue that remains following normal budgetary allocations and the financing of a minimum income as part of the negative–positive income tax arrangement. Assuming that enough revenue would remain to permit income tax reductions then, because most of these revenue raising measures will themselves be gradually implemented, the tax cuts would initially be small and perhaps take three to five years to introduce.

8. *Wealth taxes.* Wealth taxes should be introduced over the same time period it takes to reduce income tax rates.

9. *A corporate charter.* A corporate charter embodying collectively determined standards and principles, in particular, community service-type obligations, should be gradually phased in over a 5 to 10 year period.

10. *Dealing with sunk costs to increase the contestability of markets.* To recall, the contestability status of markets depends largely on the magnitude of the sunk costs associated with operating in a particular industry. Increasing the contestability of markets requires sunk costs to be detached from utilising firms which can be achieved by having sunk costs borne by a government instrumentality or a private consortium. In order to have governments bear some of the sunk costs, it is necessary that they purchase and subsequently lease existing privately owned sunk cost facilities. Given budgetary restrictions, governments can only buy so many sunk cost facilities at any point in time. For this reason, governments should purchase the facilities where exclusive access most affects the contestability of markets. They should then aim to purchase the most critical sunk costs facilities within 20 to 30 years.

11. *Intellectual royalty rights.* A system of intellectual royalty rights to cover all newly created knowledge can be introduced immediately. As for intellectual property rights presently in existence under the current system, they should be phased out over a 5 to 10 year period.

12. *International trade measures.* Of the two prescribed international trade initiatives, the introduction of the IMPEX system of foreign exchange

management is likely to be the easiest to implement. The system, therefore should be introduced over a 5 to 10 year period. The speed with which compensatory tariffs can be imposed depends on how quickly an international consensus can be achieved in relation to the proposed compensating tariff structure.

13. *Central bank control of the nation's money supply.* Reserves held by banks at a central bank should be gradually increased from their current meagre levels to somewhere between 25 to 50 percent of deposits over a 5 to ten 10 period.

chapter sixteen

Toward sustainable development: the need for moral growth

16.1 Why sustainable development requires moral growth

The final chapter of the book focuses on the need for *moral growth*, or at least the growth of morality beyond its present level. The need for moral growth arises for three main reasons of which two have already been discussed in previous chapters. To recall, moral growth is required to regenerate the moral capital eroded by the market's individualistic ethos and to ensure human needs and wants are ranked in accordance with how well their satisfaction serves the ultimate end. There is, however, a third good reason for believing in the necessity of moral growth and it arises because the preaching of the policies to achieve SD is not enough to guarantee their implementation (Boulding, 1991). Indeed, no matter how necessary certain policies appear in the minds of their advocates, necessity is unlikely to elicit their implementation while morality is in a state of decline (Daly, 1980). As Haught (1993, page 7) explains,

> "It is hard to imagine how any thorough transformation of the habits of humans will occur without a corporate human confidence in the ultimate worthwhileness of our moral endeavours."

To put the need for moral growth in another light, consider what Daly has to say about Hardin's (1968) solution to the so-called "Tragedy of the Commons." Hardin contends that the Tragedy of the Commons is a human management problem and, as such, cannot be resolved by searching for a technical solution. Instead, Hardin believes the Tragedy of the Commons requires a political solution — more specifically, "mutual coercion mutually agreed upon" (Harding, 1968). Interestingly, Hardin rejects any moral-based solution arguing that any recourse to morality is self-cancelling. Daly does not disagree with Hardin's view regarding the need for mutual coercion. Daly (1980, page 349) simply asks the following: "...where is the mutual agreement to come from if not from shared values, from a convincing morality?" Given, therefore, that "mutual coercion mutually agreed upon" is unlikely to be achieved through political means alone (Crowe, 1969), it would appear that the political solution being advocated by Hardin belongs to a broader class of technical solutions. This leads Daly (1980) to conclude that mutual coercion, which is necessary to legitimise the painful adjustment process required to move toward SD, is not a substitute for, but presupposes moral growth.

16.1.1 Boulding's three generalised systems of power

Claiming the need for moral growth on the basis that technical and political solutions are unworkable is insufficient. Why should anyone believe that moral growth is capable of triggering a movement toward SD just because technical/political solutions cannot? To forcefully demonstrate the need for moral growth, three aspects warrant careful consideration. First, it is necessary to be reminded of the dynamics of the co-evolutionary paradigm, in particular, the identity-preserving nature of higher-order systems. Second, given that human belief systems have become a major driving influence over the evolutionary pathway of the global system, one needs to take account of the human sources of power that either facilitate or impede the movement toward SD. Third, it needs to be carefully considered how the various human sources of power can be harnessed to permit the implementation of the types of SD policies outlined in the previous chapter.

To the first aspect, assume that the global system — the Great Economy — is moving along an evolutionary pathway incommensurate with the SD objective (not an unreasonable assumption given the growing evidence that many nations are becoming poorer rather than richer as their macroeconomies continue to grow). In view of the identity-preserving nature of the global system, any movement toward SD will require a society or nation to "jump" onto a more desirable evolutionary pathway. Given that jumping from one pathway to another is, as explained in Chapter 11, equivalent to a bifurcation of the evolutionary pathway of the global system, moving toward SD will undoubtedly require a human-induced bifurcation event as significant as the Great Transformation. To appreciate how an event of this magnitude might be precipitated, the second of the previously mentioned aspects requires consideration. That is: what are the human sources of power that either facilitate or impede the movement toward a SD future? Boulding has distinguished three major categories or systems of power —

the threat, exchange, and integrative systems of power (1970a, 1970b, 1981, 1985, 1989, and 1991). All three systems of power are important insofar as they contribute to the establishment, maintenance, and modification of a society's institutional structures and arrangements. In so doing, each system of power helps to define and establish the various role-creating relationships that characterise any particular society or nation.

The first of Boulding's systems of power, the *threat* system, originates when one person says to another, "You do something I want or I'll do something you do not want." If the threatened individual or party submits, a threat-submission system emerges which can be particularly potent in creating social roles. Exactly what potential a threat system of power has to create important roles depends on the *capability* and *credibility* of the threatener (Boulding, 1970a). The threatener, for instance, must be capable of carrying out a promised threat for the threat to have its desired effect. The capability of the threatener depends largely on two things. The first is the physical ability of the threatener to consummate a promised threat. The second is the personal or organisational cost of consummating the threat and the probability of having to carry it out. The higher is the former and the lower is the latter, the more the threatened can discount the capability of the threatener.

Capability aside, the actual behaviour of the threatened still depends very much on the perceived credibility of the threatener in the minds of the threatened (Boulding, 1970a). Should a promised threat not be carried out, the threatener's credibility rapidly declines. Thus, in order to maintain credibility, the threatener must sustain a certain level of force and, in some cases, use excessive force to deter potential transgressors. That having been said, if the level of force is too excessive, a threatener's credibility can be completely obliterated which can lead to noncompliance on the part of the threatened. It is because of the fine line between the level of force necessary to maintain credibility and that which results in the threatener's illegitimation that threat systems of power are inherently unstable.

In most contemporary societies, threat power resides mainly in political institutions (such as governments) as well as legal systems and affiliated organisations (e.g., the police and defence forces, taxation departments, and other governmental authorities and instrumentalities). To a lesser degree, threat power is also held by unions, employer and business organisations, and, in some cases, large corporations whose investment decisions have regionalised economic and employment consequences.

The *exchange* system of power emerges when someone says to another, "You give me what I want and I'll reciprocate by giving you what you want." Provided the terms of exchange are acceptable and both parties are free to trade, exchange usually takes place. Unlike the threat system, participating in the exchange system of power is largely voluntary.* It is also mutually beneficial, suggesting that exchange invariably leads to positive-sum game outcomes (i.e., where both parties benefit). For this reason, the exchange system is inherently more stable than the threat system which, conversely, is often characterised by zero or even negative-sum game outcomes (i.e., where one party gains at the other's expense). Because the exchange system also encourages specialisation, it is a very powerful social organiser with much greater human development potential than the threat system of power.

The third system of power is the *integrative* system, a system that Boulding (1970a) readily admits is more difficult to identify and articulate than the threat and exchange systems of power. The integrative system includes relationships involving the mutual acceptance of status and legitimacy that serve as a catalyst for the creation of social roles. An integrative relationship occurs when someone says to another, "You do something in particular because of what I am and who you are in relation to me." Integrative relationships create such things as a sense of community; personal group, family, and organisational identification; self-motivation; and, most particularly, the legitimisation of authority, love, affection, trust, and mutual respect (Boulding, 1979a). The integrative system is where the majority of images, beliefs, and values pertaining to a society's belief system are formed. It is also where a society's images and beliefs are generally passed on by the elderly members of a society to the incoming younger members. As explained in Chapter 11, the importance of human belief systems lies in the fact that a society's stock of beliefs, images, and values largely determine its choice of institutions, norms, conventions, routinised habits, and technologies. Thus, in a very real sense, the integrative system incorporates a society's moral capital that, by legitimising threat and exchange systems, underpins the cohesion and developmental success of any society.

16.1.2 *Moral growth — an essential feature of the integrative system of power*

Boulding's three systems of power provide a framework for dealing with the last of the three previously mentioned aspects. That is, how can the various human sources of power be appropriately harnessed to permit the implementation of painful SD policies? There is little doubt that the integrative system of power is the most fundamental of all three systems of power (Boulding, 1991). This does not mean that the integrative system of power

---

* The fact that participation in the exchange system of power is usually voluntary puzzles many observers. Most people associate power with the ability to get someone to do something against their will. According to Boulding (1989), this a very narrow view of power and reflects the confusion of power with "force." Despite the exchange system of power involving little if any force, it is a very powerful system because it is a system of inducement (the "carrot") whereas the threat system is a system of force (the "stick").

has the greatest development potential. Such a distinction clearly belongs to the exchange system of power.* Nevertheless, the integrative system is the most fundamental because the exchange system of power can never be particularly effective until such time as it is legitimated and rendered purposeful by the integrative system (and, to some degree, by the threat system because the ability of property rights to facilitate beneficial exchange depends very much on being able to enforce and protect such rights). This reinforces the fact that the market mechanism cannot function with any degree of effectiveness unless it operates in tandem with supporting social and moral capital.

Consider, therefore, Boulding's three systems of power in relation to the four macro policy goals requiring resolution to achieve SD. Because the goals of distributional equity and ecological sustainability are resolved respectively through the use of taxes/transfers and restrictions on the incoming resource flow, they are achieved essentially through means associated with the threat system of power. The goal of allocative efficiency is achieved through means associated with the exchange system of power, such as markets and other institutionalised exchange mechanisms. Finally, the goal of moral growth is achieved through means associated with the integrative system of power and is fundamentally required because the integrative system currently lacks the shared moral values needed to legitimate the painful threat-related measures required to move toward SD.

In view of the aforementioned, moral growth must occur at the international as well as the societal level. Moral growth at the international level is of great importance, despite it being the most difficult to achieve, because SD is a global problem requiring global solutions. This indicates that integrative systems of power and shared moral values are required at all levels and spatial dimensions of human organisation. Moral growth at the international level is also important because the international domain is dominated by threat systems of power (e.g., the United Nations and the World

---

* Despite the exchange system of power having the greatest development potential, it is often underestimated or, in some instances, completely overlooked. There are a few reasons for this. First, as a system of inducement rather than force, the exchange system is essentially a nondialectical system of power. Consequently, the progress facilitated by exchange involves incremental rather than rapid change. The threat system, which gets much greater attention than it deserves, is a dialectical system. Because dialectical events involve conflict as well as readily identifiable winners and losers, they are inherently more noticeable and receive considerably more attention than nondialectical events. In addition, much of the publicity accorded to the exchange system of power is focussed on isolated negative-sum game outcomes, such as a financial crisis or a stock market crash, people being "ripped off" by an unscrupulous seller (which says more about the information possessed by buyers or the dishonesty of the seller than it does about exchange itself), people being adversely affected by the spillover costs of activities arising from the exchange between two consenting parties (which says more about the current pervasiveness of negative externalities than it does about exchange), or the excessive use of exchange power by the privileged rich over the disadvantaged poor (which says more about the current distribution of income and wealth than it does about exchange). The countless, even if small, positive-sum game outcomes occurring each day through exchange often go completely unrecognised.

Trade Organisation) and an exchange system of power that, as a result of the increasing mobility of capital, is gradually undermining national exchange systems and, to a growing extent, national threat and integrative systems of power. Because the international domain lacks a comprehensive integrative system of power, the ability to erect one may prove to be the most important task in moving toward SD.

16.2 How is moral growth achieved?

16.2.1 A moral consensus gained through the integrative power of the truth

If moral growth is to be achieved and human belief systems are to be appropriately revised to trigger a movement toward SD, where is the moral growth itself to come from? The first thing that must be recognised is that moral growth cannot be achieved unless there is some semblance of a moral consensus among the citizens of a nation. Eventually, to move as a global community toward SD, there will need to be a global moral consensus. This suggests that as important as it is to maintain cultural diversity, the various human belief systems across the world will need to embody a great number of commonly held beliefs, images, and values.

   Given the degree of pluralism and value relativism found in most societies, attaining some semblance of a moral consensus will undoubtedly be difficult even at the national level. The task is made easier by acknowledging that only *real* objective values are capable of withstanding genuine scrutiny and commanding consensus in a sophisticated and self-analytical society (Daly, 1980). Of course, this still leaves the problem of determining what real objective values are. Nonetheless, it provides a starting point from which to ascertain objective values because, if nothing else, objective values are *true* values. This, it seems, is the essence of the matter, for if there is one thing with the potential to bring about a shared sense of and dogmatic belief in objective value, it is the integrative power of the truth. While there is bound to be considerable disagreement as to what constitutes true values in a pluralistic society, consensus can still be achieved by demonstrating that certain "truths" cannot be avoided forever — that, despite one's current images and beliefs, some truths are becoming impossibly difficult to deny. For instance, it is becoming increasingly difficult to deny that the macroeconomy is a subsystem of the sociosphere which, in turn, is a subsystem of the ecosphere; that human well-being requires a balanced system of physiological and psychological need satisfaction of which only the latter has any lasting or enduring qualities; that the physical scale of a nation's macroeconomy can expand only so much before it cannot be ecologically sustained; and that there is a point beyond which the extra benefits of a growing macroeconomy are exceeded by the additional costs (in which case growth or quantitative expansion should give way to development or qualitative

improvement). It is around these and other truths, even if they are empirically derived, that a process of moral growth must originate. If it eventually
leads to a widespread belief or faith in religiously conceived objective values,
so be it. However, in a sea of pluralism and value relativism, it is difficult
to imagine that pure faith, alone, will serve as a launching pad for a much
needed process of moral growth.

16.2.2 *The integrative power of truth — created via a learning/teaching process*

Should the integrative power of truth offer the greatest potential to bring
about a dogmatic belief in objective value, how can the integrative power
of the truth be created if many truths are not immediately obvious from
one's everyday experiences? Quite simply, emerging truths must be communicated in such a way as to become the centrepiece of the mainstream
stock of knowledge where, currently, they are found at the periphery. Only
then will they become an integral component of a society's stock of beliefs,
images, and values necessary to legitimate the many painful SD policies.

Boulding has shown that integrative relationships arising out of the
formation of shared beliefs and images are best created via a learning process
(Boulding, 1970a, 1985, 1989, and 1991). Thus, in order for truths rather than
falsehoods to become an integral part of society's belief system, a dual
learning and teaching process is required. At a societal level, a learning
process involves the revision of human belief systems via the constant updat-
ing of a society's stock of beliefs, images, and knowledge-based truths.
Provided the learning process is successful, erroneously held truths can be
gradually weeded out, existing truths reinforced, and new and previously
undiscovered truths embraced. Of course, how rapidly the discovery of
falsehoods and emerging truths leads to the progressive revision of human
belief systems depends on how well they are conveyed to individuals har-
bouring the pre-existing but erroneously held images, beliefs, and values. It
is for this reason that the teaching of new knowledge is likely to play a
considerable role in the overall learning process, for unless they can be
successfully taught, they are unlikely to be readily learned.

There are, however, a number of potential problems associated with a
reliance on a dual learning/teaching process. The first is that, as explained
in Chapter 3, belief systems are conservatively resistant to change. This is
an important feature of belief systems and is something to be valued not
despised. After all, a dogmatic belief in objective value depends on the
ability of belief systems to withstand conflicting signals yet to be supported
by convincing evidence. However, provided the learning/teaching process
is an effective one, there is no reason to believe that conservatism should
prevent the wholesale elimination of error and the eventual emergence of
the truth. While the degree of conservatism will ultimately determine the
rate at which errors are eliminated and genuine truths are embraced, it is

unlikely that conservatism can continue to repel the integrative power of the truth.

The second potential difficulty lies in the fact that as many, if not more, falsehoods are likely to be conveyed during the learning/teaching process as are genuine truths. Falsehoods, by biasing the manner in which messages and signals are received and interpreted, can perpetuate and, in some instances, reinforce existing falsehoods. For instance, the false view of the macroeconomy as an independent circular flow of physical commodities leads one to interpret a rise in real GDP as evidence of an increase in a nation's sustainable economic welfare.

Finally, with the potential for as many falsehoods to be preached as genuine truths, is there any guarantee that truths will prevail over falsehoods when so many of the latter continue to occupy the current mainstream stock of knowledge? In the short term, perhaps not. In the long run, and provided the learning/teaching process is characterised by an unbiased testing of existing images, the answer is probably yes, at least where human ignorance can be reduced through the application of the scientific method. Of course, there have undoubtedly been instances in the past where threat systems of power have successfully prevented emerging truths from becoming a part of the mainstream stock of knowledge. But their success is always short lived because, under the weight of a growing appreciation of self-evident truths, truth-subversive threat systems have little or no underlying credibility and, as such, become too difficult and costly to maintain.

16.2.3 The need for nondialectical institutions, organisations, and processes

Although history points to the eventual domination of the truth, of even greater importance is the emergence of the truth in the first place. How easily the truth is able to emerge depends not only on the state of the knowledge already acquired, but whether the much needed learning/teaching process is a genuine truth-seeking one. This, in turn, depends on whether the institutions associated with the conveyance of information and knowledge are amenable to the learning/teaching process. To appreciate whether they are, one requires a better understanding of the relationship between a society's institutions and the learning process itself.

Learning is an incremental, knowledge-building process characterised by Lamarkian evolution — the ability of acquired images, beliefs, and values to be transmitted by a society's institutional and organisational structures and arrangements. Whether institutions facilitate or impede a progressive learning process depends very much on whether institutions are, in the Hegelian sense, dialectical or nondialectical in nature. Dialectical processes are those involving systems with a contradiction implicit in their own development (Boulding, 1970a). The contradiction usually entails the negation of the present system and the eventual emergence of its opposite or antithesis. Because the antithesis also possesses a contradiction in its development, a third system eventually emerges — the synthesis. Over time, dialectical processes lead to a cyclical pattern of thesis–antithesis–synthesis where the synthesis is little if at all advanced from the original thesis (Boulding, 1970a).* Human-related dialectical processes exist whenever social systems or elements within them are dominated by conflict, separatism, and suspicion. A contradiction arises within the development of these systems because dialectical processes often involve the struggle for ideological, economic, or political supremacy. Rarely do they involve an open dialogue between opposing forces and rarely is the overriding objective the resolution of existing problems and disputes. It is for this reason that dialectical processes are most prevalent in social systems dominated by threat systems of power.

Human-related nondialectical processes are considerably different. While it is certainly true that nondialectical processes are characterised by the emergence of a new macrostate over and above a previously existing one, the new system does not emerge because of a conflict between two or more systems or, in the case of a social system, two or more elements within it. It arises because of a discontinuity in the evolutionary pathway of a once existing system caused by the co-evolution of its constituent elements pushing its parameters beyond a bifurcation threshold. Co-evolution, of course, involves the interdependent relationship between two or more closely interacting species or social elements and, as such, is not a conflictual process. From a social systems perspective, the emergence of new macrostates from nondialectical processes is usually brought about by changes arising from the integrative and exchange systems of power. Rarely are they the product of changes arising from the threat system of power.

Taking into account the fundamental difference between dialectical and nondialectical processes, there are good reasons for believing that the latter facilitate learning and the emergence of the truth while the former do not. To explain why, Boulding (1970a, page 62) is worth quoting at length:

> "The growth of knowledge is not advanced by the dialectical process at all; it is hindered by it. Dialectical struggles prevent the testing of knowledge by direct or scientific methods, and they represent a backward step towards the testing of images by the success of whole cultures, not by the testing of knowledge itself.

---

* A continual pattern of thesis–antithesis–synthesis is not inconsistent with the principle of coevolutionary balance and the identity-preserving nature of higher-order systems. Although each successive system is very much the opposite to its predecessor, most of the assumed values that dominate the system and influence its dynamics remain largely unchanged. Consider an example, such as a two party-dominated political system. Despite both parties having opposing ideological positions, they invariably share the same falsely conceived ends (e.g., continuous growth is a desirable macroeconomic objective). This is because party politics is essentially about succeeding at the ballot box which requires the objectives of political parties to closely reflect the assumed values dominating the co-evolutionary process. It is for this reason that political parties move toward the centre of the political spectrum and closely resemble each other (Downs, 1957).

The dialectical process results in severe impairments of the knowledge-building process on both sides of the conflict, for when questions are posed in dialectical terms, that is, in terms of the conflict between two cultures or two power centres, each centre gradually loses the ability to learn from the other. The value system of each begins to operate as a filter to distort the process of information input, to filter out information which is contradictory to the particular ideology held, and, hence, to prevent learning. Beyond a certain point, conflict processes are not conducive to learning. They tend rather to reinforce old images, no matter what they are and no matter how valid they are. The kind of testing of images that goes on in conflict is entirely different from the kind of testing that goes on in the scientific (nondialectical) process" (parenthesis added).

Clearly, in order to facilitate the emergence of genuine truths, a nation must have in place a set of institutional and organisational structures and arrangements that are predominantly nondialectical in nature. One would expect nations dominated by dialectical processes and ideologies to also be dominated by threat systems. This is indeed the case, as countries like China, Cuba, and the former Soviet Union have shown. Although Western societies have long had market economies, they possess fewer nondialectical institutions than most people think. This is because the depletion of moral capital and the subsequent erosion of integrative relationships caused by a successful rise of the individualistic market ethos has led to the proliferation of counteracting dialectical elements that are almost entirely contained within the threat system of power (Boulding, 1970a). For instance, almost all systems of governance are based on a dialectical form of representative democracy. In addition, industrial relations systems have traditionally involved the conflict between the representatives of workers (unions) and the representatives of the owners of the human-made means of production (business chambers). While the industrial relations environment is rapidly changing in most countries, the legal nature of the corporation maintains a dialectical relationship between workers, managers, and stockholders. Finally, the erosion of important integrative relationships has led to an increasing array of legislative non-price rules and regulations, many of which are impediments to the efficient operation of the market — one of society's important nondialectical institutions.

To facilitate a dual learning/teaching process necessary to create and harness the integrative power of the truth, the following is required. First, there is a need for more participatory forms of social, political, and industrial democracy and a reduced reliance on representative democracy. Second, organisational structures of all kinds must become less hierarchical, particularly in relation to decision-making processes. Third, commitments to free speech and freedom of expression must be maintained or established if they have not been already. Finally, it is important to recognise that societies which stress the importance of community, openness, communication, love, peace, and tolerance are those most capable of establishing and maintaining fruitful integrative relationships. They are, as a consequence, the societies that are most likely to make the greatest gains in terms of advancing the stock of knowledge and attaining the moral growth required to trigger the movement toward SD.

16.2.4 Conclusion

Unless there is a growth in morality beyond its present level, the many painful policies required to achieve SD are unlikely to be implemented. The fact that real objective values are true values suggests the one thing with the greatest potential to bring about a moral consensus is the integrative power of the truth — something that is only likely to emerge following a widespread recognition that some truths are becoming impossibly difficult to deny. Of all the emerging truths, the most glaringly obvious is that growth is ecologically unsustainable in the long run and existentially undesirable once a nation has accumulated a sufficient per capita stock of human-made capital. Sustainable development only requires growth in the early stages of a nation's development process. Beyond the initial need for growth, economic activity should be directed toward the qualitative improvement in the stock of human-made capital, the minimisation of the resource throughput that keeps the stock intact, and the responsible stewardship of the natural capital that sustains the low entropy resource flow and provides a haven for all other earthly creatures.

Creating the integrative power of the truth requires a dual learning/teaching process. It is best facilitated by institutional and organisational structures and arrangements that are nondialectical in nature. Provided a progressive learning process is successfully achieved and moral growth is attained at all levels and spatial dimensions of human organisation, the painful policies required to achieve SD should eventually see the light of day. Let us indeed hope so. More importantly, let us hope it is not too far away because any failure on humankind's part is likely to result in a feedback response of the global system that will undoubtedly have dire consequences for humankind's well-being, if not its long-term survival.

V

Appendices

appendix one

The uncancelled benefit account for Australia, 1966–1967 to 1994–1995

A1.1 Introduction

The uncancelled benefit account seeks to measure the annual net psychic income enjoyed by the Australian population. Appendix 1 reveals the methods and data sources used to calculate monetary values for the items appearing in the uncancelled benefit account. Tables revealing how the majority of the items have been calculated are located in this appendix. All items are measured in millions of Australian dollars at 1989–1990 prices. The uncancelled benefit account appears in Table 14.1 in Chapter 14. The items appearing in the uncancelled benefit account are

   a. *Psychic income*

- Private consumption expenditure.
- Index of distributional inequality.
- Distributional weighting of private consumption expenditure.
- Services yielded by consumer durables, by public dwellings, and by roads and highways.
- Services provided by nonpaid household labour and volunteer labour.
- Public expenditure on health and education counted as consumption expenditure.
- Net producer good growth.
- Change in net international position.
- Imputed value of leisure time.

   b. *Psychic outgo*

- Expenditure on consumer durables.
- Defensive private health and education expenditure.
- Cost of private vehicle accidents.

- Cost of noise pollution.
- Direct disamenity cost of air pollution.
- Cost of unemployment and underemployment.
- Cost of commuting.
- Cost of crime and cost of family breakdown.

A1.2 Psychic income

Item A — *Private consumption expenditure.*

Item A is a measure of all private consumption expenditure undertaken during each financial year measured in millions of dollars at 1989–1990 prices. The data source is Foster (1996, Table 5.2a, page 220).

Item B — *Index of distributional inequality* (Table A1.a).

Table A1.a   Item B — Index of Distributional Inequality

Year	Gini Coefficient *a*	Index of Distributional Inequality *B*
1966–1967	0.330	100.0
1967–1968	0.330	100.0
1968–1969	0.330	100.0
1969–1970	0.326	98.8
1970–1971	0.322	97.6
1971–1972	0.318	96.4
1972–1973	0.314	95.2
1973–1974	0.310	93.9
1974–1975	0.326	98.8
1975–1976	0.342	103.6
1976–1977	0.358	108.5
1977–1978	0.374	113.3
1978–1979	0.390	118.2
1979–1980	0.393	119.1
1980–1981	0.397	120.3
1981–1982	0.400	121.2
1982–1983	0.403	122.1
1983–1984	0.405	122.7
1984–1985	0.408	123.6
1985–1986	0.410	124.2
1986–1987	0.415	125.8
1987–1988	0.420	127.3
1988–1989	0.425	128.8
1989–1990	0.430	130.3
1990–1991	0.434	131.5
1991–1992	0.438	132.7
1992–1993	0.442	133.9
1993–1994	0.446	135.2
1994–1995	0.450	136.4

A total of seven different income distribution studies have been under-taken by the Australian Bureau of Statistics (ABS) during the study period (1968–1969, 1973–1974, 1978–1979, 1981–1982, 1985–1986, 1989–1990, and 1994–1995). The index of distributional inequality was constructed by making use of the estimated Gini coefficient from each of the respective studies. It was assumed that the distribution of income in the years 1966–1967 and 1967–1968 was the same as in 1968–1969. For the remaining intervening years, the Gini coefficient was estimated by way of straightline interpolation.

To calculate the index, the initial year, 1966–1967, was given a base index value of 100.0. The base index value was adjusted in accordance with changes in the Gini coefficient to reflect any improvement/deterioration in the distribution of income. The index of distributional inequality shows that Australia's income distribution improved between the years 1966–1967 and 1973–1974 (falling from 100.0 to 93.9) but deteriorated between 1973–1974 and 1994–1995 (increasing from 93.9 to 136.4). The data sources are the Australian Bureau of Statistics, Catalogue No. 6523.0 and Economic Planning Advisory Council (1995, pages 46–64).

Item C — *Distributional weighting of private consumption expenditure.*

Private consumption expenditure (Item A) was weighted by the index of distributional inequality by dividing consumption expenditure for each year by the respective index value, and multiplying by 100.

Item D — *Services yielded by the stock of consumer durables* (Table A1.b).

To calculate the service yielded by the stock of consumer durables, it is necessary to determine the rate at which the entire stock of previously bought consumer durables is "consumed." This was assumed to be 15 percent per annum on the basis that this approximates the rate at which the average consumer durable item depreciates. Thus, in order to obtain an approximate value of the annual services yielded by the stock of consumer durables, the annual value of the stock was multiplied by 0.15. The data source is from the Commonwealth Treasury of Australia, (1996), Canberra, Table 1(a), page 49 and Table 1(b), page 50

Item E — *Services yielded by the stock of public dwellings* (Table A1.c).

The imputed rental (service) rate of public dwellings for each financial year was based on the imputed rental rate for private dwellings as used in the national accounts by the ABS. The data sources are the Australian Bureau of Statistics, Catalogue No. 5204.0 and Foster (1996, Table 5.23, page 274).

Table A1.b Item D — Service from Consumer Durables

Year	Stock of Consumer Durables at Current Prices ($m) *a*	Stock at Average 1989–1990 Prices ($m) *b*	Service from Consumer Durables ($0.15 \times b$) ($m) *D*
1966–1967	6.798	45,023	6,753
1967–1968	7,247	46,758	7,014
1968–1969	7,897	49,048	7,357
1969–1970	8,579	50,762	7,614
1970–1971	9,505	53,399	8,010
1971–1972	10,517	55,351	8,303
1972–1973	11,704	56,273	8,441
1973–1974	14,078	59,395	8,909
1974–1975	17,726	63,087	9,463
1975–1976	21,315	65,991	9,899
1976–1977	24,557	68,416	10,262
1977–1978	26,969	69,688	10,453
1978–1979	28,918	69,345	10,402
1979–1980	32,251	69,501	10,425
1980–1981	35,855	70,025	10,504
1981–1982	40,677	71,998	10,800
1982–1983	45,298	72,613	10,892
1983–1984	48,368	72,407	10,861
1984–1985	52,933	74,953	11,243
1985–1986	61,035	81,054	12,158
1986–1987	68,698	85,117	12,768
1987–1988	73,937	85,175	12,776
1988–1989	79,668	84,368	12,655
1989–1990	85,828	85,828	12,874
1990–1991	89,601	86,913	13,037
1991–1992	92,297	87,867	13,180
1992–1993	96,767	90,961	13,644
1993–1994	191,306	94,113	14,117
1994–1995	106,646	97,261	14,589

Item F — *Services yielded by roads and highways* (Table A1.d).

To calculate the services provided by roads and highways, it was assumed that the imputed service rate was equal to the average service rate on dwellings over the study period, which was 11.5 percent. The total value of all publicly owned roads was multiplied by 0.115 to obtain an estimate of the annual services yielded. The data source is the Australian Bureau of Statistics, Catalogue No. 5204.0.

Item G — *Services provided by nonpaid household labour* (Table A1.e).

Nonpaid household labour includes the value of all domestic activities (household cleaning, gardening, pet care, household maintenance, etc.), child care, and shopping. Using the gross opportunity cost method, where the

Table A1.c   Item E — Service from Public Dwellings

Year	Stock of Public Dwellings (1989–1990 Prices) ($m)	Public Dwellings Contributing to a Welfare (0.5 × a) ($m)	Dwelling Imputed Rental (Service) Rate (%)	Service from Public Dwellings (b × c) ($m)
	a	b	c	E
1966–1967	6,910	3,455	10.8	373
1967–1968	7,135	3,568	11.1	396
1968–1969	7,371	3,686	11.2	413
1969–1970	7,648	3,824	11.4	436
1970–1971	8,011	4,006	11.7	469
1971–1972	8,215	4,108	11.6	476
1972–1973	8,334	4,167	10.6	442
1973–1974	8,647	4,324	9.6	415
1974–1975	9,454	4,727	9.6	454
1975–1976	10,079	5,040	9.9	499
1976–1977	10,503	5,252	10.5	551
1977–1978	10,987	5,494	11.3	621
1978–1979	11,321	5,661	11.9	674
1979–1980	11,626	5,813	11.9	692
1980–1981	12,012	6,006	11.6	697
1981–1982	12,387	6,194	11.5	712
1982–1983	12,986	6,493	11.9	773
1983–1984	13,918	6,959	11.9	828
1984–1985	14,968	7,484	11.8	883
1985–1986	16,111	8,056	11.9	959
1986–1987	17,174	8,587	12.3	1,056
1987–1988	17,890	9,945	12.1	1,082
1988–1989	18,439	9,220	11.7	1,079
1989–1990	19,342	9,671	11.7	1,132
1990–1991	20,039	10,020	12.2	1,222
1991–1992	20,671	10,336	12.5	1,292
1992–1993	21,500	10,750	12.5	1,344
1993–1994	21,723	10,862	12.4	1,347
1994–1995	22,594	11,297	12.1	1,367

gross opportunity cost of an hour of unpaid housework is equivalent to the average hourly real wage, the value of household labour was estimated by the ABS at $251 billion in 1991–1992, or $238.9 billion at 1989–1990 prices. This value was based on an opportunity cost value of $13.65 per hour (1989–1990 prices) and an estimated 17.5 billion hours of nonpaid household labour ($238.9 billion equals $13.65 per hour multiplied by 17.5 billion hours of non-paid household labour).

To estimate the value of household work in other years, the following method was used. First, the 1991–1992 estimate of total household hours worked was divided by the number of households in the year to obtain an estimate of the household labour hours worked per household. Second, it was assumed that the number of yearly hours worked per household declined over the entire period at a rate of 2 percent per year. This assumption was based on the following.

Table A1.d   Item F — Service from Roads and Highways

Year	Stock of Roads and Highways (1989–1990 Prices) ($m) *a*	Roads Contributing to Welfare (0.75 × *a*) ($m) *b*	Service from Roads and Highways (0.11 × *b*) ($m) *F*
1966–1967	22,708	17,031	1,959
1967–1968	23,852	17,889	2,057
1968–1969	25,091	18,818	2,164
1969–1970	26,383	19,787	2,276
1970–1971	27,418	20,564	2,365
1971–1972	28,882	21,662	2,491
1972–1973	31,438	23,579	2,712
1973–1974	35,667	26,750	3,076
1974–1975	37,730	28,298	3,254
1975–1976	36,852	27,639	3,178
1976–1977	36,247	27,185	3,126
1977–1978	36,527	27,395	3,150
1978–1979	37,956	28,467	3,274
1979–1980	39,714	29,786	3,425
1980–1981	41,887	31,415	3,613
1981–1982	43,288	32,466	3,734
1982–1983	43,731	32,798	3,772
1983–1984	43,588	32,691	3,759
1984–1985	43,820	32,865	3,779
1985–1986	43,811	32,858	3,779
1986–1987	43,585	32,689	3,759
1987–1988	43,204	32,403	3,726
1988–1989	43,092	32,319	3,717
1989–1990	42,562	31,922	3,671
1990–1991	44,027	33,020	3,797
1991–1992	44,335	33,251	3,824
1992–1993	44,306	33,230	3,821
1993–1994	44,357	33,268	3,826
1994–1995	44,443	33,332	3,833

- A reduction in the time devoted to housework labour owing to the increased productivity and subsequent time-saving nature of household appliances over time.
- A reduction in the number of persons per household.
- An increase in the use of other paid services, precooked food, takeaway food, and other conveniences.

Multiplying the preceding items by the number of households each year provided an annual estimate of the total yearly hours of household work. This was then multiplied by the average hourly real wage to obtain the total real value of household work. The data sources are the Australian Bureau of Statistics, Catalogue No. 5204.0 and Foster, (1996, Table 4.18, page 209).

Table A1.e   Item G — Service from Nonpaid Household Labour

Year	Yearly Hours of Household Work per Household	Number of Households (Thousands)	Total Yearly Hours of Household Work ($a \times b$) (Millions)	Average Hourly Real Opportunity Cost ($ at 1989–1990 Prices)	Service from Nonpaid Household Labour ($c \times d$) ($m)
	a	b	c	d	G
1966–1967	4,822.7	3,258.3	15,713.8	7.58	119,111
1967–1968	4,728.2	3,361.4	15,893.4	8.40	133,504
1968–1969	4,635.5	3,464.5	16,059.7	8.54	137,150
1969–1970	4,544.6	3,567.6	16,213.3	8.53	138,300
1970–1971	4,455.5	3,670.6	16,354.4	9.19	150,297
1971–1972	4,368.1	3,764.6	16,444.1	9.39	154,411
1972–1973	4,282.4	3,858.6	16,524.1	9.89	163,423
1973–1974	4,198.5	3,952.5	16,594.6	10.79	179,055
1974–1975	4,116.1	4,046.5	16,655.8	11.10	184,879
1975–1976	4,035.4	4,140.5	16,708.6	11.58	193,485
1976–1977	3,956.3	4,246.2	16,799.2	11.73	197,055
1977–1978	3,878.7	4,351.9	16,879.7	11.94	201,544
1978–1979	3,802.7	4,457.5	16,950.2	11.77	199,508
1979–1980	3,728.1	4,563.2	17,012.1	11.72	199,381
1980–1981	3,655.0	4,668.9	17,064.8	12.04	205,461
1981–1982	3,583.4	4.772.6	17,102.1	12.12	207,278
1982–1983	3,513.1	4,876.3	17,130.9	12.20	208,997
1983–1984	3,444.2	4,980.0	17,152.1	12.35	211,829
1984–1985	3,376.7	5,083.7	17,166.1	12.50	214,577
1985–1986	3,311.2	5,199.4	17,216.3	12.48	214,859
1986–1987	3,245.6	5,315.0	17,250.4	13.31	229,602
1987–1988	3,181.9	5,442.6	17,317.8	13.19	228,422
1988–1989	3,119.5	5,570.3	17,376.6	13.08	227,285
1989–1990	3,058.4	5,697.9	17,426.5	13.13	228,809
1990–1991	2,978.4	5,825.5	17,350.7	13.41	232,672
1991–1992	*2,939.6*	*5,953.1*	*17,499.7*	*13.65*	*238,871*
1992–1993	2,881.6	6,080.7	17,522.1	13.88	243,207
1993–1994	2,825.1	6,208.4	17,539.4	14.16	248,357
1994–1995	2,769.7	6,336.0	17,548.8	14.23	249,720

Item H — *Services provided by volunteer labour* (Table A1.f).

A recent ABS survey estimated the total number of voluntary hours of work at 433.9 million in 1994–1995. At a gross opportunity cost of $14.23 per hour (at 1989–1990 prices), the total value of voluntary work for 1994–1995 was estimated to be $6,174 million at 1989–1990 prices ($6,174 million equals $14.23 per hour multiplied by 433.9 hours of volunteer work). It was then assumed that the total number of voluntary hours had increased over the 1966–1967 to 1994–1995 period by 2 percent. The rationale for the 2 percent increase was based on an increase over time in the total population, particularly in the older age cohorts who are more likely to engage in volunteer work activities, and a steady increase in the number of people either unemployed, homeless, and/or below the absolute poverty line. Like the elderly,

Table A1.f   Item H — Service from Volunteer Labour

Year	Total Hours of Voluntary Work (Millions) *a*	Average Yearly Hourly Wage ($) *b*	Service from Volunteer Labour (*a* × *b*) ($m) H
1966–1967	157.9	7.58	1.197
1967–1968	161.1	8.40	1,353
1968–1969	164.3	8.54	1.403
1969–1970	167.6	8.53	1,430
1970–1971	171.0	9.19	1,571
1971–1972	174.4	9.39	1,638
1972–1973	177.9	9.89	1,759
1973–1974	181,4	11.79	2,139
1974–1975	185.1	11.10	2,055
1975–1976	190.6	11.58	2.207
1976–1977	196.3	11.73	2.303
1977–1978	202.2	11.94	2,414
1978–1979	208.3	11.77	2,452
1979–1980	214.5	11.72	2,514
1980–1981	221.0	12.04	2.661
1981–1982	227.6	12.12	2,759
1982–1983	239.2	12.20	2,918
1983–1984	251.4	12.35	3,105
1984–1985	264.1	12.50	3,301
1985–1986	277.6	12.48	3,464
1986–1987	291.7	13.31	3,883
1987–1988	306.6	13.19	4,044
1988–1989	322.2	13.08	4,214
1989–1990	338.5	13.13	4,445
1990–1991	355.8	13.41	4,771
1991–1992	373.9	13.65	5,104
1992–1993	392.9	13.88	5,453
1993–1994	412.9	14.16	5,847
1994–1995	433.9	14.23	6,174

people belonging to these categories rely heavily on the assistance provided by voluntary workers.

To calculate the total number of voluntary work hours in each year from 1966–1967 to 1993–1994, a running 2% deduction was made from the 1994–1995 figure back to 1966–1967. The estimated number of voluntary hours was then multiplied by the real average wage for each year to ascertain the total annual value of volunteer labour. The data sources are the Australian Bureau of Statistics, Catalogue No. 4441.0 and Foster, (1996, Table 4.18, page 209).

Item I — *Public expenditure on health and education counted as consumption* (Table A1.g).

The method adopted here is similar to that employed by Daly and Cobb (1989). It was assumed that one-half of all public education expenditure was

Table A1.g Item I — Public Expenditure on Health and Education Counted as Consumption

Year	Public Expenditure on Education ($m) _a_	Expenditure on Education Counted as Consumption (0.5 × _a_) ($m) _b_	Public Expenditure on Health ($m) _c_	Expenditure on Health Conributing to Improved Health (_c_ − 4349.3) × 0.5 ($m) _d_	Public Expenditure on Health and Education Counted as Consumption (_b_ + _d_) ($m) _I_
1966–1967	5,297.4	2,648.7	4,349.3	0.0	2,649
1967–1968	5,711.3	2,855.7	4,506.2	78.4	2,934
1968–1969	6,268.2	3,134.1	4,860.8	255.8	3,390
1969–1970	6,997.4	3,498.7	5,372.6	511.7	4,010
1970–1971	7,901.6	3,950.8	6,143.2	897.0	4,848
1971–1972	8,748.7	4,374.4	6,812.1	1,231.4	5,606
1972–1973	9,066.9	4,533.5	7,191.0	1,420.9	5,954
1973–1974	10,764.7	5,382.4	7,989.2	1,820.0	7,202
1974–1975	13,278.2	6,639.1	9,454.8	2,552.8	9,192
1975–1976	13,735.2	6,867.6	12,544.0	4,097.4	10,965
1976–1977	14,147.8	7,073.9	11,793.5	3,722.1	10,796
1977–1978	14,777.5	7,388.8	11,403.1	3.526.9	10,916
1978–1979	14,637.1	7,318.6	11,530.1	3,590.4	10,909
1979–1980	14,195.5	7,097.8	11,185.1	3,417.9	10,516
1980–1981	14,697.4	7,348.7	11,793.3	3,722.0	11,071
1981–1982	14,925.7	7,462.9	12,054.7	3,852.7	11,316
1982–1983	15,365.8	7,682.9	12,181.0	3,915.9	11,599
1983–1984	15,975.8	7,987.9	13,661.8	4,656.3	12,644
1984–1985	16,591.3	8,295.7	16,095.7	5,873.2	14,169
1985–1986	17,034.6	8,517.3	16,550.2	6,100.5	14,618
1986–1987	17,019.5	8,509.8	17,126.5	6,388.6	14,898
1987–1988	16,588.8	8,294.4	17,615.7	6,633.2	14,928
1988–1989	17,021.9	8,511.0	18,414.0	7,032.4	15,543
1989–1990	17,226.0	8,613.0	18,838.0	7,244.4	15,857
1990–1991	18,556.3	9,278.2	19,939.1	7,794.9	17,073
1991–1992	19,623.4	9,811.7	20,567.0	8,108.9	17,921
1992–1993	20,193.1	10,096.6	21,241.2	8,446.0	18,543
1993–1994	20,533.5	10,266.8	21,923.8	8,787.3	19,054
1994–1995	20,577.8	10,288.9	22,844.6	9,247.7	19,537

purely defensive and did not increase psychic income. As for public health expenditure, it was assumed that only one-half of all expenditure above the 1966–1967 level contributed to improved health and, thus, to increased psychic income. The data source is Foster (1996, Table 2.12, page 83).

Item J — *Net producer goods growth* (Table A1.h).

As per the "strong" sustainability concept, there is a need for the stock of producer goods to meet the minimum requirement necessary to sustain the nation's net psychic income. The net growth in producer goods was calculated in the following manner. First, the net producer goods growth requirement was calculated by multiplying the percentage increase in the labour force by the previous year's stock of producer goods. This provided an estimate of the quantity of producer goods necessary to maintain the average quantity of producer goods per worker. Next, the growth requirement was

Table A1.h Item J — Net Producer Goods Growth

Year	Australian Labour Force (Thousands) a	Percentage Change in Labour Force (%) b	5 Year Rolling Ave. % Change (%) c	Net Stock of Producer Goods (Exc. Labour) ($ at 1989–1990 Prices) d	5 Year Rolling Average of Producer Goods e	Change in 5 Year Rolling Average $(e - t_{-1})$ f	Producer Goods Requirement $(c \times e_{t-1})$ g	Net Producer Goods Growth $(f - g)$ ($m at 1989–1990 Prices) J
1961–1962	4,335.0			289,009				
1962–1963	4,396.0	1.4		305,483				
1963–1964	4,541.0	3.3		322,896				
1964–1965	4,683.0	3.1		341,301				
1965–1966	4,903.0	4.7		360,755	323,889			
1966–1967	5,006.8	2.1	2.9	381,318	342,351	18,462	9,488	8,973
1967–1968	5,132.9	2.5	3.2	403,138	361,882	19,531	10,790	8,741
1968–1969	5,255.8	2.4	3.0	429,871	383,277	21,395	10,751	10,644
1969–1970	5,464.6	4.0	3.1	457,369	406,490	23,214	12,035	11,178
1970–1971	5,634.3	3.1	2.8	486,858	431,711	25,221	11,470	13,751
1971–1972	5,701.9	1.2	2.6	514,515	458,350	26,639	11,389	15,250
1972–1973	5,852.1	2.6	2.7	556,185	488,960	30,609	12,198	18,411
1973–1974	6,041.9	3.2	2.8	622,348	527,455	38,495	13,843	24,652
1974–1975	6,139.8	1.6	2.4	648,780	565,737	38,282	12,451	25,831
1975–1976	6,258.2	1.9	2.1	657,355	599,837	34,099	12,023	22,076
1976–1977	6,378.0	1.9	2.3	674,776	631,889	32,052	13,605	18,447
1977–1978	6,436.7	0.9	1.9	698,742	660,400	28,511	12,166	16,345
1978–1979	6,475.7	0.6	1.4	723,170	680,565	20,164	9,231	10,933
1979–1980	6,690.0	3.3	1.7	742,324	699,273	18,709	11,812	6,897
1980–1981	6,791.6	1.5	1.7	787,454	725,293	26,020	11,564	14,456
1981–1982	6,894.7	1.5	1.6	818,886	754,115	28,822	11,419	17,403
1982–1983	6,993.3	1.4	1.7	831,371	780,641	26,526	12,642	13,884
1983–1984	7,127.5	1.9	1.9	847,066	805,420	24,779	15,137	9,642
1984–1985	7,267.1	2.0	1.7	901,248	837,205	31,785	13,442	18,343
1985–1986	7,573.5	4.2	2.2	942,361	868,186	30,981	18,489	12,493
1986–1987	7,387.7	2.2	2.3	966,797	897,769	29,582	20,325	9,258
1987–1988	7,947.9	2.7	2.6	1,003,004	932,095	34,327	23,303	11,023
1988–1989	8,238.2	3.7	2.9	1,054,228	973,528	41,432	27,426	14,006
1989–1990	8,457.2	2.7	3.1	1,054,757	1,004,229	30,702	30,007	694
1990–1991	8,509.2	0.6	2.4	1,066,761	1,029,109	24,880	23,721	1,159
1991–1992	8,545.3	0.4	2.0	1,067,448	1,049,240	20,130	20,692	-658
1992–1993	8,573.4	0.3	1.5	1,074,486	1,063,536	14,296	16,114	-1,862
1993–1994	8,729.1	1.8	1.2	1,093,310	1,071,352	7,816	12,427	-4,521
1994–1995	8,984.4	2.9	1.2	1,112,024	1,082,806	11,453	13,089	-1,403

subtracted from the actual change in the net stock of producer goods. This enabled the net increase/decrease in the stock of producer goods relative to the minimum requirement to be obtained. Like Daly and Cobb (1989) who also used this method, the changes in the labour force and the stock of producer goods was calculated on the basis of five year rolling averages. The data source is Foster, (1996, Table 4.3, page 180 and Table 5.23, page 274).

Item K — *Change in net international position* (Table A1.j).

Table A1.j   Item K — Change in Net International
                         Position

Year	Net Foreign Liabilities ($m) a	Change in Net International Position $(a_t - a_{i-1})$ ($m) K
1966–1967	31,857	0
1967–1968	33,570	–1,713
1966–1969	35,906	–2,336
1969–1970	36,395	–489
1970–1971	36,629	–234
1971–1972	31,925	4,704
1972–1973	24,997	6,928
1973–1974	23,816	1,181
1974–1975	24,141	–325
1975–1976	22,406	1,735
1976–1977	25,692	–3,286
1977–1978	30,669	–4,977
1978–1979	35,080	–4,411
1979–1980	56,745	–21,665
1980–1981	63,885	–7,140
1981–1982	68,280	–4,395
1982–1983	77,612	–9,332
1983–1984	82,584	–7,972
1984–1985	110,247	–27,663
1985–1986	129,220	–18,973
1986–1987	145,430	–16,210
1987–1988	145,351	79
1988–1989	161,090	–15,739
1989–1990	170,871	–9,781
1990–1991	186,223	–15,352
1991–1992	191,184	–4,961
1992–1993	205,373	–14,189
1993–1994	223,033	–17,660
1994–1995	236,044	–13,011

Item K was calculated by subtracting the previous year's net foreign liabilities from the current year's liabilities. The data source Foster, (1996), Table 1.20c, page 55).

Table A1.k Item L — Imputed Value of Leisure Time

Year	Average Weekly Hours of Paid Employment a	Total Number of Paid Employed (Thousands) b	Total Weekly Hours of Paid Employment ($a \times b$) (Millions) c	Australian Population (Aged 16–65) (Thousands) d	Australian Population (Aged 15+) (Thousands) e	Ave. Weekly Hours of Paid Employment (c/d) f	Total Weekly Hours of Unpaid Work (Millions) g	Average Weekly Hours of Unpaid Work (f/e) h	Ave. Weekly Hours Spent Driving/Commuting by Private Vehicle j	Ave. Weekly Hours Spent Commuting by Public/Commercial Transport k	Average Weekly Hours Spent Commuting ($j+k$) m
1966–1967	39.2	4.933	193.4	7,342	8,342	26.3	305.2	36.6	3.06	0.36	3.42
1967–1968	38.6	5.056	195.2	7,512	8,526	26.0	308.7	36.2	3.13	0.35	3.48
1968–1969	38.4	5.183	199.0	7,692	8,719	25.9	312.0	35.8	3.20	0.34	3.54
1969–1970	38.5	5.396	207.7	7,861	8,905	26.4	315.0	35.4	3.27	0.32	3.59
1970–1971	38.9	5.516	214.6	8,227	9,317	26.1	317.7	34.1	3.27	0.32	3.59
1971–1972	38.4	5.610	215.4	8,394	9,512	25.7	319.5	33.6	3.34	0.33	3.67
1972–1973	38.1	5.783	220.3	8,548	9,697	25.8	321.1	33.1	3.42	0.32	3.74
1973–1974	37.4	5.855	219.0	8,713	9,894	25.1	322.5	32.6	3.51	0.33	3.84
1974–1975	36.9	5.841	215.5	8,860	10,072	24.3	323.9	32.2	3.60	0.28	3.88
1975–1976	36.7	5.898	216.5	8,991	10,244	24.1	325.0	31.7	3.69	0.28	3.97
1976–1977	36.5	5.995	218.8	9,143	10,431	23.9	326.9	31.3	3.75	0.27	4.02
1977–1978	36.5	6.005	219.2	9,284	10,611	23.6	328.5	31.0	3.81	0.27	4.08
1978–1979	36.8	6.079	223.7	9,430	10,800	23.7	330.0	30.6	3.87	0.26	4.13
1979–1980	36.4	6.281	228.6	9,564	10,977	23.9	331.2	30.2	3.98	0.26	4.24
1980–1981	36.2	6.394	231.5	9,737	11,192	23.8	332.3	29.7	4.09	0.28	4.37
1981–1982	35.8	6.397	229.0	9,935	11,434	23.1	333.2	29.1	4.19	0.29	4.48
1982–1983	35.9	6.241	224.1	10,102	11,638	22.2	334.0	28.7	4.26	0.29	4.55
1983–1984	36.2	6.466	234.1	10,266	11,840	22.8	334.6	28.3	4.33	0.29	4.62
1984–1985	25.8	6.676	239.0	10,441	12,062	22.9	335.3	27.8	4.39	0.29	4.69
1985–1986	35.8	6.919	247.7	10,636	12,318	23.3	336.3	27.3	4.45	0.30	4.76
1986–1987	35.6	7.092	252.5	10,833	12,572	23.3	337.3	26.8	4.50	0.31	4.80
1987–1988	36.0	7.353	264.7	11,048	12,839	24.0	339.0	26.4	4.55	0.30	4.85
1988–1989	36.1	7.715	278.5	11,234	13,081	24.8	340.4	26.0	4.44	0.30	4.73
1989–1990	35.9	7.808	280.3	11,418	13,311	24.5	341.7	25.7	4.33	0.29	4.63
1990–1991	35.3	7.629	269.3	11,548	13,499	23.3	342.8	25.4	4.25	0.30	4.55
1991–1992	35.3	7.618	268.9	11,672	13,676	23.0	343.7	25.1	4.27	0.30	4.57
1992–1993	35.8	7.621	272.8	11,768	13,825	23.2	344.5	24.9	4.31	0.29	4.60
1993–1994	36.4	7.886	287.1	11,895	14,003	24.1	345.2	24.7	4.34	0.29	4.63
1994–1995	36.0	8.216	295.8	12,036	14,190	24.6	345.8	24.4	4.37	0.30	4.67

Table A1.k Item L — Imputed Value of Leisure Time (Continued)

Year	Average Weekly Hours Spent Resting and Sleeping — n	Average Weekly Hours Spent Getting an Education — p	Average Weekly Hours Spent Job Seeking — q	Weekly Hours of Nonleisure $(f+h+m+n+p+q)$ — r	Weekly Hours of Leisure $(168-r)$ — s	Yearly Hours of Leisure $(s \times 52)$ — t	Total Leisure Hours $(t \times$ 15+population) (Millions of Hours) — u	Real Hourly Value of Leisure ($ at 1989–1990 Prices) — v	Imputed Value of Leisure Time $(u \times v)$ ($m at 1989–1990 Prices) — L
1966–1967	70	2.49	0.18	139.0	29.0	1,507.3	12,573.7	4	50,295
1967–1968	70	2.60	0.16	138.4	29.6	1,537.8	13,111.3	4	52,445
1968–1969	70	2.68	0.15	138.0	30.0	1,558.5	13,588.7	4	54,355
1969–1970	70	2.72	0.15	138.3	29.7	1,546.4	13,771.0	4	55,084
1970–1971	70	2.69	0.17	136.6	31.4	1,631.2	15,198.0	4	60,792
1971–1972	70	2.76	0.26	135.9	32.1	1,667.0	15,856.1	4	63,424
1972–1973	70	2.83	0.19	135.6	32.4	1,682.2	16,312.7	4	65,251
1973–1974	70	3.10	0.24	134.9	33.1	1,720.8	17,025.6	4	68,102
1974–1975	70	3.44	0.47	134.3	33.7	1,753.7	17,663.3	4	70,653
1975–1976	70	3.65	0.49	133.9	34.1	1,772.6	18,159.0	4	72,636
1976–1977	70	3.81	0.59	133.7	34.3	1,784.0	18,609.0	4	74,436
1977–1978	70	3.89	0.64	133.2	34.8	1,810.8	19,214.3	4	76,857
1978–1979	70	3.92	0.60	132.9	35.1	1,823.7	19,696.2	4	78,785
1979–1980	70	3.93	0.62	132.3	35.1	1,826.9	20,053.9	4	80,216
1980–1981	70	3.91	0.59	132.3	35.7	1,854.7	20,758.0	4	83,032
1981–1982	70	3.88	0.70	131.3	36.7	1,910.9	21,849.0	4	87,396
1982–1983	70	4.02	1.02	130.5	37.5	1,951.7	22,713.4	4	90,854
1983–1984	70	3.92	0.88	130.5	37.5	1,951.0	23,100.0	4	92,400
1984–1985	70	3.73	0.82	129.9	38.1	1,979.7	23,879.3	4	95,517
1985–1986	70	3.84	0.84	130.0	38.0	1,974.4	24,320.9	4	97,284
1986–1987	70	3.97	0.83	129.7	38.3	1,989.7	25,015.1	4	100,061
1987–1988	70	4.14	0.73	130.1	37.9	1,971.6	25,314.0	4	101,256
1988–1989	70	4.15	0.62	130.3	37.7	1,959.7	25,634.3	4	102,537
1989–1990	70	4.10	0.77	129.7	38.3	1,990.6	26,496.3	4	105,985
1990–1991	70	4.23	1.04	128.5	39.5	2,052.2	27,702.5	4	110,810
1991–1992	70	4.37	1.15	128.3	39.7	2,066.4	28,260.5	4	113,042
1992–1993	70	4.49	1.17	128.4	39.6	2,061.1	28,495.2	4	113,981
1993–1994	70	4.55	1.01	129.0	39.0	2,029.4	28,417.1	4	113,638
1994–1995	70	4.62	0.90	129.1	38.9	2,021.1	28,678.8	4	114,715

Item L — *Imputed value of leisure time* (Table A1.k).

To estimate the annual services gained from leisure time, it was first necessary to estimate the annual number of leisure hours enjoyed by Australians. Because the gross opportunity cost method was used to value such services, the total number of leisure hours was calculated for Australians aged 15 years and older. Nonleisure activities included in the calculation of leisure services were

- Hours of paid employment.
- Time spent commuting and travelling by private vehicle.
- Time spent commuting by public and commercial transport.
- Hours spent resting and sleeping.
- Time spent getting an education.
- Time spent job-seeking.

a. Hours of paid employment — The data source is Foster, (1996, Table 4.12, page 198).
b. Time spent commuting and travelling by private vehicle.

A total to seven motor vehicle usage surveys conducted by the ABS were used to determine the time spent commuting and travelling by private vehicle. The ABS surveys were conducted for the years 1970–1971, 1975–1976, 1978–1979, 1981–1982, 1984–1985, 1987–1988, and 1990–1991. To obtain estimates over the study period of the time spent commuting and travelling by private vehicle, the following method was employed. First, the total kilometres travelled to and from work were added to the number of kilometres of private travel. Second, it was assumed that 25 percent of all private travel is for leisure purposes and contributes to psychic income. Hence, 25 percent of privately travelled kilometres was deducted from the preceding total in order to give the total number of nonleisure kilometres of travel. Next, it was assumed that the average speed travelled is 45 kilometres per hour (kph). Although this is below the urban speed limit of 60 kph, much travel is conducted during peak traffic periods when average speeds in some major cities can be as low as 25 kph. By dividing the total number of nonleisure kilometres travelled by 45 kph, an estimate for the total number of nonleisure hours of travel per year was obtained. This was then divided by 52 in order to provide a measure of the average number of nonleisure hours of travel per week. Last, to ascertain the average per person, the running total was divided by the number of people in Australia aged 15 years and older.

In order to obtain estimates for the intervening years, interpolation of the change in total kilometres travelled was based on a geometric progression of 4.61 percent between 1970–1971 and 1975–1976; 3.37 percent between 1975–1976 and 1978–1979; 4.41 percent between 1978–1979 and 1981–1982; 3.45 percent between 1981–1982 and 1984–1985; and 3.1 percent between 1984–1985 and 1987–1988. Between 1987–1988 and 1990–1991, a 0.77 percent decline was estimated, but a 2 percent increase was assumed after 1990–1991. As for the period 1966–1967 to 1970–1971, it was assumed that an increase occurred at the rate of 4.61 percent per annum, the same rate of increase as for the period 1970–1971 to 1975–1976.

For hours of private travel, interpolation was based on the following geometric progressions: 1966–1967 to 1975–1976 (5.7 percent); 1975–1976 to 1978–1979 (3.46 percent); 1978–1979 to 1981–1982 (2.71 percent); 1981–1982 to 1984–1985 (3.56 percent); 1984–1985 to 1987–1988 (1.53 percent); 1987–1988 to 1990–1991 (–1.79 percent); and beyond 1990–1991 (2 percent). The data source is the Australian Bureau of Statistics, Catalogue No. 9208.0.

    c. *Time spent commuting by public and commercial transport.*

To calculate this subitem, household expenditures on rail, bus, tram, and taxi fares were multiplied by 0.8 on the assumption that, of all expenditure involving people travelling on a fare-paid form of short-distance transport, 80 percent are by people 15 years of age and older. The average fare per trip was assumed to be $2.50 in 1989–1990 prices. The number of trips was then estimated by dividing the 80 percent of all fares paid by $2.50. Following this, the running total was divided by 1/3 (or, equivalently, multiplied by 3), on the assumption that the average time of each trip is twenty minutes (1/3 of 1 h). This provided the number of yearly hours spent on short-distance, fare-paid public and commercial transport. The total was then divided by 52 to get a weekly average, and again divided by the population of Australians aged 15 years and older.

For long-distance forms of travel (e.g., air and longer train and bus trips), other fares were, as above, multiplied by 0.8. The average fare per trip was assumed to be $150 in 1989–1990 prices. The total number of long-distance trips was calculated by dividing 80 percent of all trips undertaken by $150. In order to estimate the number of yearly hours spent on long-distance travel, the running total was then divided by two on the basis that an average trip is of two hours duration. The total was then divided by 52 to obtain a weekly average, and again by the 15+ year old population.

Finally, all subtotals were added to obtain an estimate of the average weekly time spent travelling and commuting by all forms of public and commercial transport. This figure was then added to the time spent commuting and travelling by private vehicle to obtain a total measure of the average weekly time of nonleisure travel. It should also be noted that the estimated total measure for the financial year 1991–1992 was 4.57 hours per week. This compares favourably with the value of 4.72 hours per week for 1991–1992 as calculated from a time-use survey conducted by the ABS (1993). The data source is the Australian Bureau of Statistics, Catalogue No. 5204.0.

    d. *Hours spent resting and sleeping.*

It was assumed that the average number of hours spent resting and sleeping was 10 per day or 70 per week.

e. *Time spent getting an education.*

To calculate the time taken getting an education, it was assumed, first, that secondary school students (aged 15 years and older), spent an average of 40 hours per week for 40 weeks of the year either at school or at home studying. Second, it was assumed that full-time, higher education students spent an average of 40 hours per week for 40 weeks of the year getting an education. For part-time students, it was assumed that they studied for the same number of weeks per year, but at 20 hours per week. As for technical education and training, the same was assumed for full and part-time enrolled students as higher education students. The data sources are the Australian Bureau of Statistics, Catalogue No. 4221.0, the Australian Bureau of Statistics, Catalogue No. 1301.0, the Australian Bureau of Statistics, Catalogue No. 4224.0, and the Department of Employment, Education, and Training, (1995, various articles).

f. *Time spent job-seeking.*

It was assumed that the average time spent by an unemployed person seeking work was 15 per week. The 15 was chosen to include the time spent reading the job section of one or more newspapers at an average of 3 times a week, visiting the local employment service office 2 or 3 times a week, writing job applications; travelling to and from job interviews, and having to visit a Social Security office to obtain unemployment benefits. While 15 per week might seem excessive, it should be pointed out that no allowance was made for the time spent by the underemployed seeking additional hours of employment. The data source is Foster (1996, Table 4.3, page 180).

Having calculated the preceding 6 subitems, the totals were added and then subtracted from 168 (the number of hours in a week). This enabled the average number of leisure hours enjoyed by each person aged 15 and older per week to be obtained. The total was then multiplied twice by 52 (the number of weeks per year), and the population of Australia aged 15 years and older. The total imputed value of leisure time was calculated by multiplying the total number of leisure hours by the opportunity cost of an hour of leisure (assumed to be a constant $4 per hour at 1989–1990 prices). Of course, this assumes that the value of a leisure hour, free from the use and/or consumption of goods, has remained unchanged over the study period. A constant $4 per hour was chosen to reflect two interdependent facts. First, to gain the full benefit from a leisure hour, it is often necessary to consume and/or use goods. Second, the enjoyment of consumption and consumer durable services, which are reflected in other items, are not independent of time (i.e., time is required to eat, to dine, to see a movie, to watch television, to play a computer game, or to play a game with a bat and ball, etc.).

It is true that an increase in available choices and an improvement in many consumer goods has probably increased the psychic income enjoyed from a leisure hour over time. However, to incorporate it into the value of leisure time would amount to double-counting because the boost to psychic income from goods with higher use value is reflected in other items (e.g., Items A, D, E, F, etc.). Hence, assuming a constant real value of a leisure hour over time seems well justified.

Item M — *Total psychic income.*

Equal to the sum of Items C to L.

Item M1 — *Australian population.*

Indicated in the thousands; the data source is Foster, (1996, Table 4.1, page 177).

Item M2 — *Per capita psychic income.*

Item M divided by Item M1.

A1.3 Psychic outgo

Item N — *Expenditure on consumer durables.*

Item N is simply a pecuniary measurement of all expenditure on consumer durables during each financial year. The data source is the Australian Bureau of Statistics, Catalogue No. 5204.0.

Item O — *Defensive private health and education expenditure* (Table A1.m).

The method employed is the same as Item I and, once again, is similar to that employed by Daly and Cobb (1989). The data source is the Australian Bureau of Statistics, Catalogue No. 5204.0.

Item P — *Cost of private vehicle accidents* (Table A1.n).

A total of four studies on the cost of vehicle accidents in Australia have been undertaken during the study period (1969, 1978, 1985, and 1988). For each of these years, the total cost of private vehicle accidents was divided by the total number of accidents to obtain a measure of the average cost per vehicle accident. By calculating the number of accidents per registered vehicle, the average accident cost per registered vehicle was estimated by multiplying the number of accidents per registered vehicle by the cost per vehicle accident. The same was done for the intervening years by interpolating a straight line change in the cost per accident. The data sources are Paterson, (1973), Atkins, (1981), Bureau of Transport and Communications (1979), Steadman and Bryan, (1991) and the Australian Bureau of Statistics, Catalogue No. 1301.0.

Table A1.m   Item O — Private Defensive Expenditure on Health and Education

Year	Private Expenditure on Education ($m) *a*	Private Defensive Expenditure on Education (0.5 × a) ($m) *b*	Private Expenditure on Health ($m) *c*	Private Defensive Expenditure on Health (c–5363.7) × 0.5 ($m) *d*	Private Defensive Expenditure on Health and Education (b+d) ($m) *O*
1966–1967	1,663	832	5,363.7	0.0	832
1967–1968	1,795	898	5,704.9	170.6	1,068
1968–1969	1,878	939	6,020.2	328.3	1,267
1969–1970	1,958	979	6,345.1	490.7	1,470
1970–1971	2,002	1,001	6,944.0	790.2	1,791
1971–1972	2,137	1,069	7,563.8	1,100.1	2,169
1972–1973	2,150	1,075	7,878.8	1,257.6	2,333
1973–1974	1,855	928	8,019.1	1,327.7	2,255
1974–1975	1,496	748	8.563.4	1,599.9	2,348
1975–1976	1,498	749	8,949.1	1,792.7	2,542
1976–1977	1,637	819	9,003.1	1,819.7	2,638
1977–1978	1,686	843	9,126.1	1,881.2	2,724
1978–1979	1,695	848	9,738.3	2,187.3	3,035
1979–1980	1,725	863	9,610.0	2,123.2	2,986
1980–1981	1,911	956	9,863.0	2,249.7	3,205
1981–1982	2,024	1,012	10,553.1	2,594.7	3,607
1982–1983	2,261	1,131	11,292.2	2,964.3	4,095
1983–1984	2,383	1,192	11,768.0	3,202.2	4,394
1984–1985	2,434	1,217	11,825.8	3,231.1	4,448
1985–1986	2,575	1,288	12,207.4	3,421.9	4,709
1986–1987	2,817	1,409	13,049.1	3,842.7	5,251
1987–1988	3,022	1,511	13,736.8	4,186.6	5,698
1988–1989	3,422	1,711	14,510.9	4,573.6	6,285
1989–1990	3,945	1,973	14,438.0	4,737.2	6,710
1990–1991	4,175	2,088	16,043.6	5,340.0	7,427
1991–1992	4,538	2,269	17,173.4	5,904.9	8,174
1992–1993	4,725	2,363	17,912.6	6,274.5	8,637
1993–1994	4,903	2,452	18,679.1	6,657.7	9,109
1994–1995	5,040	2,520	19,361.2	6,998.8	9,519

Item Q — *Cost of noise pollution* (Table A1.p).

A study by the Organisation for Economic Co-operation and Development in 1986 indicated that Australia has been as noisy as any other major industrialised nation, including the United States. In view of this finding, it is reasonable to assume that the per capita cost of noise pollution in Australia was, at the time, much the same as in the United States. The damage caused by noise pollution in the United States was estimated by the World Health Organisation to be US$4 billion in 1972 (at 1972 prices). Hence, allowing for population differences and the exchange rate in 1972, the cost of noise pollution in Australia in 1972 was approximately Aus$934.8 million at 1989–1990 prices.

To estimate the cost of noise pollution for each year over the study period, the following was assumed. First, it was assumed that noise pollution, as with most forms of pollution, is closely correlated with the rate of production and consumption or the rate of increase in real GDP. Second,

Table A1.n Item P — Cost of Private Vehicle Accidents

Year	Cost per Accident ($ Thousands) a	Accident per Registered Vehicle b	Accident Cost per Reg. Vehicle ($a \times b$) ($) c	Number of Registered Vehicles (Millions) d	Cost of Private Vehicle Accidents per Calendar Year ($c \times d$) e	Year f	Cost of Private Vehicle Accidents (Ave. Between Calendar Years) P
1966	2,693	0.110	296.2	3.9699	1,176		
1967	2,643	0.107	282.8	4.1632	1,177	1966–1967	1,177
1968	2,594	0.105	272.4	4.1965	1,143	1967–1968	1,160
1969	2,544	0.102	259.5	4.6559	1,208	1968–1969	1,176
1970	2,494	0.099	246.9	4.9107	1,213	1969–1970	1,210
1971	2,445	0,097	237.2	5.2047	1,235	1970–1971	1,224
1972	2,395	0.094	225.1	5.3251	1,199	1971–1972	1,217
1973	2,346	0.091	213.5	5.6341	1,203	1972–1973	1,201
1974	2,296	0.089	204.3	5.9861	1,223	1973–1974	1,213
1975	2,247	0.086	193.2	6.3303	1,223	1974–1975	1,223
1976	2,197	0.083	182.4	6.6660	1,216	1975–1976	1,219
1977	2,148	0.081	174.0	6.8181	1,186	1976–1977	1,201
1978	2,098	0.078	163.6	7.1181	1,165	1977–1978	1,188
1979	2,227	0.078	172.8	7.3581	1,272	1978–1979	1,230
1980	2,357	0.077	181.7	7.5735	1,376	1979–1980	1,324
1981	2,486	0.077	190.7	7.9176	1,510	1980–1981	1,443
1982	2,616	0.076	199.6	8.3461	1,666	1981–1982	1,588
1983	2,745	0.076	208.3	8.6079	1,793	1982–1983	1,730
1984	2,875	0.075	216.8	8.8328	1,915	1983–1984	1,854
1985	3,004	0.075	225.3	8.9597	2,019	1984–1985	1,967
1986	3,126	0.071	221.9	9.2906	2,062	1985–1986	2,040
1987	3,248	0.067	217.6	9.3737	2,040	1986–1987	2,051
1988	3,370	0.063	212.3	9.4180	1,991	1987–1988	2,015
1989	3,492	0.060	210.6	9.8061	2,065	1988–1989	2,028
1990	3,614	0.058	208.5	10.0806	2,102	1989–1990	2,083
1991	3,736	0.055	205.5	9.9341	2,041	1990–1991	2,072
1992	3,858	0.052	201.8	10.2369	2,066	1991–1992	2,053
1993	3,980	0.050	197.8	10.4315	2,063	1992–1993	2,064
1994	4,102	0.047	192.8	10,6992	2,063	1993–1994	2,063
1995	4,224	0.044	187.1	10.9354	2,046	1994–1995	2,109

because of significant legislative requirements introduced in relation to noise pollution in the early 1970s and late 1980s, it was assumed that the impact of noise pollution had somewhat abated. Hence, a noise pollution index was constructed on the basis that noise pollution increased between the years of 1966–1967 and 1972–1973 at the rate of increase in GDP; at 2/3 the rate of increase in GDP over the period 1972–1973 and 1988–1989; and at 1/3 the rate of increase in GDP after 1988–1989. To calculate the total cost of noise pollution in the years other than 1972–1973, the 1972–1973 figure of $934.8 million was adjusted by the noise pollution index. The data sources are the Department of Arts, Heritage and Environment (1987), Australian Bureau of Statistics, Catalogue No. 5204.0, and Fabricius, (1994).

Table A1.p Item Q — Cost of Noise Pollution

Year	Percentage Change in Real GDP (%)	Percentage Change in Real GDP (× 0.67) (%)	Percentage Change in Real GDP (× 0.33) (%)	Index of Noise Pollution	Cost of Noise Pollution (d × 34.8)/100 ($m at 1989–1990 Prices)
	a	b	c	d	Q
1966–1967	6.6			74.2	694
1967–1968	3.7			76.9	719
1968–1969	8.8			83.7	782
1969–1970	5.6			88.4	826
1970–1971	4.8			92.6	866
1971–1972	4.0			96.3	901
1972–1973	3.8			100.0	935
1973–1974	2.2	1.5		101.5	949
1974–1975	1.0	0.7		102.1	955
1975–1976	1.5	1.0		103.2	964
1976–1977	1.4	0.9		104.1	973
1977–1978	0.5	0.3		104.5	977
1978–1979	2.8	1.9		106.4	995
1979–1980	1.0	0.7		107.1	1,001
1980–1981	1.8	1.2		108.4	1,014
1981–1982	1.2	0.8		109.3	1,022
1982–1983	−0.8	−0.5		108.7	1.017
1983–1984	3.0	2.0		110.9	1,037
1984–1985	2.5	1.7		112.8	1,054
1985–1986	1.9	1.3		114.2	1,067
1986–1987	1.3	0.9		115.2	1,077
1987–1988	2.4	1.6		117.0	1,094
1988–1989	2.4	1.6		118.9	1,111
1989–1990	1.5		0.5	119.5	1,117
1990–1991	−0.3		−0.1	119.4	1,116
1991–1992	0.2		0.1	119.5	1,117
1992–1993	1.8		0.6	120.2	1,123
1993–1994	2.4		0.8	121.1	1,132
1994–1995	2.0		0.7	121.9	1,140

Item R — *Direct disamenity cost of air pollution* (Table A1.q).

Following Freeman's (1982) study of air pollution in the United States, Daly and Cobb (1989) identified six major categories of air pollution-related costs. These are

- The cost of the damage to agricultural vegetation and crops.
- The cost of materials damage.
- The cost of cleaning soiled and damaged goods.
- The cost of acid rain damage to aquatic and forest resources.
- The cost of urban disamenities such as reduced property values and compensation to affected individuals.
- The cost of aesthetic impairment.

The latter two were included in the estimation of uncancelled benefits rather than uncancelled costs (sacrificed sink function of natural capital) because they directly impact on the net psychic income of the nation. Together they incur what might be regarded as a direct disamenity cost of

Table A1.q Item R — Direct Disamenity Cost of Air Pollution

Year	Index of Real GDP a	Percent. Change in Real GDP (%) b	Percent. Change in Real GDP (×0.5) (%) c	Air Pollution Index d	Air Pollution Control Costs $(504.4 \times d)/100$ ($m) e	Air Pollution Damage Costs $(e \times 10)$ ($m) f	Total cost of Air Pollution $(e+f)$ ($m at 1989–1990 Prices) g	Direct Disamenity Cost of Air Pollution $(0.4 \times g)$ ($m at 1989–1990) R
1966–1967	56.5	6.6		75.5	380.9	3,808.6	4,189	1,676
1967–1968	58.6	3.7		78.3	395.0	3,950.1	4,345	1,738
1968–1969	63.8	8.9		85.3	430.1	4,300.6	4,731	1,892
1969–1970	67.3	5.5		89.9	453.7	4,536.6	4,990	1,996
1970–1971	70.6	4.9		94.3	475.9	4,759.0	5,235	2,094
1971–1972	73.4	4.0		98.1	494.8	4,947.8	5,443	2,177
1972–1973	76.2	3.8		101.8	513.6	5,136.5	5,650	2,260
1973–1974	77.9	2.2	1.1	103.0	519.4	5,193.8	5,713	2,285
1974–1975	78.7	1.0	0.5	103.5	522.0	5,220.5	5,743	2,297
1975–1976	79.9	1.5	0.8	104.3	526.0	5,260.3	5,786	2,315
1976–1977	81.0	1.4	0.7	105.0	529.6	5,296.5	5,826	2,330
1977–1978	81.4	0.5	0.2	105.3	531.0	5,309.6	5,841	2,336
1978–1979	83.7	2.8	1.4	106.8	538.5	5,384.6	5,923	2,369
1979–1980	84.5	1.0	0.5	107.3	541.0	5,410.3	5,951	2,381
1980–1981	86.0			106.2	535.7	5,356.7	5,892	2,357
1981–1982	87.0			106.2	535.7	5,356.7	5,892	2,357
1982–1983	86.3			106.2	535.7	5,356.7	5,892	2,357
1983–1984	88.9			106.2	535.7	5,356.7	5,892	2,357
1984–1985	91.1			106.2	535.7	5,356.7	5,892	2,357
1985–1986	92.8			106.2	535.4	5,354.3	5,890	2,356
1986–1987	94.0			105.1	530.1	5,301.3	5,831	2,333
1987–1988	96.3			104.1	524.9	5,248.8	5,774	2,309
1988–1989	98.6			103.0	519.7	5,196.8	5,717	2,287
1989–1990	100.1			102.0	514.5	5,145.4	5,660	2,264
1990–1991	99.8			101.0	509.4	5,094.4	5,604	2,242
1991–1992	100.0			100.0	504.4	5,044.0	5,548	2,219
1992–1993	101.8			99.0	499.4	4,994.1	5,493	2,197
1993–1994	104.2			98.0	494.5	4,944.6	5,439	2,176
1994–1995	106.3			97.1	489.6	4,895.7	5,385	2,154

air pollution. Based on Freeman's study, the urban disamenity and aesthetic impairment caused by air pollution constituted approximately 40 percent of the total cost of air pollution. Although it is likely that such a percentage would vary between 1966–1967 and 1994–1995, the 40 percent share was assumed to have remained constant over the study period.

There has, unfortunately, been very little previous work on the cost of air pollution in Australia. Even data on air pollution emissions is limited in both specificity and scope. Hence, to estimate the direct disamenity cost of air pollution, an air pollution index was constructed on the basis that air pollution is very much a function of production and, therefore, closely related to the rate of increase in GDP. Because of important legislation concerning air pollution in Australia being enacted in the early 1970s, the mid-1980s, and again in the early 1990s, it was assumed that air pollution increased at the rate of growth in GDP until 1972–1973. Following 1972–1973, it was assumed that air pollution increased at half the rate of increase in GDP between 1972–1973 and 1979–1980.

According to a recent State of the Environment Report for Australia (State of the Environment Advisory Council, 1996), air quality remained steady in the first half of the 1980s and, since the introduction of more stringent legislative controls in the mid-1980s, has gradually improved. It was therefore, assumed that the air pollution index remained constant between 1980–1981 and 1985–1986 and improved at a rate of 1 percent per year from 1985–1986 onwards.

To calculate the total cost of air pollution in Australia, the ABS estimated air pollution control costs, in other words, the amount Australians spent defending themselves against air pollution, at $504.4 million in 1991–1992 at 1989–1990 prices. Based on estimates of the ratio of air pollution control to damage costs by Zolotas (1980), the air pollution damage cost was assumed to be 10 times the control cost. As a consequence, the total cost of air pollution in Australia was conservatively estimated at $5.5 billion in 1991–1992 at 1989–1990 prices. To calculate the total cost of air pollution in other years, the 1991–1992 value of $5.5 billion was adjusted by the air pollution index. Of course, to calculate the direct disamenity cost of air pollution, it was necessary to multiply the total air pollution cost by 0.4 (40 percent) for reasons given previously. The data sources are Freeman, (1982), Australian Bureau of Statistics, Catalogue No. 4603.0, Fabricius, (1994), and State of the Environment Advisory Council, (1996, Chapter 5).

Item S — *Cost of unemployment* (Table A1.r).

To calculate the cost of unemployment, the total number of unemployed persons were divided into two groups — those seeking either full-time or part-time employment. In the former category, the number of unemployed was multiplied by the average weekly hours of full-time employment and again by 52 weeks for the year. This yielded the total yearly number of unemployed hours suffered by people seeking full-time work. Because an

Table A1.r   Item S — Cost of Unemployment

Year	Number of Unemployed (Thousands) a	Average Weekly Hours of Paid Employment b	Total Weekly u/ed Hours (a × b) (Thousands) c	Total Yearly u/ed Hours (c × 52) (Millions) d	Seeking Full-Time Work — Real Hourly Cost of u/e ($) e	Gross Cost of u/e (d × e) ($m) f	Unemp. Benefit Recipients (Thousands) g	Yearly Benefit Received per Recipient ($ at 1989–1990 Prices) h	Total u/e Benefits Received ($m) j	Cost of u/e (Full-Time) ($m at 1989–1990 Prices) k
1966–1967	70	41.5	2,905.0	1,510.6	4	6,042.4	19.3	2,732.6	52.8	5,900
1967–1968	61	41.1	2,507.1	1,303.7	4	5,214.8	16.0	2,631.7	42.2	5,173
1968–1969	58	41.0	2,378.0	1,236.6	4	4,946.2	11.7	2,568.8	30.0	4,916
1969–1970	58	41.1	2,383.8	1,239.6	4	4,958.3	9.7	3,023.3	29.2	4,929
1970–1971	70	41.4	2,898.0	1,507.0	4	6,027.8	14.5	2,889.1	41.8	5,986
1971–1972	111	41.2	4,573.2	2,378.1	4	9,512.3	32.1	4,580.2	146.9	9,365
1972–1973	67	41.1	2,753.7	1,431.9	4	5,727.7	24.0	5,480.3	131.3	5,596
1973–1974	106	40.5	4,293.0	2,232.4	4	8,929.4	24.1	5,852.6	140.8	8,789
1974–1975	216	40.2	8,683.2	4,515.3	4	18,061.1	124.4	6,959.2	865.8	17,195
1975–1976	237	40.3	9,551.1	4,966.6	4	19,866.3	152.4	7,055.9	1075.3	18,791
1976–1977	283	40.3	11,404.9	5,930.5	4	23,722.2	192.3	7,078.8	1361.0	22,361
1977–1978	334	40.5	13,527.0	7,034.0	4	28,136.2	236.8	7,059.0	1671.7	26,464
1978–1979	316	40.8	12,892.8	6,704.3	4	26,817.0	261.6	6,525.5	1706.9	25,110
1979–1980	335	40.5	13,567.5	7,055.1	4	28,220.4	262.1	5,919.2	1551.7	26,669
1980–1981	324	40.2	13,024.8	6,772.9	4	27,091.6	266.3	5,626.4	1498.6	25,593
1981–1982	388	39.9	15,481.2	8,050.2	4	32,200.9	314.5	5,533.3	1740.3	30,461
1982–1983	607	40.2	24,401.4	12,688.7	4	50,754.9	559.6	5,862.0	3280.1	47,475
1983–1984	524	40.6	21,274.4	11,062.7	4	44,250.8	509.4	6,287.8	3202.8	41,048
1984–1985	488	40.3	19,666.4	10,226.5	4	40,906.1	478.9	6,534.3	3129.2	37,777
1985–1986	492	40.5	19,926.0	10,361.5	4	41,446.1	467.9	6,749.6	3158.1	38,288
1986–1987	495	40.6	20,097.0	10,450.4	4	41,801.8	455.3	6775.1	3084.6	38,717
1987–1988	445	41.2	18,334.0	9,533.7	4	38,134.7	393.0	6,754.8	2654.5	35,480
1988–1989	369	41.6	15,350.4	7,982.2	4	31,928.8	307.3	6,775.6	2082.4	29,846
1989–1990	476	41.4	19,706.4	10,247.3	4	40,989.3	341.6	6,760.0	2309.1	38,680
1990–1991	680	41.1	27,948.0	14,533.0	4	58,131.8	575.9	6,846.8	3943.2	54,189
1991–1992	758	41.8	31,684.4	16,475.9	4	65,903.6	718.2	6,830.7	4905.8	60,998
1992–1993	774	42.1	32,585.4	16,944.4	4	67,777.6	772.1	6,780.8	5235.7	62,542
1993–1994	656	43.0	28,208.0	14,668.2	4	58,672.6	722.0	6,926.4	5000.9	53,672
1994–1995	583	42.6	24,835.8	12,914.6	4	51,658.5	664.2	6,882.2	4571.4	47,087

Table A1.r   Item S — Cost of Unemployment (Continued)

Year	Number of Unemployed (Thousands) m	Average Weekly Hours of Paid Employment n	Total Weekly u/ed Hours ($m \times n$) (Thousands) p	Total Yearly u/ed Hours ($o \times 52$) (Millions) q	Real Hourly Cost of u/e ($) r	Gross Cost of u/e ($d \times e$) ($m) s	Seeking Part-Time Work Unemp. Benefit Recipients (Thousands) t	Yearly Benefit Received per Recipient ($ at 1989–1990 Prices) u	Total u/e Benefits Received ($m) v	Cost of u/e (Part-Time) ($m at 1989–1990 Prices) w	Total Cost of Unemployment ($k+w$) ($m at 1989–1990 Prices) s
1966–1967	17	19.3	328.1	170.6	4	682	4.7	1,366.3	6.4	676.0	6,666
1967–1968	20	18.1	362.0	188.2	4	753	5.3	1,315.9	6.9	746.0	5,919
1968–1969	21	17.4	365.4	190.0	4	760	4.2	1,284.4	5.4	754.6	5,671
1969–1970	20	16.4	328.0	170.6	4	682	3.3	1,511.6	5.0	677.2	5,606
1970–1971	23	17.5	402.5	209.3	4	837	4.7	1,444.6	6.9	830.3	6,816
1971–1972	33	15.9	524.7	272.8	4	1,091	9.5	2,290.1	21.8	1,069.5	10,435
1972–1973	39	15.9	620.1	322.5	4	1,290	13.9	2,740.1	38.2	1,251.6	6,848
1973–1974	35	15.6	546.0	283.9	4	1,136	7.9	2,926.3	23.2	1,112.4	9,901
1974–1975	63	15.9	1,001.7	520.9	4	2,084	36.3	3,479.6	126.3	1,957.3	19,153
1975–1976	56	15.7	879.2	457.2	4	1,829	36.0	3,527.9	127.0	1,701.7	20,493
1976–1977	76	15.0	1,140.0	592.8	4	2,371	51.6	3,539.4	182.8	2,188.4	24,550
1977–1978	64	15.5	992.0	515.8	4	2,063	45.4	3,529.5	160.2	1,903.2	28,368
1978–1979	62	15.6	967.2	502.9	4	2,012	51.3	3,262.7	167.5	1,844.3	26,954
1979–1980	60	15.5	930.0	483.6	4	1,934	47.0	2,959.6	139.0	1,795.4	28,464
1980–1981	57	16.0	912.0	474.2	4	1,897	46.9	2,813.2	131.8	1,765.1	27,358
1981–1982	74	16.5	1,221.0	634.9	4	2,540	60.0	2,766.7	166.0	2,373.7	32,834
1982–1983	80	15.6	1,248.0	649.0	4	2,596	73.7	2,931.0	216.2	2,379.7	49,855
1983–1984	81	15.8	1,279.8	665.5	4	2,662	78.7	3,143.9	247.5	2,414.4	43,462
1984–1985	85	15.5	1,317.5	685.1	4	2,740	83.4	3,267.2	272.5	2,467.9	40,245
1985–1986	106	15.7	1,664.2	865.4	4	3,462	100.8	3,374.8	340.2	3,121.3	41,409
1986–1987	107	15.7	1,679.9	873.5	4	3,494	98.4	3,387.5	333.4	3,160.8	41,878
1987–1988	93	15.3	1,422.9	739.9	3	2,960	82.1	3,377.4	277.4	2,682.3	38,163
1988–1989	99	15.4	1,524.6	792.8	4	3,171	82.5	3,387.8	279.4	2,891.8	32,738
1989–1990	109	15.6	1,700.4	884.2	4	3,537	78.2	3,380.0	264.4	3,272.5	41,953
1990–1991	119	15.4	1,832.6	953.0	4	3,812	100.8	3,423.4	345.0	3,466.8	57,655
1991–1992	141	15.3	2,157.3	1,121.8	4	4,487	133.6	3,415.4	456.3	4,030.9	65,029
1992–1993	142	15.7	2,229.4	1,159.3	4	4,637	141.7	3,390.4	480.3	4,156.9	66,699
1993–1994	142	16.0	2,272.0	1,181.4	4	4,726	156.3	3,463.2	541.3	4,184.5	57,856
1994–1995	139	15.9	2,210.1	1,149.3	4	4,597	158.4	3,441.1	545.0	4,052.0	51,139

hour of unemployment can be treated as a cost in the same way an hour of leisure can be considered a benefit, the total number of unemployed hours was multiplied by $4 per hour. This enabled a measure of the gross cost of unemployment to be obtained. Next, the 1989–1990 real price value of unemployment benefits received was subtracted from the gross cost to ascertain the net cost of unemployment. The previous method was again used for people seeking part-time employment. The only variation here was the difference in the average weekly hours spent working by those employed part-time. The data source is Foster, (1996, Table 4.3, page 180; Table 4.12, page 198; Table 4.15, page 202; and Table 2.25, page 108),

Item T — *Cost of underemployment* (Table A1.s).

Table A1.s   Item T — Cost of Underemployment

Year	Total Number of Underemployed (Thousands) *a*	Total Number of Underemployed Hours (*a* × 260) *b*	Real Hourly Cost of Underemployment ($ at 1989–1990 Prices) *c*	Cost of Underemployment (*b* × *c*) ($m at 1989–1990 Prices) T
1966–1967	46.0	11.96	4	48
1967–1968	42.0	10.92	4	44
1968–1969	40.0	10.40	4	42
1969–1970	43.0	11.18	4	45
1970–1971	48.0	12.48	4	50
1971–1972	55.0	14.30	4	57
1972–1973	43.0	11.18	4	45
1973–1974	64.0	16.64	4	67
1974–1975	78.0	20.28	4	81
1975–1976	94.0	24.44	4	98
1976–1977	121.0	31.46	4	126
1977–1978	171.7	44.64	4	179
1978–1979	168.2	43.73	4	175
1979–1980	195.6	50.86	4	203
1980–1981	178.5	46.41	4	186
1981–1982	232.0	60.32	4	241
1982–1983	266.5	69.29	4	277
1983–1984	242.8	63.13	4	253
1984–1985	233.1	60.61	4	242
1985–1986	272.5	70.85	4	283
1986–1987	314.0	81.64	4	327
1987–1988	281.5	73.19	4	293
1988–1989	321.0	83.46	4	334
1989–1990	372.8	96.93	4	388
1990–1991	512.8	133.33	4	533
1991–1992	603.5	156.91	4	628
1992–1993	586.2	152.41	4	610
1993–1994	561.7	146.04	4	584
1994–1995	580.0	150.80	4	603

It was assumed that the average underemployed person is working five hours less than what they would like. To estimate the cost of underemployment, the number of underemployed was multiplied by 5 and then by $4 per hour (as was the case in calculating the cost of unemployment). The data

Table A1.t   Item U — Cost of Commuting

Year	Cost of Car Operation (at 1989–1990 Prices) ($m) a	Rail, Tram and Bus Fares (at 1989–1990 Prices) ($m) b	Cost of Commuting $(0.21a + 0.3b)$ ($m) U
1966–1967	4,932.7	1,405.6	1,458
1967–1968	5,448.5	1,397.4	1,563
1968–1969	6,094.6	1,407.4	1,702
1969–1970	6,588.2	1,423.2	1,810
1970–1971	7,173.6	1,411.2	1,930
1971–1972	7,480.2	1,487.7	2,017
1972–1973	7,508.4	1,438.2	2,008
1973–1974	7,715.9	1,387.8	2,037
1974–1975	7,922.5	1,238.7	2,035
1975–1976	8,525.4	1,209.6	2,153
1976–1977	8,365.6	1,126.1	2,095
1977–1978	8,563.2	1,159.1	2,146
1978–1979	9,181.4	1,142.3	2,271
1979–1980	10,341.4	1,187.2	2,528
1980–1981	10,613.7	1,309.5	2,622
1981–1982	10,779.3	1,365.9	2,673
1982–1983	11,529.1	1,461.2	2,859
1983–1984	11,623.5	1,513.8	2,895
1984–1985	12,230.4	1,596.1	3,047
1985–1986	12,503.0	1,628.9	3,114
1986–1987	12,518.9	2,044.3	3,242
1987–1988	12,959.4	2,087.3	3,348
1988–1989	12,982.7	2,175.1	3,379
1989–1990	13,851.0	2,284.0	3,594
1990–1991	15,438.9	2,382.4	3,957
1991–1992	15,645.1	2,448.0	4,020
1992–1993	15,470.5	2,425.2	3,976
1993–1994	15,748.6	2,467.3	4.047
1994–1995	15,880.4	2,690.0	4,142

sources are the Australian Bureau of Statistics, Catalogue No. 6246.0, and Kryger, (1993).

Item U — *Cost of commuting* (Table A1.t).

The method employed for calculating the cost of commuting is the same the method used by Daly and Cobb (1989). The cost of commuting was calculated on the basis of the following formula.

$$C = 0.3.(A - 0.3A) + 0.3B$$

$$= 0.3.(0.7A) + 0.3B$$

$$= 0.21A + 0.3B \qquad \text{(A1.1)}$$

where

- *C* equals the cost of commuting.
- *A* equals the cost of private car operation.
- 0.3*A* equals the estimated cost of the depreciation of private motor vehicles.
- 0.3 equals the estimated portion of total non-commercial vehicle kilometres used in commuting.
- *B* equals the price paid for the use of public transport.
- 0.3*B* = is the estimated portion of passenger kilometres on local public transport used for commuting.

The data source is the Australian Bureau of Statistics, Catalogue No. 5204.0.

Item V — *Cost of crime* (Table A1.u).

The cost of crime was assumed to be the sum of the theft of privately owned property, consumer durables, and vehicles, and the psychic outgo cost from being confined indoors at particular times of the day or to particular places owing to the fear of being a victim of crime (i.e., the indirect psychic outgo cost of crime).

a. *Cost of theft.*

The cost of various elements of crime have been estimated for Australia by the Australian Institute of Criminology. Those which are directly applicable to the nation's psychic income include the cost of

- Robbery in 1988 — valued at $111 million at 1989–1990 prices.
- Stealing from a person in 1988 — valued at $318 million at 1989–1990 prices.
- Breaking and entering in 1988 — valued at $608 million at 1989–1990 prices.
- Motor vehicle theft in 1988 — valued at $737 million at 1989–1990 prices.
- Bicycle theft in 1988 — valued at $33 million at 1989–1990 prices.
- Stealing from motor vehicle in 1988 — valued at $123 million at 1989–1990 prices.
- Damage from vehicle-related offences in 1988 — valued at $265 million at 1989–1990 prices.
- Crime prevention/security costs in 1991-92 — valued at $476 million at 1989–1990 prices.

To estimate the cost of household property and vehicle-related offences in other years, three indexes were constructed — a robbery index (used to

Table A1.u Item V — Cost of Crime

Year	Robbery Index (1987-1988 = 100.0) a	Robbery Cost (429 × a/100) ($m) b	Breaking and Entering Index c	Breaking and Entering Cost (608 × c/100) ($m) d	Motor Vehicle Theft Index e	Motor Vehicle Theft Cost (2,195 × e/100) ($m) f	Breaking and Entering Index (1991-1992 = 100.0) g	Crime Prev./Security Exp. (476 × g/100) ($m) h	Cost of Private Property-Related Theft ($m) j	Indirect Cost of Crime k	Cost of Crime (j+k) ($m at 1989-1990 Prices) V
1966-1967	12.0	51.5	12.9	78.4	21.7	476.3	10.4	49.5	656	1,165	656
1967-1968	13.8	59.2	15.4	93.6	22.7	498.3	12.4	59.0	710	1,265	710
1968-1969	17.7	75.9	18.6	113.1	25.0	548.8	15.0	71.4	809	1,366	809
1969-1970	22.2	95.2	21.5	130.7	28.9	634.4	17.3	82.3	943	1,462	943
1970-1971	29.7	127.4	25.7	156.3	34.3	752.9	20.8	99.0	1,136	1,584	1,136
1971-1972	38.2	163.9	34.2	207.9	38.0	834.1	27.6	131.4	1,337	1,784	1,337
1972-1973	36.5	156.6	33.6	204.3	36.0	790.2	27.2	129.5	1,281	1,904	1,281
1973-1974	37.4	160.4	35.2	214.0	38.7	849.5	28.5	135.7	1,360	2,046	1,360
1974-1975	42.7	183.2	37.9	230.4	41.7	915.3	30.6	145.7	1,475	2,385	1,475
1975-1976	36.6	157.0	36.8	223.7	40.6	891.2	29.7	141.4	1,413	2,582	1,413
1976-1977	37.8	162.2	38.0	231.0	43.5	954.8	30.7	146.1	1,494	2,701	1,494
1977-1978	44.1	189.2	42.8	260.2	50.4	1,106.3	34.6	164.7	1,720	2,918	1,720
1978-1979	45.4	194.8	49.3	299.7	54.8	1,202.9	39.8	189.4	1,887	3,060	1,887
1979-1980	52.4	224.8	55.2	335.6	57.1	1,253.3	44.6	212.3	2,026	3,254	2,026
1980-1981	56.4	242.0	65.6	398.8	61.5	1,349.9	53.0	252.3	2,243	3,433	2,243
1981-1982	63.1	270.7	72.7	442.0	70.4	1,545.3	58.7	279.4	2,537	3,542	2,537
1982-1983	81.4	349.2	87.3	530.8	77.5	1,701.1	70.5	335.6	2,917	3,731	2,917
1983-1984	81.6	350.1	96.0	583.7	80.2	1,760.4	77.5	368.9	3,063	3,816	3,063
1984-1985	82.1	352.2	95.8	582.5	83.5	1,832.8	77.4	368.4	3,136	3,960	3,136
1985-1986	119.0	510.5	84.9	516.2	97.8	2,146.7	68.5	326.1	3,499	4,252	3,499
1986-1987	103.1	442.3	95.8	582.5	108.8	2,388.2	77.4	368.4	3,781	4,379	3,781
1987-1988	100.0	429.0	100.0	608.0	100.0	2,195.0	80.8	384.6	3,617	4,433	3,617
1988-1989	104.5	448.3	104.0	632.3	102.7	2,254.3	84.0	399.8	3,735	4,718	3,735
1989-1990	113.3	486.1	110.2	670.0	100.4	2,203.8	89.0	423.6	3,783	5,030	3,783
1990-1991	123.0	527.7	116.8	710.1	98.9	2,170.9	94.3	448.9	3,858	5,198	3,858
1991-1992	133.4	572.3	123.8	752.7	97.5	2,140.1	100.0	476.0	3,941	5,370	3,941
1992-1993	144.8	621.2	131.2	797.7	96.0	2,107.2	106.0	504.6	4,031	5,327	4,031
1993-1994	158.6	680.4	139.7	849.4	94.6	2,076.5	112.8	536.9	4,143	5,401	4,143
1994-1995	189.0	810.8	142.7	867.6	100.2	2,199.4	115.3	548.8	4,427	5,703	4,427

estimate the cost of robbery and stealing from a person), a break and enter index, a motor vehicle theft index (used to estimate the cost of motor vehicle theft, bicycle theft, stealing from motor vehicles, and damage from vehicle-related offences). Each index was based on the number of reported robbery, breaking and entering, and motor vehicle offences for each respective year. It was assumed that the real cost of offences remained constant throughout the period.

Because the cost of the preceding forms of crime was estimated for 1988, all 3 indexes used were given an index value of 100.0 for the year 1987–1988. As for the cost of crime prevention/security expenditure, the breaking and entering index was used once more but, because the cost was estimated for the financial year 1991–1992, the index was recalculated with a base index value of 100.0 applied to 1991–1992. The data sources are Walker, (1992), the Australian Bureau of Statistics, Catalogue No. 1301.0 (various articles), and the Australian Bureau of Statistics, Catalogue No. 4510.0.

b. *Indirect psychic outgo cost of crime.*

The indirect psychic outgo cost of crime was assumed to be equivalent to the amount spent by all forms of government on public order and safety. The data source is Foster, (1986, Table 2.12, page 83).

Item W — *Cost of family breakdown* (Table A1.v).

The cost of family breakdown was calculated by using a similar approach to that employed by Redefining Progress (1995). The approach involved two steps. In the first instance, the number of divorces in Australia was multiplied by $5000. It was assumed that, following divorce, the $5000 at 1989–1990 prices approximated the average cost of legal fees, counselling services, and establishment of separate residences. Second, the number of divorces was multiplied by the average number of affected children to calculate the total number of children directly affected by divorce each year. This figure was multiplied by $10,000 to account for the counselling costs, health costs, the cost of disruption at school, and the difficulties children have in establishing and maintaining healthy personal relationships.

While $5,000 and $10,000 might seem a little excessive, it is less, in real terms, than the figures used by Redefining Progress. In addition, the total figure does not include the cost of disharmonious family relationships in cases where divorce has not ensued. The data source is the Australian Bureau of Statistics, Catalogue No. 1301.0 (various articles).

Item X — *Total psychic outgo,*

Equal to the sum of Items N to W.

Item X1 — *Australian population.*

Table A1.v   Item W — Cost of Family Breakdown

Year	Number of Divorces (Thousands) a	Direct Cost of Divorce (a × \$5,000) (\$m) b	Average No. of Children Affected per Divorce c	Number of Directly Affected Children (a × c) d	Indirect Cost of Affect of Divorce on Children (d × \$10,000) (\$m) e	Cost of Family Breakdown (b+c) (\$m) W
1966–1967	9,803	49.02	2.10	20,586	205.86	255
1967–1968	10,268	51.34	2.10	21,563	215.63	267
1968–1969	10,887	54.44	2.10	22,863	228.63	283
1969–1970	11,617	58.09	2.10	24,396	243.96	302
1970–1971	12,598	62.99	2.10	26,456	264.56	328
1971–1972	14,327	71.64	2.05	29,370	293.70	365
1972–1973	15,987	79.94	2.05	32,773	327.73	408
1973–1974	16,977	84.89	2.05	34,803	348.03	433
1974–1975	20,973	104.87	2.05	42,995	429.95	535
1975–1976	35,884	179.42	2.00	71,768	717,68	897
1976–1977	46,330	231.65	2.00	92,660	926.60	1,158
1977–1978	42,879	214.40	2.00	85,758	857.58	1,072
1978–1979	39,231	196.16	2.00	78,462	784.62	981
1979–1980	38,556	192.78	2.00	77,112	771.12	964
1980–1981	40,335	201.68	1.95	78,653	786.53	988
1981–1982	42,750	213.75	1.95	83,363	833.63	1,047
1982–1983	43,807	219.04	1.95	85,424	854.24	1,073
1983–1984	43,325	216.63	1.90	82,318	823.18	1,040
1984–1985	41,477	207.39	1.90	78,806	788.06	995
1985–1986	39,624	198.12	1.90	75,286	752.86	951
1986–1987	39,571	197.86	1.90	75,185	751.85	950
1987–1988	40,366	201.83	1.90	76,695	766.95	969
1988–1989	41,195	205.98	1.90	78,271	782.71	989
1989–1990	42,009	210.05	1.90	79,817	798.17	1,008
1990–1991	44,133	220.67	1.90	83,853	838.53	1,059
1991–1992	45,648	228.24	1.90	86,731	867.31	1,096
1992–1993	46,995	234.98	1.90	89,291	892.91	1,128
1993–1994	48,290	241.45	1.90	91,751	917.51	1,159
1994–1995	48,978	244.89	1.90	93,058	930,58	1,175

As per Item M1.

Item X2 — *Per capita psychic outgo.*

Item X divided by Item X1.

Column AA — *Net psychic income.*

Item X less Item M.

Column BB — *Australian population.*

The data source is Foster, R. (1996, Table 4.1, page 177).

Column CC — *Per capita net psychic income.*

Item AA divided by Item BB.

appendix two

The uncancelled cost account for Australia, 1966–1967 to 1994–1995

A2.1 Introduction

The uncancelled cost account measures the source, sink, and life-support services of natural capital annually sacrificed in the process of converting natural capital to human-made capital. Appendix 2 reveals the methods and data used to calculate monetary values for the items appearing in the uncancelled cost account. Tables revealing how the items have been calculated are located in this appendix. All items are measured in millions of Australian dollars at 1989–1990 prices. The uncancelled cost account appears in Table 14.2 in Chapter 14. The items appearing in the uncancelled cost account are

a. *Sacrificed source function.*

- User cost of nonrenewable resources (metallic minerals, coal, oil and gas, and non-metallic minerals).
- Cost of lost agricultural land.
- User cost of timber resources.
- User cost of fishery resources.
- Cost of lost and degraded wetlands, mangroves, and saltmarshes.

b. *Sacrificed sink function.*

- Cost of water pollution.
- Cost of air pollution.
- Cost of solid waste pollution.
- Cost of ozone depletion.

c. *Sacrificed life-support function.*

- Cost of long-term environmental damage.
- An ecosystem health index.

A2.2 Sacrificed source function of natural capital

Item 1 — *User cost of metallic minerals* (Table A2.a).

The method employed to calculate the cost of metallic minerals is a variation on El Serafy's user cost approach (Lawn, 1998). El Serafy's (1989) method attempts to overcome the current error of treating the entire proceeds from nonrenewable resource depletion as income. This error arises because a proper measure of income must take into account the need to maintain intact the stock of income-generating capital (Hicks, 1946). Because, by definition, nonrenewable resource use involves the diminution of the stock of income-generating natural capital, at least some of the proceeds from nonrenewable resource depletion must be set aside to establish a suitable replacement asset. The implications of this, according to El Serafy, is that any estimation of sustainable income requires a nonrenewable resource earmarked for depletion to be converted into a perpetual income stream. This, he has explained, requires a finite series of earnings from the sale of the resource to be converted to an infinite series of true income such that the capitalised values of the two series are equal (El Serafy, 1989). To do this, an income and capital component of the finite series of earnings must be identified. If correctly estimated, the income component constitutes an amount that can be consumed without any fear of undermining the capacity to sustain the same level of consumption over time. The capital component, on the other hand, is an amount that needs to be set aside each year to ensure a perpetual income stream of constant value, both during the life of the resource as well as following its exhaustion. It is this capital component that constitutes the true "user cost" of resource depletion. To identify the income and capital components, El Serafy has suggested the use of the following formula

$$X/R = 1 = -\frac{1}{\left(1+r\right)^{n+1}} \tag{A2.1}$$

where

- X equals the true money income.
- R equals net receipts (gross receipts minus extraction and other operating costs).
- r equals the rate of discount.
- n equals the number of periods over which the resource is to be liquidated, or equivalently, the mine life of the resource in question.

Table A2.a Item 1, 2, 3, 4, and 4A — User Cost of Nonrenewable Resources

Year	Total Value of Metallic Minerals Mined — a	User Cost of Metallic Minerals (a × 0.56) — 1	Total Value of Coal Mined — b	User Cost of Coal (b × 0.56) — 2	Total Value of Oil and Gas Mined — c	User Cost of Oil and Gas (c × 0.56) — 3	Total Value of Nonmetallic Minerals Mined — d	User Cost of Nonmetallic Minerals (d × 0.56) — 4	Total Value of Mineral Production ($m at 1989–1990 Prices) — e	User Cost of Nonrenewable Resources ($m at 1989–1990 Prices) — 4A
1966–1967	2,459.0	1,377.0	1,198.6	671.2	141.1	79.0	820.4	459.4	4,619	2,587
1967–1968	3,001.0	1,680.6	1,348.3	755.0	252.0	141.1	857.3	480.1	5,459	3,057
1968–1969	3,490.5	1,954.7	1,361.5	762.4	251.2	140.7	973.7	545.3	6,077	3,403
1969–1970	4,519.5	2,530.9	1,560.7	874.0	516.0	289.0	1,098.7	615.3	7,695	4,309
1970–1971	4,734.4	2,651.3	1,752.3	981.3	1,213.6	679.6	1,162.4	650.9	8,863	4,963
1971–1972	4,811.0	2,694.2	1,897.7	1,062.7	1,419.7	795.0	1,217.7	681.9	9,346	5,234
1972–1973	4,788.0	2,681.3	2,043.6	1,144.4	1,500.3	840.2	1,281.5	717.6	9,613	5,384
1973–1974	5,473.2	3,065.0	2,084.3	1,167.2	1,617.3	905.7	1,335.1	747.7	10,510	5,886
1974–1975	5,631.7	3,153.8	3,318.0	1,858.1	1,597.7	894.7	1,283.2	718.6	11,831	6,625
1975–1976	5,146.2	2,881.9	3,903.6	2,186.0	1,499.4	839.7	1,245.6	697.5	11,795	6,605
1976–1977	5,483.2	3,070.6	4,165.2	2,332.5	1,476.1	826.6	1,271.9	712.3	12,396	6,942
1977–1978	5,293.5	2,964.4	4,262.0	2,386.7	1,725.1	966.1	1,319.2	738.8	12,600	7,056
1978–1979	5,705.8	3,195.2	4,150.4	2,324.2	2,179.9	1,220.7	1,347.1	754.4	13,383	7,495
1979–1980	7,275.6	4,074.3	3,978.9	2,228.2	2,523.4	1,413.1	1,501.4	840.8	15,279	8,556
1980–1981	6,178.0	3,459.7	4,892.2	2,739.6	3,107.1	1,740.0	1,525.2	854.1	15,703	8,793
1981–1982	5,795.0	3,245.2	5,371.0	3,007.8	3,453.2	1,933.8	1,551.9	869.1	16,171	9,056
1982–1983	6,090.2	3,410.5	5,866.5	3,285.2	3,635.6	2,035.9	1,447.5	810.6	17,040	9,542
1983–1984	5,686.6	3,184.5	5,536.5	3,100.4	4,505.8	2,523.2	1,565.5	876.7	17,294	9,685
1984–1985	6,579.9	3,684.7	6,466.0	3,621.0	5,728.8	3,208.1	1,656.7	927.8	20,431	11,442
1985–1986	6,983.5	3,910.8	7,166.2	4,013.1	5,141.4	2,879.2	1,857.9	1,040.4	21,149	11,843
1986–1987	7,872.1	4,408.4	6,642.0	3,719.5	4,554.0	2,550.2	1,912.8	1,071.2	20,981	11,749
1987–1988	9,009.5	5,045.3	5,568.3	3,118.2	4,759.5	2,665.3	1,788.8	1,001.7	21,126	11,831
1988–1989	—	—	—	—	—	—	—	—	21,063	11,795
1989–1990	—	—	—	—	—	—	—	—	26,133	14,634
1990–1991	—	—	—	—	—	—	—	—	26,293	14,724
1991–1992	—	—	—	—	—	—	—	—	25,985	14,552
1992–1993	—	—	—	—	—	—	—	—	29,191	16,347
1993–1994	—	—	—	—	—	—	—	—	27,108	15,180
1994–1995	—	—	—	—	—	—	—	—	26,962	15,099

- $R - X$ equals the user cost or the amount that must be set aside to ensure a perpetual income stream. This is the amount included in the uncancelled cost account.

The two key parameters in the determination of the income and capital components are the number of periods over which the resource is to be liquidated and the chosen discount rate. The greater are the values of these two parameters, the more substantial is the income component and that much smaller is the amount that needs to be set aside each period to ensure a perpetual income stream. The number of periods over which the resource is to be liquidated is relatively easy to calculate and, in the majority of cases, will usually depend on the size of a nonrenewable resource deposit as well as expected future resource prices. The major difficulty lies in determining the most appropriate discount rate. According to El Serafy, the chosen discount rate should approximate an available market parameter that would indicate prudent behaviour on the part of the resource liquidator.

It is at this point where the user cost approach used here differs from that of El Serafy. El Serafy is of the view that the replacement asset, that is, the asset established to replace the depleted resource, can be in any particular form, whether it be a human-made or a natural capital asset or even a financial asset (El Serafy, 1996). Because this view assumes the substitutability of human-made for natural capital, El Serafy's method of measuring Hicksian income is one of the "weak sustainability" kind. The standard ecological economic view is that human-made and natural capital are complements and, as such, the only true measure of Hicksian income is one of the "strong sustainability" kind. Because the strong sustainability approach requires both natural and human-made capital to be kept intact, a nonrenewable resource must be replaced by a suitable *renewable* resource substitute.

What does this mean in terms of the preceding user cost formula? It does not mean a great deal, except in relation to the choice of discount rate. A weak sustainability advocate, such as El Serafy, would probably suggest a discount rate equal to the real interest rate earned on alternative assets, whether it be human-made capital or financial assets, because the real interest rate reflects the likely real returns on the re-investment. It is the view here that the chosen discount rate should be equivalent to the real interest rate on the cultivated renewable resource. This happens to be its natural regenerative rate (Lawn, 1998). Thus, in order to put the strong sustainability approach into practice, the discount rate used in the user cost formula must be replaced with the regenerative rate of the proposed renewable resource substitute.

To calculate the user cost of nonrenewable resources, a discount rate of 5 percent was used on the assumption that it approximates the regeneration rate of renewable resources. It was also assumed that the average mine life of nonrenewable resources is 25 years. This means a set-aside ratio of 28 percent is required, that is, for every dollar of net receipts received, 72 cents constitutes money income while 28 cents is equal to the user cost of the resource.

It was felt that it was necessary to make two further adjustments in calculating the user cost of nonrenewable resource depletion. The first was deemed necessary because the setting aside of 28 percent of the net receipts from nonrenewable resource depletion assumes that nonrenewable resource prices will remain constant over time. This is unlikely. Indeed, increased resource scarcity is likely to lead to a doubling of nonrenewable resource prices over a 25 year period. This makes it necessary to set aside 56 percent of the net receipts from nonrenewable resource depletion. It also means that 56 percent of the net receipts constitute the user cost of nonrenewable resource depletion and only 44 percent legitimate income. Second, resource receipts were treated in the same manner as Daly and Cobb (1989). Daly and Cobb argue that the value of the net receipts from resource depletion (R) understates the figure that ought to be used in the calculation of user costs. Their argument is based on the belief that nonrenewable resource availability is not only a function of the relative and absolute scarcity of resources, but is also a function of the cost of exploration and extraction activities. Thus, according to Daly and Cobb, neither cost should be excluded from the calculation of user costs on the assumption that they both constitute "regrettable necessities" associated with the depletion of nonrenewable resources. In order to include the cost of exploration and extraction activities, the user cost of metallic minerals and remaining nonrenewable resources was calculated at 56 percent of the total dollar value of all production, not 56 percent of the net receipts from resource depletion as implied by Eq. (A2.1). The data sources are the Australian Bureau of Statistics, Catalogue No. 8405.0 (various articles) and the Australian Bureau of Agricultural and Resource Economics, *Commodity Statistical Bulletin* (various articles).

Item 2 — *User cost of coal* (Table A2.a). As per Item 1.

Item 3 — *User cost of oil and gas* (Table A2.a). As per Items 1 and 2.

Item 4 — *User cost of non-metallic minerals* (Table A2.a). As per Items 1, 2, and 3.

Item 4A — *User cost of nonrenewable resources* (Table A2.a). The user cost of nonrenewable resources is the sum of Items 1 to 4.

Item 5 — *Cost of lost agricultural land* (Table A2.b).

Agricultural land is lost in two main ways — through productivity losses arising from erosion, acidification, compaction, soil structure decline, and growing salinity; and through encroachment onto already existing agricultural land from increasing levels of urbanisation.

a. *Productivity loss.*

Very little work has been done on the actual cost of productivity losses on agricultural land. According to an Industry Commission Staff Information Paper, the production equivalent of land degradation in Australia was approximately $1.59 billion in 1994–1995 (at 1989–1990 prices). This works

Table A2.b   Item 5 — Cost of Lost Agricultural Land

Year	Index of Volume of Agricultural Production a	Productivity Loss: Productivity Loss of Soil $[1592 \times (a/100)]$ ($m) b	Productivity Loss: Cumulative Productivity Loss ($m) c	Area of Residential and Commercial Land (Hectares) d	Urbanisation Loss: Total Loss of Agricultural Land $(0.75 \times d)$ (Hectares) e	Urbanisation Loss: Cumulative Cost of Lost Ag. Land from Urbanisation ($10,000 \times e$) ($m) f	Cost of Lost Agricultural Land $(c + f)$ ($m at 1989–1990 Prices) 5
Up to 1966–1967			50,000				
1966–1967	75	1,194	51,194	842,785	632,089	6,321	57,515
1967–1968	66	1,051	52,245	857,785	643,339	6,433	58,678
1968–1969	80	1,274	53,518	875,927	656,945	6,569	60,088
1969–1970	77	1,226	54,744	893,357	670,018	6,700	61,444
1970–1971	79	1,258	56,002	933,356	700,017	7,000	62,002
1971–1972	83	1,321	57,323	950,246	712,685	7,127	64,450
1972–1973	76	1,210	58,533	964,643	723,482	7,235	65,768
1973–1974	80	1,274	59,807	980,215	735,161	7,352	67,158
1974–1975	84	1,337	61,144	992,357	744,268	7,443	68,587
1975–1976	88	1,401	62,545	1,002,357	751,768	7,518	70,063
1976–1977	89	1,417	63,962	1,013,713	760,285	7,603	71,565
1977–1978	80	1,274	65,235	1,025,643	769,232	7,692	72,928
1978–1979	101	1,608	66,843	1,036,857	777,643	7,776	74,620
1979–1980	95	1,512	68,356	1,049,643	787,232	7,872	76,228
1980–1981	86	1,369	69,725	1,065,928	799,446	7,994	77,719
1981–1982	96	1,528	71,253	1,084,572	813,429	8,134	79,387
1982–1983	83	1,321	72,575	1,099,546	824,660	8,247	80,821
1983–1984	104	1,656	74,230	1,112,786	834,590	8,346	82,576
1984–1985	105	1,672	75,902	1,127,715	845,786	8,458	84,360
1985–1986	102	1,624	77,526	1,144,143	858,107	8,581	86,107
1986–1987	106	1,688	79,213	1,161,715	871,286	8,713	87,926
1987–1988	104	1,656	80,869	1,180,858	885,644	8,856	89,725
1988–1989	105	1,672	82,540	1,200,285	900,214	9,002	91,543
1989–1990	106	1,688	84,228	1,218,929	914,197	9,142	93,370
1990–1991	112	1,783	86,011	1,234,572	925,929	9,259	95,270
1991–1992	106	1,688	87,699	1,249,215	936,911	9,369	97,068
1992–1993	112	1,783	89,482	1,261,143	945,857	9,459	98,940
1993–1994	116	1,847	91,328	1,274,143	955,607	9,556	100,884
1994–1995	100	1,592	92,920	1,289,571	967,178	9,672	102,592

out to around 6 percent of the value of agricultural production. To estimate the cost of productivity losses in other years, the index of volume of agricultural production was used. This assumes that productivity losses are a function of the volume of agricultural production. Because the total cost of lost soil productivity is a cumulative rather than annual loss, it was conservatively estimated that the cumulative productivity loss at the beginning of the 1966–1967 financial year was $50 billion. By 1994–1995, this had increased to just under $93 billion.

b. *Urbanisation loss.*

Required here is an estimation of the cumulative total area of residential and commercial land in Australia for each year. To achieve this, three distinct types of regions were assumed to exist in Australia: major urban centres (which include 13 major urban cities), minor urban centres, and rural areas. It was assumed that the average population density per hectare of residential and commercial land is 14 persons per hectare for each of the three types of regions. The figure of 14 was used because it is the approximate average population density for major urban centres over the study period. The population of the three categories of regions was divided by the population density of 14 in order to obtain an estimate of the area of residential and commercial land in each type of region.

The estimated area of residential and commercial land in each type of region was added to give an estimate of the total area of residential and commercial land in Australia. Because not all land being used for residential and/or commercial purposes was necessarily fit for agriculture prior to its new use, the total area was multiplied by 0.75 (75 percent) to ascertain a measure of the total cumulative loss of agricultural land (i.e., it was assumed that only 75 percent of the land resumed for residential and/or commercial purposes was suitable for agricultural production). The total cumulative loss of agricultural land was then multiplied by a capitalised value of $10,000 per hectare on the assumption that, on average, each hectare of current or potential agricultural land lost from increasing urbanisation was worth $10,000 per hectare. This is higher than the average value of agricultural land in Australia. However, land lost from urbanisation tends to be of much higher quality than the average hectare of Australian agricultural land which, on the whole, is fit for little else than low density stock grazing.

Finally, the productivity and urbanisation losses were summed in order to obtain a final estimate of the cost of agricultural land lost. The data sources are the Australian Bureau of Statistics, Catalogue No. 2822.0 (various articles), the Australian Bureau of Statistics, Catalogue No. 4601.0, Australian Bureau of Statistics, Catalogue No. 1301.0 (various articles), and the Australian Bureau of Agricultural and Resource Economics, *Commodity Statistical Bulletin 1993*, Table 18, page 18.

Item 6 — *User cost of timber resources* (Table A2.c).

TableA2.c   Item 6 — User Cost of Timber Resources

Year	Native Forest					Woodland				
	Area of Native Forest (km²) a	Decline in Forest Area (km²) b	Value of Timber per km² ($) c	Net Decline in Value of Native Forest Timber ($m) d	User Cost (d × 0.56 Where d Is Positive) ($m) e	Area of Woodland (km²) f	Decline in Woodland Area (km²) g	Value of Timber per km² ($) h	Net Decline in Value of Woodland Timber ($m) j	User Cost (j × 0.56 where j Is Positive) ($m) k
1965–1966	413,410					715,000				
1966–1967	403,410	10,000	39,091	391	219	710,000	5,000	19,546	98	55
1967–1968	393,410	10,000	39,091	391	219	705,000	5,000	19,546	98	55
1968–1969	383,410	10,000	39,091	391	219	700,000	5,000	19,546	98	55
1969–1970	373,410	10,000	39,091	391	219	695,000	5,000	19,546	98	55
1970–1971	390,620	-17,210	39,091	-673	-673	690,000	5,000	19,546	98	55
1971–1972	407,830	-17,210	39,091	-673	-673	685,000	5,000	19,546	98	55
1972–1973	425,040	-17,210	39,091	-673	-673	680,000	5,000	19,546	98	55
1973–1974	426,810	-1,770	39,091	-69	-69	675,000	5,000	19,546	98	55
1974–1975	428,580	-1,770	39,091	-69	-69	670,000	5,000	19,546	98	55
1975–1976	430,370	-1,790	39,091	-70	-70	665,000	5,000	19,546	98	55
1976–1977	431,000	-630	39,091	-25	-25	660,000	5,000	19,546	98	55
1977–1978	438,300	-7,300	39,091	-285	-285	655,000	5,000	19,546	98	55
1978–1979	419,932	18,368	39,091	718	402	650,000	5,000	19,546	98	55
1979–1980	408,840	11,092	39,091	434	243	650,000	0	19,546	0	0
1980–1981	408,670	170	39,091	7	4	650,000	0	19,546	0	0
1981–1982	408,590	80	39,091	3	2	650,000	0	19,546	0	0
1982–1983	408,410	180	39,091	7	4	650,000	0	19,546	0	0
1983–1984	409,380	-970	39,091	-38	-38	650,000	0	19,546	0	0
1984–1985	413,100	-3,720	39,091	-145	-145	640,160	9,840	19,546	192	108
1985–1986	412,820	280	39,091	11	6	640,060	100	19,546	2	1
1986–1987	408,410	4,410	39,091	172	97	639,960	100	19,546	2	1
1987–1988	408,280	130	39,091	5	3	639,860	100	19,546	2	1
1988–1989	409,720	-1,440	39,091	-56	-56	634,660	5,200	19,546	102	57
1989–1990	408,190	1,530	39,091	60	33	634,560	100	19,546	2	1
1990–1991	408,180	10	39,091	0	0	634,460	100	19,546	2	1
1991–1992	410,570	-2,390	39,091	-93	-93	634,360	100	19,546	2	1
1992–1993	407,190	3,380	39,091	132	74	634,260	100	19,546	2	1
1993–1994	407,190	0	39,091	0	0	634,260	0	19,546	0	0
1994–1995	407,090	100	39,091	4	2	634,160	100	19,546	2	1

Table A2.c   Item 6 — User Cost of Timber Resources (Continued)

	Plantation - Broadleaved					Plantation - Coniferous					
Year	Area of Broadleaved Forest (km²)	Decline in Forest Area (km²)	Value of Timber per km² ($)	Net Decline in Value of Timber in Broadleaved Forest ($m)	User Cost (q × 0.56 Where q Is Positive) ($m)	Area of Coniferous Forest (km²)	Decline in Forest Area (km²)	Value of Timber per km² ($)	Net Decline in Value of Timber in Coniferous Forest ($m)	User Cost (v × 0.56 Where v Is Positive) ($m)	User Cost of Timber Resources ($m at 1989–1990 Prices)
	m	n	p	q	r	s	t	u	v	w	6
1965-1966	202					4,750					
1966-1967	222	-20	171,039	-3	-3	4,900	-150	556,976	84	47	317
1967-1968	242	-20	171,039	-3	-3	5,050	-150	556,976	84	47	317
1968-1969	262	-20	171,039	-3	-3	5,200	-150	556,976	84	47	317
1969-1970	282	-20	171,039	-3	-3	5,350	-150	556,976	84	47	317
1970-1971	302	-20	171,039	-3	-3	5,500	-150	556,976	84	47	-575
1971-1972	322	-20	171,039	-3	-3	5,650	-150	556,976	84	47	-575
1972-1973	342	-20	171,039	-3	-3	5,800	-150	556,976	84	47	-575
1973-1974	362	-20	171,039	-3	-3	5,950	-150	556,976	84	47	29
1974-1975	382	-20	171,039	-3	-3	6,100	-150	556,976	84	47	29
1975-1976	402	-20	171,039	-3	-3	6,250	-150	556,976	84	47	28
1976-1977	422	-20	171,039	-3	-3	6,400	-150	556,976	84	47	73
1977-1978	442	-20	171,039	-3	-3	6,550	-150	556,976	84	47	-187
1978-1979	458	-16	171,039	-3	-3	6,816	-266	556,976	148	83	537
1979-1980	479	-21	171,039	-4	-4	7,183	-367	556,976	204	114	354
1980-1981	502	-23	171,039	-4	-4	7,411	-228	556,976	127	71	71
1981-1982	423	79	171,039	14	8	7,773	-362	556,976	202	113	122
1982-1983	430	-7	171,039	-1	-1	7,745	28	556,976	-16	-16	-13
1983-1984	420	10	171,039	2	1	7,927	-182	556,976	101	57	20
1984-1985	390	30	171,039	5	3	8,030	-103	556,976	57	32	-3
1985-1986	410	-20	171,039	-3	-3	8,360	-330	556,976	184	103	107
1986-1987	470	-60	171,039	-10	-10	8,580	-220	556,976	123	69	156
1987-1988	520	-50	171,039	-9	-9	8,900	-320	556,976	178	100	95
1988-1989	600	-80	171,039	-14	-14	9,140	-240	556,976	134	75	62
1989-1990	960	-360	171,039	-62	-62	9,260	-120	556,976	67	37	10
1990-1991	1,060	-100	171,039	-17	-17	9,400	-140	556,976	78	44	28
1991-1992	1,170	-110	171,039	-19	-19	9,560	-160	556,976	89	50	-61
1992-1993	1,310	-140	171,039	-24	-24	9,570	-10	556,976	6	3	54
1993-1994	1,460	-150	171,039	-26	-26	9,590	-20	556,976	11	6	-19
1994-1995	1,550	-90	171,039	-15	-15	9,640	-50	556,976	28	16	3

Unlike a nonrenewable resource, the exploitation of timber resources results in a "user" cost only if there is a decline in the source function of natural capital, that is, if the rate at which timber stocks are harvested (h_{timber}) exceeds the natural regeneration rate (y_{timber}). Should this occur, the proceeds from the portion of all extracted timber exceeding the sustainable harvest rate (i.e., $h_{timber} - y_{timber}$) must be treated the same way as the proceeds from the sale of nonrenewable resources. Because this requires the user cost formula to be applied [Eq. (A2.1)], a discount rate must be decided upon as well as the approximate life of an unsustainably harvested forest/plantation. A decision is also required in relation to the likely price effect should timber resources be exhausted.

First, a 5 percent discount rate was chosen to reflect the regeneration rate of the timber stocks that would have to be cultivated elsewhere to maintain intact stocks. Second, it was assumed that, on average, a forest or timber plantation that was continuously harvested at an unsustainable rate would be depleted in 25 years. Finally, it was assumed that the prices of timber resources would double if they were harvested unsustainably. Based on these assumptions, the set-aside rate or user cost of timber resource depletion (should it occur) comes to 56 percent of net receipts.

In addition, it was decided that, in a similar manner to nonrenewable resources, the cost of timber extraction activities should not be excluded when calculating the user cost of timber resources or any other renewable resource. To include such costs, the user cost of all renewable resources was calculated at 56 percent of all production costs in the years where timber stocks declined.

Of course, one of the major differences between renewable and nonrenewable resources is that the stocks of the former can be augmented and, in doing so, can increase the source function of natural capital. In instances where this has occurred, the value of the entire expansion in the stock of timber resources was counted in the uncancelled cost account as a positive figure or "negative" uncancelled cost. The reason for this is as follows. The annual benefit or negative uncancelled cost of an increase in the source function of natural capital is equal to the value of the maximum amount of timber that can be sustainably harvested from the larger stock. It is not equal to the total increase in the stock of harvestable timber. Thus, the annual benefit (B) equals the value of the increase in the stock of timber resources (ΔS) multiplied by the annual regeneration rate (y). This is given by the following:

$$B = \Delta S \times y \qquad (A2.2)$$

where

- B equals the negative uncancelled cost from increasing the ecosphere's source function.

- ΔS equals the increase in the stock of timber resources or any other renewable resource.
- y equals the natural regeneration rate.

If the timber resource is harvested sustainably (i.e., if $h = y$), the benefits from having increased the stock of timber are enjoyed indefinitely. The net present value (NPV) of an infinite series of benefits is equal to the benefits received each year divided by the discount rate (r). That is:

$$\text{NPV of benefits} = \frac{B}{r} \qquad\qquad\text{(A2.3)}$$

$$= \frac{\Delta S \times y}{r} \qquad\qquad\text{(A2.4)}$$

If it is assumed that the discount rate is approximately equal to the regeneration rate of timber stocks (a reasonable one given the neoclassical economic view that a renewable resource with a regeneration rate less than the discount rate can be justifiably be harvested to extinction), then the NPV of benefits is equal to the value of the increased stock of timber itself. That is,

$$\text{NPV of benefits} = \Delta S \qquad\qquad\text{(A2.5)}$$

In determining the user cost or otherwise of timber resources, Australia's timber stocks were classified into the following four categories.

- Native forest timber.
- Woodland forest timber.
- Plantation timber — broadleaved.
- Plantation timber — coniferous.

The respective 1989–1990 net present values for each category of timber are $39,091 per average km² for native forest timber, $171,039 per average km² for plantation timber — broadleaved, and $556,976 per average km² for plantation timber — coniferous. It was assumed that the value of woodland forest timber is one-half that of native forest timber. The preceding values were multiplied by the net increase/decrease in timber resource stocks to ascertain the overall net change (positive/negative) in the value of all timber stocks. The appropriate user cost adjustment, as discussed previously, was made in cases where timber stocks declined. The data sources are the Australian Bureau of Statistics, Catalogue No. 5241.0, Table 4.2, page 72; Table 4.3, page 73; Table 4.4, page 73; Table 4.5, page 74; Table 4.7, page 79; Table 4.8, page 80; Table 4.9, page 81; and Table 4.9, page 82 the Australian Bureau of Statistics, Catalogue No. 4601.0, Table 6.1.12, page 137; the Department of Primary Industries, *Australian Forest Resources* (various articles); and the

Australian Bureau of Agricultural and Resource Economics, *Commodity Statistical Bulletin* (various articles).

Item 7 — *User cost of fishery resources* (Table A2.d).

Being a renewable resource, the proceeds from the sale of fishery resources must be treated in the same way as the proceeds from timber resources. That is, where stocks decline, the El Serafy user cost formula must be invoked [Eq. (A2.1)]. All assumptions in relation to timber stocks regarding the choice of discount rate, the average life expectancy of unsustainably harvested stocks, and the increase in prices owing to the increasing scarcity of stocks were applied to fishery stocks. Hence, the set-aside rate or user cost of depleted fishery stocks was set at 56 percent of all production costs in the years where fishery stocks declined.

In all, there are seven major fishery categories which make up the total stock of fishery resources. These include prawns, rock lobster, abalone, scallops, oysters, fish (excluding tuna), and tuna.

There is, regrettably, no accurate measure of Australia's fishery stocks nor any knowledge of the actual net change in fishery stocks. In spite of this, the following status has been given to each of the preceding seven listed categories (State of the Environment Advisory Council, 1996, pages 8–29). They are,

- Prawns — fully exploited.
- Rock lobster — fully exploited.
- Abalone — fully exploited.
- Scallops — over exploited.
- Oysters — increasing, largely owing to aquaculture.
- Fish (excluding tuna) — fully exploited;
- Tuna — over exploited.

Since the mid-1960s, when the level of fisheries production began to increase significantly, Australia's fishery stocks have unquestionably declined (State of the Environment Advisory Council, 1996, page 8–14). In order to estimate the net change in fishery stocks, the following was assumed. Where a fishery stock category has been fully exploited, it was assumed the stock had declined by an amount equivalent to 25 percent of the annual production tonnage. Where a category has been over exploited, the stock was assumed to have fallen by an amount equivalent to 50 percent of the annual production tonnage. As for the stock of oysters, which has been increasing in recent times, it was assumed to have declined by 25 percent of the annual production tonnage between 1966–1967 and 1974–1975 (prior to extensive aquaculture); to have remained unchanged between 1974–1975 and 1984–1985; and to have increased at the rate of 12.5 percent of the annual production tonnage since 1984–1985.

Table A2.d Item 7 — User Cost of Fishery Resources

	Prawns (Fully Exploited)					Rock Lobster (Fully Exploited)				
Year	Annual Harvest (Kilotonnes) *a*	Net Decline in Prawn Stocks (*a* × 0.25) (Kilotonnes) *b*	Value of Prawns per Tonne ($) *c*	Net Decline in Value of Prawn Stocks ($m at 1989–1990 Prices) *d*	User Cost of Prawn Stocks (*d* × 0.56) ($m) *e*	Annual Harvest (Kilotonnes) *f*	Net Decline in Rock Lobster Stocks (*f* × 0.25) (Kilotonnes) *g*	Value of Rock Lobster per Tonne ($) *h*	Net Decline in Value of Rock Lobster Stocks ($m at 1989–1990 Prices) *j*	User Cost of Rock Lobster Stocks (*j* × 0.56) ($m) *k*
1966–1967	10.8	2.7	9,717	26.2	14.7	9.0	2.3	15,586	35.1	19.6
1967–1968	12.8	3.2	9,717	31.1	17.4	10.0	2.5	15,586	39.0	21.8
1968–1969	14.8	3.7	9,717	36.0	20.1	11.0	2.8	15,586	42.9	24.0
1969–1970	16.8	4.2	9,717	40.8	22.9	12.0	3.0	15,586	46.8	26.2
1970–1971	18.8	4.7	9,717	45.7	25.6	13.0	3.3	15,586	50.7	28.4
1971–1972	17.5	4.4	9,717	42.5	23.8	13.1	3.3	15,586	51.0	28.6
1972–1973	16.8	4.2	9,717	40.8	22.9	13.0	3.3	15,586	50.7	28.4
1973–1974	24.5	6.1	9,717	59.5	33.3	11.8	3.0	15,586	46.0	25.7
1974–1975	16.3	4.1	9,717	39.6	22.2	12.3	3.1	15,586	47.9	26.8
1975–1976	19.5	4.9	9,717	47.4	26.5	12.9	3.2	15,586	50.3	28.1
1976–1977	23.1	5.8	9,717	56.1	31.4	12.9	3.2	15,586	50.3	28.1
1977–1978	19.3	4.8	9,717	46.9	26.3	14.5	3.6	15,586	56.5	31.6
1978–1979	21.7	5.4	9,717	52.7	29.5	15.4	3.9	15,586	60.0	33.6
1979–1980	22.0	5.5	9,717	53.4	29.9	14.5	3.6	15,586	56.5	31.6
1980–1981	26.9	6.7	9,717	65.3	36.6	15.6	3.9	15,586	60.8	34.0
1981–1982	21.8	5.5	9,717	53.0	29.7	16.0	4.0	15,586	62.3	34.9
1982–1983	22.2	5.6	9,717	53.9	30.2	18.0	4.5	15,586	70.1	39.3
1983–1984	22.7	5.7	9,717	55.1	30.9	16.0	4.0	15,586	62.3	34.9
1984–1985	20.5	5.1	9,717	49.8	27.9	14.1	3.5	15,586	54.9	30.8
1985–1986	18.5	4.6	9,717	44.9	25.2	11.8	3.0	15,586	46.0	25.7
1986–1987	20.8	5.2	9,717	50.5	28.3	12.3	3.1	15,586	47.9	26.8
1987–1988	22.6	5.7	9,717	54.9	30.7	15.9	4.0	15,586	62.0	34.7
1988–1989	28.4	7.1	9,717	69.0	38.6	17.7	4.4	15,586	69.0	38.6
1989–1990	23.3	5.8	9,717	56.6	31.7	15.7	3.9	15,586	61.2	34.3
1990–1991	29.1	7.3	9,717	70.7	39.6	14.7	3.7	15,586	57.3	32.1
1991–1992	24.5	6.1	9,717	59.5	33.3	18.5	4.6	15,586	72.1	40.4
1992–1993	24.8	6.2	9,717	60.2	33.7	18.4	4.6	15,586	71.7	40.1
1993–1994	22.8	5.7	9,717	55.4	31.0	16.8	4.2	15,586	65.5	36.7
1994–1995	25.3	6.3	9,717	61.5	34.4	16.3	4.1	15,586	63.5	35.6

Table A2.d Item 7 — User Cost of Fishery Resources (Continued)

| | Abelone (Fully Exploited) | | | | | Scallops (Overexploited) | | | | |
Year	Annual Harvest (Kilotonnes) m	Net Decline in Abelone Stocks ($m \times 0.25$) (Kilotonnes) n	Value of Abelone per Tonne ($) p	Net Decline in Value of Abelone Stocks ($m at 1989–1990 Prices) q	User Cost of Rock Abelone Stocks ($q \times 0.56$) ($m) r	Annual Harvest (Kilotonnes) s	Net Decline in Scallops Stocks ($s \times 0.5$) (Kilotonnes) t	Value of Scallops per Tonne ($) u	Net Decline in Value of Scallops Stocks ($m at 1989–1990 Prices) v	User Cost of Scallop Stocks ($v \times 0.56$) ($m) w
1966–1967	7.0	1.8	17,902	31.3	17.5	6.0	3.0	4,048	12.1	6.8
1967–1968	7.0	1.8	17,902	31.3	17.5	7.0	3.5	4,048	14.2	7.9
1968–1969	7.0	1.8	17,902	31.3	17.5	8.0	4.0	4,048	16.2	9.1
1969–1970	7.0	1.8	17,902	31.3	17.5	9.0	4.5	4,048	18.2	10.2
1970–1971	7.7	1.9	17,902	34.5	19.3	9.3	4.7	4,048	18.8	10.5
1971–1972	8.0	2.0	17,902	35.8	20.1	10.1	5.1	4,048	20.4	11.4
1972–1973	6.4	1.6	17,902	28.6	16.0	17.0	8.5	4,048	34.4	19.3
1973–1974	6.0	1.5	17,902	26.9	15.0	12.4	6.2	4,048	25.1	14.1
1974–1975	5.0	1.3	17,902	22.4	12.5	6.1	3.1	4,048	12.3	6.9
1975–1976	5.3	1.3	17,902	23.7	13.3	4.6	2.3	4,048	9.3	5.2
1976–1977	6.9	1.7	17,902	30.9	17.3	4.4	2.2	4,048	8.9	5.0
1977–1978	5.1	1.3	17,902	22.8	12.8	9.1	4.6	4,048	18.4	10.3
1978–1979	6.2	1.6	17,902	27.7	15.5	10.5	5.3	4,048	21.3	11.9
1979–1980	6.4	1.6	17,902	28.6	16.0	16.4	8.2	4,048	33.2	18.6
1980–1981	6.9	1.7	17,902	30.9	17.3	9.8	4.9	4,048	19.8	11.1
1981–1982	7.7	1.9	17,902	34.5	19.3	15.4	7.7	4,048	31.2	17.5
1982–1983	6.1	1.5	17,902	27.3	15.3	30.5	15.3	4,048	61.7	34.6
1983–1984	8.5	2.1	17,902	38.0	21.3	29.7	14.9	4,048	60.1	33.7
1984–1985	7.7	1.9	17,902	34.5	19.3	19.2	9.6	4,048	38.9	21.8
1985–1986	7.0	1.8	17,902	31.3	17.5	16.2	8.1	4,048	32.8	18.4
1986–1987	6.7	1.7	17,902	30.0	16.8	11.7	5.9	4,048	23.7	13.3
1987–1988	6.8	1.7	17,902	30.4	17.0	6.0	3.0	4,048	12.1	6.8
1988–1989	5.5	1.4	17,902	24.6	13.8	5.4	2.7	4,048	10.9	6.1
1989–1990	5.1	1.3	17,902	22.8	12.8	6.3	3.2	4,048	12.8	7.1
1990–1991	5.2	1.3	17,902	23.3	13.0	14.5	7.3	4,048	29.3	16.4
1991–1992	5.0	1.3	17,902	22.4	12.5	29.8	14.9	4,048	60.3	33.8
1992–1993	4.7	1.2	17,902	21.0	11.8	33.6	16.8	4,048	68.0	38.1
1993–1994	4.7	1.2	17,902	21.0	11.8	24.6	12.3	4,048	49.8	27.9
1994–1995	5.1	1.3	17,902	22.8	12.8	12.7	6.4	4,048	25.7	14.4

Table A2.d Item 7 — User Cost of Fishery Resources (Continued)

	Oysters (Increasing)					Fish — Excluding Tuna (Fully Exploited) Tuna (overexploited)				
Year	Annual Harvest (Kilotonnes)	Net Decline in Oyster Stocks (Kilotonnes)	Value of Oysters per Tonne ($)	Net Decline in Value of Oyster Stocks ($m at 1989-1990 Prices)	User Cost ($aa \times 0.56$ Where aa Is Positive) ($m)	Annual Harvest (Kilotonnes)	Net Decline in Fish Stocks ($cc \times 0.25$) (Kilotonnes)	Value of Fish per Tonne ($)	Net Decline in Value of Fish Stocks ($m at 1989-1990 Prices)	User Cost of Fish Stocks ($ff \times 0.56$) ($m)
	x	y	z	aa	bb	cc	dd	ee	ff	gg
1966–1967	6.0	1.5	5,258	7.9	4.4	36.8	9.2	2,297	21.1	11.8
1967–1968	7.0	1.8	5,258	9.2	5.2	38.8	9.7	2,297	22.3	12.5
1968–1969	8.0	2.0	5,258	10.5	5.9	40.8	10.2	2,297	23.4	13.1
1969–1970	9.0	2.3	5,258	11.8	6.6	42.8	10.7	2,297	24.6	13.8
1970–1971	9.8	2.5	5,258	12.9	7.2	44.8	11.2	2,297	25.7	14.4
1971–1972	10.4	2.6	5,258	13.7	7.7	46.8	11.7	2,297	26.9	15.0
1972–1973	9.2	2.3	5,258	12.1	6.8	45.8	11.5	2,297	26.3	14.7
1973–1974	10.5	2.6	5,258	13.8	7.7	56.0	14.0	2,297	32.2	18.0
1974–1975	8.9	2.2	5,258	11.7	6.6	46.3	11.6	2,297	26.6	14.9
1975–1976	10.3	0	5,258	0.0	0	44.3	11.1	2,297	25.4	14.2
1976–1977	10.8	0	5,258	0.0	0	48.5	12.1	2,297	27.9	15.6
1977–1978	9.8	0	5,258	0.0	0	50.8	12.7	2,297	29.2	16.3
1978–1979	6.7	0	5,258	0.0	0	50.2	12.6	2,297	28.8	16.1
1979–1980	8.4	0	5,258	0.0	0	43.6	10.9	2,297	25.0	14.0
1980–1981	8.3	0	5,258	0.0	0	58.0	14.5	2,297	33.3	18.7
1981–1982	7.8	0	5,258	0.0	0	57.9	14.5	2,297	33.2	18.6
1982–1983	7.8	0	5,258	0.0	0	58.5	14.6	2,297	33.6	18.8
1983–1984	7.0	0	5,258	0.0	0	64.0	16.0	2,297	36.8	20.6
1984–1985	7.5	0	5,258	0.0	0	73.0	18.3	2,297	41.9	23.5
1985–1986	6.8	-0.9	5,258	-4.5	-4.5	76.0	19.0	2,297	43.6	24.4
1986–1987	6.9	-0.9	5,258	-4.5	-4.5	85.0	21.3	2,297	48.8	27.3
1987–1988	7.5	-0.9	5,258	-4.9	-4.9	94.0	23.5	2,297	54.0	30.2
1988–1989	8.0	-1.0	5,258	-5.3	-5.3	97.6	24.4	2,297	56.0	31.4
1989–1990	7.5	-0.9	5,258	-4.9	-4.9	120.9	30.2	2,297	69.4	38.9
1990–1991	8.1	-1.0	5,258	-5.3	-5.3	145.4	36.4	2,297	83.5	46.8
1991–1992	8.2	-1.0	5,258	-5.4	-5.4	139.8	35.0	2,297	80.3	45.0
1992–1993	8.6	-1.1	5,258	-5.7	-5.7	135.0	33.8	2,297	77.5	43.4
1993–1994	8.8	-1.1	5,258	-5.8	-5.8	123.6	30.9	2,297	71.0	39.7
1994–1995	9.3	-1.2	5,258	-6.1	-6.1	129.3	32.3	2,297	74.3	41.6

Table A2.d Item 7 — User Cost of Fishery Resources (Continued)

			Tuna (Overexploited)			
Year	Annual Harvest (Kilotonnes) *hh*	Net Decline in Tuna Stocks (*hh* × 0.5) (Kilotonnes) *jj*	Value of Tuna per Tonne ($) *kk*	Net Decline in Value of Tuna Stocks ($m at 1989–1990 Prices) *mm*	User Cost of Fish Stocks (*mm* × 0.56) ($m) *nn*	User Cost of Fishery Resources ($m at 1989–1990 Prices) 7
1966–1967	6.0	3.0	8,013	24.04	13.5	88
1967–1968	6.0	3.0	8,013	24.04	13.5	96
1968–1969	6.0	3.0	8,013	24.04	13.5	103
1969–1970	6.0	3.0	8,013	24.04	13.5	111
1970–1971	6.8	3.4	8,013	27.24	15.3	121
1971–1972	10.2	5.1	8,013	40.87	22.9	129
1972–1973	13.6	6.8	8,013	54.49	30.5	139
1973–1974	9.7	4.9	8,013	38.86	21.8	136
1974–1975	11.1	5.6	8,013	44.47	24.9	115
1975–1976	10.7	5.4	8,013	42.87	24.0	111
1976–1977	10.1	5.1	8,013	40.47	22.7	120
1977–1978	12.3	6.2	8,013	49.28	27.6	125
1978–1979	11.3	5.7	8,013	45.27	25.4	132
1979–1980	13.6	6.8	8,013	54.49	30.5	141
1980–1981	18.2	9.1	8,013	72.92	40.8	159
1981–1982	21.7	10.9	8,013	86.94	48.7	169
1982–1983	21.3	10.7	8,013	85.34	47.8	186
1983–1984	15.8	7.9	8,013	63.30	35.4	177
1984–1985	13.0	6.5	8,013	52.08	29.2	152
1985–1986	13.5	6.8	8,013	54.09	30.3	137
1986–1987	11.8	5.9	8,013	47.28	26.5	134
1987–1988	10.5	5.3	8,013	42.07	23.6	138
1988–1989	8.6	4.3	8,013	34.46	19.3	143
1989–1990	8.2	4.1	8,013	32.85	18.4	138
1990–1991	11.7	5.9	8,013	46.88	26.3	169
1991–1992	13.2	6.6	8,013	52.89	29.6	189
1992–1993	10.2	5.1	8,013	40.87	22.9	184
1993–1994	7.7	3.9	8,013	30.85	17.3	159
1994–1995	7.9	4.0	8,013	31.65	17.7	150

Based on ABS estimates, the real 1989–1990 net present values of each category of fish stock are $9,717 per tonne of prawns, $15,586 per tonne of rock lobster, $17,902 per tonne of abalone, $4,048 per tonne of scallops, $5,258 per tonne of oysters, $2,297 per tonne of fish, and $8,013 per tonne of tuna.

These values were subsequently multiplied by the net increase/decrease in fishery stocks to determine the overall net change (positive/negative) in the value of total fishery stocks. As with timber stocks, an appropriate user cost adjustment was made in cases where fishery stocks declined. The data sources are the State of the Environment Advisory Council, (1996, Figure 8.4, page 8–14, and Table 8.11, page 8–29) and the Australian Bureau of Agricultural and Resource Economics, *Commodity Statistical Bulletin* (various articles).

Item 8 — *Cost of lost wetlands, mangroves, and saltmarshes* (Table A2.e).

Of all the different land types, wetlands, mangroves, and saltmarshes provide some of the most important natural capital services. Unfortunately,

Table A2.e   Item 8 — Cost of Lost Wetlands, Mangroves, and Saltmarshes

Year	Wetlands and Inland Waters			Mangroves			Saltmarshes			Total Cost of Lost Wetlands, Mangroves, and Saltmarshes $(c + f + j)$ ($m at 1989–1990 Prices)
	Annual Loss and/or Degradation of Wetlands (Hectares)	Cumulative Loss of Wetlands (Hectares)	Cumulative Cost of Lost Wetlands $(b \times \$3000)$ ($m)	Annual Loss of Mangroves (Hectares)	Cumulative Loss of Mangroves (Hectares)	Cumulative Cost of Lost Mangroves $(e \times \$10,000)$ ($m)	Annual Loss of Saltmarshes (Hectares)	Cumulative Loss of Saltmarshes (Hectares)	Cumulative Cost of Lost Saltmarshes $(h \times \$6,500)$ ($m)	
	a	b	c	d	e	f	g	h	j	8
Up to 1966–1967		7,057,000			108,422			216,511		
1966–1967	109,400	7,166,400	21,499	2,096	110,518	1,105	5,438	221,949	1,443	24,047
1967–1968	109,400	7,275,800	21,827	2,096	112,615	1,126	5,438	227,387	1,478	24,431
1968–1969	109,400	7,385,200	22,156	2,096	114,711	1,147	5,438	232,825	1,513	24,816
1969–1970	109,400	7,494,600	22,484	2,096	116,807	1,168	5,438	238,263	1,549	25,201
1970–1971	109,400	7,604,000	22,812	2,096	118,904	1,189	5,438	243,701	1,584	25,585
1971–1972	109,400	7,713,400	23,140	2,096	121,000	1,210	5,438	249,139	1,619	25,969
1972–1973	109,400	7,822,800	23,468	2,096	123,096	1,231	5,438	254,577	1,655	26,354
1973–1974	109,400	7,932,200	23,797	2,096	125,192	1,252	5,438	260,015	1,690	26,739
1974–1975	109,400	8,041,600	24,125	2,096	127,289	1,273	5,438	265,453	1,725	27,123
1975–1976	109,400	8,151,000	24,453	2,096	129,385	1,294	5,438	270,891	1,761	27,508
1976–1977	109,400	8,260,400	24,781	2,096	131,481	1,315	5,438	276,329	1,796	27,892
1977–1978	109,400	8,369,800	25,109	2,096	133,578	1,336	5,438	281,767	1,831	28,276
1978–1979	109,400	8,479,200	25,438	2,096	135,674	1,357	5,438	287,205	1,867	28,662
1979–1980	109,400	8,588,600	25,766	2,096	137,770	1,378	5,438	292,643	1,902	29,046
1980–1981	109,400	8,698,000	26,094	2,096	139,867	1,399	5,438	298,081	1,938	29,431
1981–1982	109,400	8,807,400	26,422	2,096	141,963	1,420	5,438	303,519	1,973	29,815
1982–1983	54,700	8,862,100	26,586	2,834	144,797	1,448	5,841	309,360	2,011	30,045
1983–1984	54,700	8,916,800	26,750	2,834	147,631	1,476	5,841	315,201	2,049	30,275
1984–1985	54,700	8,971,500	26,915	2,834	150,465	1,505	5,841	321,042	2,087	30,507
1985–1986	54,700	9,026,200	27,079	2,834	153,298	1,533	5,841	326,883	2,125	30,737
1986–1987	54,700	9,080,900	27,243	2,834	156,132	1,561	5,841	332,724	2,163	30,967
1987–1988	54,700	9,135,600	27,407	2,834	158,966	1,590	5,841	338,565	2,201	31,198
1988–1989	54,700	9,190,300	27,571	2,834	161,800	1,618	5,841	344,406	2,239	31,428
1989–1990	54,700	9,245,000	27,735	2,834	164,634	1,646	5,841	350,247	2,277	31,658
1990–1991	54,700	9,299,700	27,899	2,834	167,468	1,675	5,841	356,088	2,315	31,889
1991–1992	54,700	9,354,400	28,063	2,834	170,302	1,703	5,841	361,929	2,353	32,119
1992–1993	54,700	9,409,100	28,227	2,834	173,136	1,731	5,841	367,770	2,391	32,349
1993–1994	54,700	9,463,800	28,391	2,834	175,970	1,760	5,841	373,611	2,428	32,579
1994–1995	54,700	9,518,500	28,556	2,834	178,804	1,788	5,841	379,452	2,466	32,810

along with other forms of native vegetation, they have suffered from high rates of clearance, drainage, and/or severe degradation from increased sedimentation and nutrient insertion.

a. *Wetlands.*

Wetlands are defined as "areas of land that are flooded naturally or are inundated or waterlogged on a permanent, seasonal, or intermittent basis" (State of the Environment Advisory Council, 1996, page 7–26). According to Usback and James (1993), both the extent and condition of Australia's wetlands have deteriorated greatly since European settlement.

To calculate the cumulative cost of lost wetlands, use was made of a recent environmental report by the ABS (Catalogue No. 4601.0). This report indicates that the area of wetlands prior to European settlement was in the order of 190,370 km² or, equivalently, 19.037 million hectares. It also indicates that approximately 50 percent of all Australian wetlands have been lost since European settlement — an area of approximately 9,518,500 hectares. Of this area, it was assumed that one-half as much again had been lost since 1945. The assumption was made in line with the belief that one-half of all native vegetation clearance and other forms of significant vegetation disturbance in Australia had occurred since World War 2 (Graetz et al., 1995). It was assumed that the rate of loss between 1945 and 1981–1982 was twice that of the period 1981–1982 to 1994–1995. This amounts to an annual loss of wetlands in the order of 109,400 hectares between 1966–1967 and 1981–1982, and 54,700 hectares between 1981–1982 and 1994–1995. Based on the ABS Report's findings and our assumptions, the cumulative loss of wetlands was then calculated by working back from the estimated cumulative impact as of 1994–1995.

It was assumed that the value of each hectare of wetlands was $3000 at 1989–1990 prices. This value compares very conservatively with the estimated $6000 to $7000 per hectare value for wetlands in the United States by Daly and Cobb (1989) and by Redefining Progress (1995) in the calculation of the Genuine Progress Indicator (GPI). The data sources are the Australian Bureau of Statistics, Catalogue No. 4601.0, Table 12.2.2.9, page 356, Daly and Cobb (1989, pages 432 and 433), Redefining Progress (1995, pages 28 and 29), and Paijmans et al. (1995).

b. *Mangroves.*

To calculate the cumulative cost of lost mangroves, use was made of Table 8.7 of the recent State of the Environment Report for Australia (1996, p. 8–22). The table indicates that, as of 1982, the total area of mangroves in Australia stood at around 1,157,600 hectares. Of this total, approximately one million hectares remained in northern bioregions where human settlement and manipulation of natural capital has been less extensive than in southern bioregions. It was then assumed that the loss of mangroves prior to

1994–1995 in northern bioregions was 100,000 hectares (10 percent of the 1982 total), of which 75 percent of the total loss (75,000 hectares) had occurred since 1945. The 75 percent assumption since 1945 reflects the rapid growth of human settlement and activity in the northern parts of Australia since World War II. More particularly, it was assumed that the rate of loss of mangroves in these regions from 1981–1982 onward occurred at twice the 1945 to 1981–1982 rate. All told, the annual loss of mangroves in northern bioregions was estimated to be 1190.5 hectares between 1966–1967 and 1981–1982 and 2381 hectares since 1981–1982.

As for mangroves in the southern bioregions, it was conservatively estimated that one-third of all mangroves in existence prior to white settlement have been lost (approximately 78,800 hectares). Of this, it was assumed that, as in the case of native bushland, most of the mangroves lost since 1945 were lost before 1981–1982. Indeed, we assumed that mangroves were lost in the southern bioregions at a rate of 905.8 hectares per year between 1945 and 1981–1982 and at a rate of 452.9 hectares per year since 1981–1982.

In the final analysis, the annual rate of loss of all mangroves in Australia was estimated at 2096 hectares between 1966–1967 and 1981–1982 and 2834 hectares between 1981–1982 and 1994–1995. The cumulative loss of mangroves then was determined by working back from the estimated cumulative impact as of 1994–1995.

Because of the critical role played by mangroves in aquatic ecosystems, the value of each hectare of mangroves was assumed to be higher than that of wetlands, indeed, $10,000 per hectare at 1989–1990 prices. The data sources are the State of the Environment Advisory Council, (1996, Table 8.7, page 8–22) and Galloway et al. (1984).

c. *Saltmarshes.*

The calculation of the cumulative cost of lost saltmarshes was based on Table 8.6 of the State of the Environment Report for Australia (1996, page 8–22). Along similar lines to mangroves, the distribution of saltmarshes was divided into northern and southern bioregions. Once again, the majority of Australia's saltmarshes are located in northern bioregions, in fact, 1,748,100 of Australia's 2,332,200 hectares.

The percentage rates of saltmarsh lost in both bioregions was assumed to be the same as with mangroves. This meant an approximate annual loss of saltmarshes in northern bioregions of 2081.1 hectares between 1966–1967 and 1981–1982 and of 4162.1 hectares between 1981–1982 and 1994–1995 and, for the southern bioregions, 3356.9 hectares between 1966–1967 and 1981–1982 and 1678.5 hectares thereafter. The combined annual loss of Australian saltmarshes, therefore, amounted to 5438 hectares during the period 1966–1967 to 1981–1982 and 5841 hectares between 1981–1982 and 1994–1995. As with wetlands and mangroves, the cumulative loss of saltmarshes was determined by working back from the estimated cumulative impact as of 1994–1995.

It was assumed that the value of a hectare of saltmarsh was greater than that of the average wetland but less than that of mangroves. Hence, the average value of a hectare of saltmarsh was estimated to be $6500 at 1989–1990 prices. The data sources are State of the Environment Advisory Council, (1996, Table 8.6, page 8–22) and Galloway et al. (1984).

Item 9 — *Cost of sacrificed source function of natural capital.*

Equal to the sum of Items 4A to 8.

A2.3 Sacrificed sink function of natural capital

Item 10 — Cost of water pollution (Table A2.f).

To estimate the cost of water pollution, a water pollution index was constructed along similar lines to the noise and air pollution indexes previously calculated. Like air and noise pollution, water pollution is closely related to the rate of increase in GDP. Critical legislation concerning water pollution in Australia was enacted at all levels of government in the early 1970s, in the early 1980s, and again in the early 1990s. It was, therefore, assumed that water pollution increased at the rate of growth in GDP until 1972–1973. Following 1972–1973, it was assumed that water pollution increased at three-fourths the rate of increase in GDP between 1972–1973 and 1982–1983, one-half the rate between 1982–1983 and 1991–1992, and one-fourth the rate thereafter.

In a study by the ABS on water pollution control costs, it was estimated that, in 1991–1992, Australians spent $1515 million at 1989–1990 prices defending themselves against the impact of water pollution. To estimate the full cost of water pollution, the water pollution damage cost was assumed to be three times the control cost. This made the total cost of water pollution in 1991–1992 a conservative $6060 million at 1989–1990 prices. To calculate the total cost of water pollution in other years, the $6060 million was adjusted by the water pollution index. The data sources are the Australian Bureau of Statistics, Catalogue No. 4603.0 and Fabricius (1994).

Item 11 — *Cost of air pollution* (Table A2.g).

The total cost of air pollution has been estimated previously in the calculation of the direct disamenity cost of air pollution (see Item R, Appendix 1). To recall, six major categories of air pollution-related costs were identified, four of which can be associated with the loss of the ecosphere's sink function. These were

- The cost of the damage to agricultural vegetation and crops.
- The cost of materials damage.
- The cost of cleaning soiled and damaged goods.
- The cost of acid rain damage to aquatic and forest resources.

Table A2.f Item 10 — Cost of Water Pollution

Year	Index of Real GDP \a	Percentage Change in Real GDP (%) \b	Percentage Change in Real GDP (×0.75) (%) \c	Percentage Change in Real GDP (×0.5) (%) \d	Percentage Change in Real GDP (×0.25) (%) \e	Pollution Index \f	Water Pollution Control Costs (1515×f)/100 ($m) \g	Water Pollution Damage Costs (g×3) ($m) \h	Cost of Water Pollution (g+h) ($m at 1989–1990 Prices) \10
1966–1967	56.5	6.6				62.7	950	2,851	3,801
1967–1968	58.6	3.7				65.1	986	2,957	3,943
1968–1969	63.8	8.9				70.8	1,073	3,219	4,293
1969–1970	67.3	5.5				74.7	1,132	3,396	4,528
1970–1971	70.6	4.9				78.4	1,188	3,563	4,750
1971–1972	73.4	4.0				81.5	1,235	3,704	4,939
1972–1973	76.2	3.8				84.6	1,282	3,845	5,127
1973–1974	77.9	2.2	1.7			86.0	1,303	3,910	5,213
1974–1975	78.7	1.0	0.8			86.7	1,313	3,940	5,253
1975–1976	79.9	1.5	1.1			87.7	1,328	3,985	5,313
1976–1977	81.0	1.4	1.0			88.6	1,342	4,026	5,368
1977–1978	81.4	0.5	0.4			88.9	1,347	4,041	5,388
1978–1979	83.7	2.8	2.1			90.8	1,375	4,126	5,502
1979–1980	84.5	1.0	0.7			91.4	1,385	4,156	5,541
1980–1981	86.0	1.8	1.3			92.7	1,404	4,211	5,615
1981–1982	87.0	1.2	0.9			93.5	1,416	4,248	5,664
1982–1983	86.3	-0.8	-0.6			92.9	1,408	4,223	5,630
1983–1984	88.9	3.0		1.5		94.3	1,429	4,286	5,715
1984–1985	91.1	2.5		1.2		95.5	1,446	4,339	5,786
1985–1986	92.8	1.9		0.9		96.4	1,460	4,380	5,840
1986–1987	94.0	1.3		0.6		97.0	1,469	4,408	5,877
1987–1988	96.3	2.4		1.2		98.2	1,487	4,462	5,949
1988–1989	98.6	2.4		1.2		99.3	1,505	4,515	6,020
1989–1990	100.1	1.5		0.8		100.1	1,517	4,550	6,066
1990–1991	99.8	-0.3		-0.2		99.9	1,513	4,540	6,054
1991–1992	100.0	0.2		0.1		100.0	1,515	4,545	6,060
1992–1993	101.8	1.8			0.4	100.5	1,522	4,565	6,087
1993–1994	104.2	2.4			0.6	101.0	1,531	4,592	6,123
1994–1995	106.3	2.0			0.5	101.6	1,538	4,615	6,154

Table A2-g Item 11 — Cost of Air Pollution (Sacrificed Sink Function)

Year	Index of Real GDP *a*	Percent. Change in Real GDP (%) *b*	Percent. Change in Real GDP (× 0.5) (%) *c*	Air Pollution Index *d*	Air Pollution Control Costs (504.4 × d)/100 ($m) *e*	Air Pollution Damage Costs (e × 10) ($m) *f*	Total Cost of Air Pollution (e + f) ($m at 1989–1990 Prices) *g*	Cost of Air Pollution (Degraded Sink Function of Ecosphere) (0.6 × g) ($m at 1989–1990 Prices) 11
1966–1967	56.5	6.6		75.5	380.9	3,808.6	4,189	2,514
1967–1968	58.6	3.7		78.3	395.0	3,950.1	4,345	2,607
1968–1969	63.8	8.9		85.3	430.1	4,300.6	4,731	2,838
1969–1970	67.3	5.5		89.9	453.7	4,536.6	4,990	2,994
1970–1971	70.6	4.9		94.3	475.9	4,759.0	5,235	3,141
1971–1972	73.4	4.0		98.1	494.8	4,947.8	5,443	3,266
1972–1973	76.2	3.8		101.8	513.6	5,136.5	5,650	3,390
1973–1974	77.9	2.2	1.1	103.0	519.4	5,193.8	5,713	3,428
1974–1975	78.7	1.0	0.5	103.5	522.0	5,220.5	5,743	3,446
1975–1976	79.9	1.5	0.8	104.3	526.0	5,260.3	5,786	3,472
1976–1977	81.0	1.4	0.7	105.0	529.6	5,296.5	5,826	3,496
1977–1978	81.4	0.5	0.2	105.3	531.0	5,309.6	5,841	3,504
1978–1979	83.7	2.8	1.4	106.8	538.5	5,384.6	5,923	3,554
1979–1980	84.5	1.0	0.5	107.3	541.0	5,410.3	5,951	3,571
1980–1981	86.0			106.2	535.7	5,356.7	5,892	3,535
1981–1982	87.0			106.2	535.7	5,356.7	5,892	3,535
1982–1983	86.3			106.2	535.7	5,356.7	5,892	3,535
1983–1984	88.9			106.2	535.7	5,356.7	5,892	3,535
1984–1985	91.1			106.2	535.7	5,356.7	5,892	3,535
1985–1986	92.8			106.2	535.4	5,354.3	5,890	3,534
1986–1987	94.0			105.1	530.1	5,301.3	5,831	3,499
1987–1988	96.3			104.1	524.9	5,248.8	5,774	3,464
1988–1989	98.6			103.0	519.7	5,196.8	5,717	3,430
1989–1990	100.1			102.0	514.5	5,145.4	5,660	3,396
1990–1991	99.8			101.0	509.4	5,094.4	5,604	3,362
1991–1992	100.0			100.0	504.4	5,044.0	5,548	3,329
1992–1993	101.8			99.0	499.4	4,994.1	5,493	3,296
1993–1994	104.2			98.0	494.5	4,944.6	5,439	3,263
1994–1995	106.3			97.1	489.6	4,895.7	5,385	3,231

According to Freeman (1982), approximately 60 percent of the total cost of air pollution can be apportioned to the four previously mentioned air pollution-related costs. Although it is likely that such a percentage would vary between 1966–1967 and 1994–1995, the 60 percent share was assumed to have remained constant over the study period. The data sources are Item R, Appendix 1 and Freeman (1982).

Item 12 — *Cost of solid waste pollution* (Table A2.h).

Table A2.h Item 12 — Cost of Solid Waste Pollution

Year	Total Waste to Landfill (Teragram per Year) *a*	Solid Waste Pollution Index *b*	Solid Waste Pollution Control Costs (1017.6×*b*)/100 ($m) *c*	Solid Waste Pollution Damage Costs (*c* × 3) ($m) *d*	Cost of Solid Waste Pollution (*c* + *d*) ($m at 1989–1990 Prices) 12
1966–1967	7.61	57.0	580.5	1,741.5	2,322
1967–1968	7.83	58.7	597.3	1,791.9	2,389
1968–1969	8.04	60.3	613.3	1,839.9	2,453
1969–1970	8.32	62.4	634.7	1,904.0	2,539
1970–1971	8.71	65.3	664.4	1,993.2	2,658
1971–1972	8.78	65.8	669.8	2,009.3	2,679
1972–1973	8.98	67.3	685.0	2,055.0	2,740
1973–1974	9.18	68.8	700.3	2,100.8	2,801
1974–1975	9.53	71.4	727.0	2,180.9	2,908
1975–1976	9.53	71.4	727.0	2,180.9	2,908
1976–1977	9.71	72.8	740.7	2,222.1	2,963
1977–1978	9.89	74.1	754.4	2,263.3	3,018
1978–1979	10.08	75.6	768.9	2,306.8	3,076
1979–1980	10.28	77.1	784.2	2,352.5	3,137
1980–1981	10.50	78.7	801.0	2,402.9	3,204
1981–1982	10.78	80.8	822.3	2,467.0	3,289
1982–1983	11.00	82.5	839.1	2,517.3	3,356
1983–1984	11.22	84.1	855.9	2,567.6	3,424
1984–1985	11.45	85.8	873.4	2,620.3	3,494
1985–1986	11.71	87.8	893.3	2,679.8	3,573
1986–1987	11.98	89.8	913.9	2,741.6	3,655
1987–1988	12.26	91.9	935.2	2,805.6	3,741
1988–1989	12.56	94.2	958.1	2,874.3	3,832
1989–1990	12.84	96.3	979.5	2,938.4	3,918
1990–1991	13.10	98.2	999.3	2,997.9	3,997
1991–1992	13.34	100.0	1,017.6	3,052.8	4,070
1992–1993	13.58	101.8	1,035.9	3,107.7	4,144
1993–1994	13.81	103.5	1,053.5	3,160.4	4,214
1994–1995	14.00	104.9	1,067.9	3,203.8	4,272

To calculate the cost of solid waste pollution, it was first necessary to devise a solid waste pollution index. This was achieved on the basis that the full impact of solid waste pollution is a function of the total quantity of solid waste that makes its way to landfill sites. Second, the ABS estimate of the control cost of solid waste pollution, $1.02 billion in 1991–1992 at 1989–1990 prices, was used as a partial estimate of its total uncancelled cost. The solid waste pollution damage cost was assumed to be three times the control cost. This made the total cost of solid waste pollution $4.07 billion in 1991–1992.

To calculate the cost of solid waste pollution in other years, the $4.07 billion was adjusted by the solid waste pollution index. The data sources are the National Greenhouse Gas Inventory Committee, (1996, Table A2, page 73), the Australian Bureau of Statistics, Catalogue No. 4603.0, and Fabricius (1994).

Item 13 — *Cost of ozone depletion* (Table A2.j).

Table A2.j Item 13 — Cost of Ozone Depletion

Year	Cumulative World Production of CFC-11 and CFC-12 (Millions of Kilograms) *a*	Cumulative Cost to Australia from Ozone Depletion ($0.75 × *a*) ($m at 1989–1990 Prices) 13
1966–1967	2,880	2,160
1967–1968	3,320	2,490
1968–1969	3,800	2,850
1969–1970	4,350	3,263
1970–1971	4,880	3,660
1971–1972	5,610	4,208
1972–1973	6,340	4,755
1973–1974	7,170	5,738
1974–1975	8,050	6,038
1975–1976	8,800	6,600
1976–1977	9,600	7,200
1977–1978	10,360	7,770
1978–1979	11,090	8,318
1979–1980	11,790	8,843
1980–1981	12,470	9,353
1981–1982	13,160	9,870
1982–1983	13,800	10,350
1983–1984	14,500	10,875
1984–1985	15,250	11,438
1985–1986	16,000	12,000
1986–1987	16,810	12,608
1987–1988	17,680	13,260
1988–1989	18,530	13,898
1989–1990	19,270	14,453
1990–1991	19,770	14,828
1991–1992	20,240	15,180
1992–1993	20,670	15,503
1993–1994	21,060	15,795
1994–1995	21,380	16,035

The method used here is that employed by Redefining Progress (1995). To calculate the cost of ozone depletion, it is necessary to estimate the amount that needs to be raised to compensate future generations for the long-term impact of declining ozone. To do this, the cumulative world production of CFC-11 and CFC-12 was multiplied by $1 per kilogram. The world production level was used in recognition of the fact that ozone depletion is a global problem. The data sources are Redefining Progress (1995) and the Subcommittee on Health and Environment (1987).

Item 14 — *Cost of sacrificed sink function of natural capital.*

Equal to the sum of Items 10 to 13.

A2.4 Sacrificed life-support function of natural capital

Item 15 — *Cost of long-term environmental damage* (Table A2.k).

Table A2.k   Item 15 — Cost of Long-Term Environmental Damage

Year	Total Energy Consumption (Petajoules) *a*	Total Energy Consumption — Crude Oil Barrel Equivalent (a/6.12 × 10-6) (Millions of Barrels) *b*	Environmental Damage Cost ($2.00 × b) ($m at 1989–1990 prices) *c*	Cumulative Cost of Long-Term Environmental Damage ($m at 1989–1990 Prices) 15
Up to 1966–1967		6,500.0		13,000
1966–1967	1,805.8	295.1	590	13,590
1967–1968	1,898.9	310.3	621	14,211
1968–1969	2,025.9	331.0	662	14,873
1969–1970	2,137.6	349.3	699	15,572
1970–1971	2,210.3	361.2	722	16,294
1971–1972	2,331.2	380.9	762	17,056
1972–1973	2,447.8	400.0	800	17,856
1973–1974	2,615.1	427.3	855	18,711
1974–1975	2,694.5	440.3	881	19,592
1975–1976	2,730.6	446.2	892	20,484
1976–1977	2,905.6	474.8	950	21,434
1977–1978	2,982.7	487.4	975	22,409
1978–1979	3,050.9	498.5	997	23,406
1979–1980	3,130.2	511.5	1,023	24,429
1980–1981	3,146.1	514.1	1,028	25,457
1981–1982	3,236.5	528.8	1,058	26,515
1982–1983	3,122.9	510.3	1,021	27,536
1983–1984	3,220.4	526.2	1,052	28,588
1984–1985	3,369.6	550.6	1,101	29,689
1985–1986	3,403.0	556.0	1,112	30,801
1986–1987	3,514.8	574.3	1,149	31,950
1987–1988	3,622.3	591.9	1,184	33,134
1988–1989	3,832.1	626.2	1,252	34,386
1989–1990	3,945.2	644.6	1,289	35,675
1990–1991	3,946.6	644.9	1,290	36,965
1991–1992	4,003.2	654.1	1,308	38,273
1992–1993	4,079.2	666.5	1,333	39,606
1993–1994	4,174.2	682.1	1,364	40,970
1994–1995	4,269.2	697.6	1,395	42,365

Over time, humankind has disrupted ecological systems, exacted resources from the ecosphere's stocks of low entropy, and inserted highly durable and toxic wastes into its waste absorbing sinks. However, much of the cost of these previous actions has yet to fully materialise. Hence, the long-term cost will be inevitably borne by future generations. To gain an approximate measure of the cost of long-term environmental damage, Daly and Cobb's (1989) method was adopted. Daly and Cobb assumed that the long-term damage of ecological disruption is directly proportional to the consumption of energy resources.

To ascertain the cost of long-term environmental damage, the annual consumption of energy for Australia (total energy consumption) was converted to a crude oil barrel equivalent and then multiplied by $2.50 per barrel. As with the previous item, the eventual cost represents the amount that needs to be accumulated to compensate future generations for the long-term impact of environmental damage.

Because the long-term environmental damage of human activity is a cumulative process, two additional steps were required. First, it was necessary to add the annual environmental impact cost to the cost previously incurred in order to compile a cumulative running total. Second, it was necessary to make an assumption regarding the quantity of energy consumed up to 1966–1967. This was conservatively estimated to be 6.5 billion barrels of crude oil meaning, therefore, a cumulative cost up to 1966–1967 of $16.25 billion. The data sources are the Australian Bureau of Statistics, Catalogue No. 4604.0, Table 3.4, page 21, and the Bureau of Resource Economics, (1987, page 73).

Item 16 — *Final uncancelled costs* (unweighted).

Item 16 is the sum of Items 9, 14 and 15.

Item 17 — *Ecosystem health index* (Table A2.m).

The life-support function of the ecosphere and, thus, its overall state of health, is very much a function of the biodiversity found within it. While items 1 to 16 are able to capture the loss of the ecosphere's source and sink services, they do not capture the impact of the loss of biodiversity on the ecosphere's life-support function. As a consequence, an ecosystem health index was calculated as a means of weighting the annual uncancelled costs of human exploitation of the ecosphere.

Unfortunately, there is no inventory of biodiversity levels for the years 1966–1967 to 1994–1995. Attempts to measure and record Australia's biodiversity are still in the infancy stage. To overcome this deficiency, a ecosystem health index was constructed on the basis that habitat loss and, more particularly, remnant vegetation loss, remains the "greatest threat to biodiversity" (Biodiversity Unit, 1995, page 8). It is believed that around one-half of all native vegetation clearance and other forms of significant vegetation disturbance in Australia has occurred since 1945 [Graetz et al. (1995)]. Because the ability to accurately estimate the clearance of native vegetation is confined to the period since 1981–1982, it was necessary to approximate the average loss of native vegetation between 1945 and 1981–1982. Based on

Table A2.m Item 17 — Ecosystem Health Index

Year	Area of Remnant Vegetation Significantly Disturbed (Annually) (km²) a	Remaining Area of Native Veg. Insignificantly Disturbed (km²) b	Ecosystem Health Index (1966–1967 = 100.0) 17
1966–1967	30,000	5,534,530	100.0
1967–1968	30,000	5,504,530	99.5
1968–1969	30,000	5,474,530	98.9
1969–1970	30,000	5,444,530	98.4
1970–1971	30,000	5,414,530	97.8
1971–1972	30,000	5,384,530	97.3
1972–1973	30,000	5,354,530	96.7
1973–1974	30,000	5,324,530	96.2
1974–1975	30,000	5,294,530	95.7
1975–1976	30,000	5,264,530	95.1
1976–1977	30,000	5,234,530	94.6
1977–1978	30,000	5,204,530	94.0
1978–1979	30,000	5,174,530	93.5
1979–1980	30,000	5,144,530	93.0
1980–1981	30,000	5,114,530	92.4
1981–1982	30,000	5,084,530	91.9
1982–1983	7,500	5,077,030	91.7
1983–1984	7,500	5,069,530	91.6
1984–1985	7,500	5,062,030	91.5
1985–1986	7,500	5,054,530	91.3
1986–1987	7,500	5,047,030	91.2
1987–1988	10,500	5,036,530	91.0
1988–1989	7,500	5,029,030	90.9
1989–1990	9,750	5,019,280	90.7
1990–1991	7,500	5,011,780	90.6
1991–1992	7,500	5,004,280	90.4
1992–1993	7,500	4,996,780	90.3
1993–1994	7,500	4,989,280	90.1
1994–1995	9,750	4,979,530	90.0

data gathered by Graetz et al. (1995, page 37), this amounted to an annual loss of 20,000 km² of native vegetation per year. Because vegetation thinning also has an effect on biodiversity, it was assumed that the rate of vegetation thinning was one-half as much as that of wholesale clearance (10,000 km² of native vegetation per year). All told, this meant that in each year between 1945 and 1981–1982, an area of approximately 30,000 km² of vegetation was significantly disturbed.

Thanks to the introduction of vegetation clearance legislation in some states in the early 1980s, the extent of "significant" vegetation disturbance has declined markedly since 1981–1982. Nevertheless, it still remains extremely high by international standards — an indictment on the rate of vegetation disturbance in Australia (Biodiversity Unit, 1995).

The ecosystem health index was finally constructed by giving the financial year 1966–1967 a base index value of 100.0 and adjusting it in accordance with the annual change in the area of Australia which remained relatively undisturbed. Over the study period, the ecosystem health index fell from the base index value of 100.0 to a value of 90.0. This represented a 10 percent reduction in the life-support potential of Australia's natural capital. The data source is the Biodiversity Unit, (1995) and Graetz, R. et al. (1995).

Item 18 — *Final uncancelled costs.*

Item 16 was weighted by the ecosystem health index (Item 17). This was achieved by dividing the uncancelled costs for each year by the index and multiplying by 100.

Item 19 — *Australian population.*

Indicated in the thousands. The data source is Foster (1996, Table 4.1, page 177).

Item 20 — *Per capita uncancelled costs.*

Item 18 divided by Item 19.

appendix three

The human-made capital account for Australia, 1966–1967 to 1994–1995

A3.1 Introduction

Appendix 3 reveals the methods used to calculate monetary values for the items appearing in the human-made capital account. Tables revealing how some of the items have been calculated are located in this appendix. All items are measured in millions of Australian dollars at 1989–1990 prices. The human-made capital account for Australia appears in Table 14.4 in Chapter 14 and includes the following human-made producer and consumer goods.

- Public sector dwellings, nondwelling construction, and equipment.
- Inventories (private nonfarm, farm, and public marketing and other public authorities).
- Inventories (livestock).
- Land (residential and commercial).
- Labour (capitalised value).
- Stock of household dwellings.
- Stock of consumer durables.

A3.2 Human-made producer and consumer goods

Items A, B and C — *Private business sector dwellings, nondwelling construction, and equipment.*

The data source is Foster (1996, Table 5.23, page 274).

Items D, E and F — *Public sector dwellings, nondwelling construction, and equipment.*

The data source is Foster (1996, Table 5.23, page 274).

Items G, H and I — *Inventories (private nonfarm, farm, and public marketing and other public authorities).*

The data sources are Foster (1996, Table 5.2a, page 221) and the Australian Bureau of Statistics, Catalogue No. 5204.0, (various articles).

Item J — *Inventories (statistical discrepancy).*

The data sources are Foster (1996, Table 5.2a, page 221) and the Australian Bureau of Statistics, Catalogue No. 5204.0, (various articles).

Item K — *Inventories (livestock)* (Table A3.a).

Table A3.a   Item K — Livestock (Inventories)

Year	Cattle (Thousands)	Per Head Value of Cattle ($)	Total Value of Beef Cattle ($m)	Lambs (Thousands)	Per Head Value of Lambs ($)	Total Value of Lambs ($m)	Value of Livestock — Inventories (c + f) ($m at 1989–1990 Prices)
	a	b	c	d	e	f	K
1966–1967	4,914	978.35	4,808	11,580	71.00	822	5,630
1967–1968	5,303	978.35	5,188	12,075	71.00	857	6,046
1968–1969	5,861	978.35	5,734	12,225	71.00	868	6,602
1969–1970	6,455	978.35	6,315	12,675	71.00	900	7,215
1970–1971	7,294	978.35	7,136	13,073	71.00	928	8,064
1971–1972	8,410	978.35	8,228	12,878	71.00	914	9,142
1972–1973	9,043	978.35	8,847	11,708	71.00	831	9,678
1973–1974	9,749	978.35	9,538	10,058	71.00	714	10,252
1974–1975	10,465	978.35	10,238	10,463	71.00	743	10,981
1975–1976	10,739	978.35	10,507	10,935	71.00	776	11,283
1976–1977	10,177	978.35	9,957	10,808	71.00	767	10,724
1977–1978	9,457	978.35	9,252	9,848	71.00	699	9,951
1978–1979	8,730	978.35	8,541	9,563	71.00	679	9,220
1979–1980	8,413	978.35	8,231	9,788	71.00	695	8,926
1980–1981	8,053	978.35	7,879	9,953	71.00	707	8,585
1981–1982	7,837	978.35	7,667	9,855	71.00	700	8,367
1982–1983	7,099	978.35	6,945	10,125	71.00	719	7,664
1983–1984	6,970	978.35	6,819	9,788	71.00	695	7,514
1984–1985	7,193	978.35	7,037	10,133	71.00	719	7,757
1985–1986	6,880	978.35	6,731	10,928	71.00	776	7,507
1986–1987	6,941	978.35	6,791	11,280	71.00	801	7,592
1987–1988	6,937	978.35	6,787	11,490	71.00	816	7,603
1988–1989	7,160	978.35	7,005	11,430	71.00	812	7,817
1989–1990	7,441	978.35	7,280	12,120	71.00	861	8,140
1990–1991	7,643	978.35	7,478	12,773	71.00	907	8,384
1991–1992	7,722	978.35	7,555	12,240	71.00	869	8,424
1992–1993	7,654	978.35	7,488	11,115	71.00	789	8,277
1993–1994	7,585	978.35	7,421	10,875	71.00	772	8,193
1994–1995	7,517	978.35	7,354	10,500	71.00	746	8,100

Livestock can be divided into two categories. The first category is livestock raised specifically for slaughter, in other words, livestock that is in the process of production and therefore considered as work in progress. The second category is livestock used repeatedly or continuously over periods of more than one year in order to produce other goods (such as wool, milk, etc.). The former category has been chosen to be part of the human-made capital account. The latter category, which can be regarded as fixed natural capital assets, has been included in the natural capital account.

Because of the lack of appropriate data, estimates of livestock inventories were confined to beef and dairy cattle, as well as sheep and lambs. Pigs, poultry, deer, goats, horses and greyhounds were excluded. It should be noted that cattle and sheep constitute the majority of total livestock numbers and livestock value in Australia.

The real 1989–1990 (inventory) price values of beef cattle and lambs were derived from a recent study by the ABS on livestock. To calculate the total value of such inventories, the quantities of beef cattle and lambs were multiplied by their respective per head value. The data sources are Australian Bureau of Statistics, Catalogue No. 7221.0 (various articles), the Australian Bureau of Statistics, Catalogue No. 7102.0 (various articles), the Australian Bureau of Agricultural and Resource Economics, *Commodity Statistical Bulletin* (various articles), and the Australian Bureau of Statistics, Catalogue No. 5241.0, Table 5.2 and Table 5.3, page 107, and Table 5.4 and 5.5, page 108.

Items L and M — *Land (residential and commercial)* (Table A3.b).

The calculation of the real 1989–1990 value of residential and commercial land begins with an estimation of the combined residential and commercial land area in Australia for each year over the study period. This has already been achieved in the estimation of lost agricultural land arising from increasing urbanisation (see Item 5 in Appendix 2). It then was necessary to determine the total area of both residential and commercial land separately.

In the derivation of Item 5 in Appendix 2, it was assumed that three distinct types of regions exist in Australia — major urban centres (which includes 13 major urban cities), minor urban centres, and rural areas. To estimate the area of residential and commercial land in the three major regions, it was assumed that the following ratio of residential to commercial land exists in each.

- Major urban centres — 70 percent residential and 30 percent commercial.
- Minor urban centres — 82.5 percent residential and 17.5 percent commercial.
- Rural areas — 95 percent residential and 5 percent commercial.

Each of the preceding percentages were multiplied by the annual combined residential and commercial land area in the three regions to ascertain

Table A3.b Items L and M — Real Value of Residential and Commercial Land

Year	Population of Major Urban Centres	Population Density (Persons Per Hectare)	Land Area of Major Urban Centres (Hectares) (a/b)	Resid. Land Area of Major Urban Centres (Hectares) (c × 0.7)	Real Ave. Value of Resid. Land per Hectare at 1989–1990 Prices	Total Value of Resid. Land in Major Urban Centres (d × e) ($m)	Comm. Land Area of Major Urban Centres (c × 0.3) (Hectares)	Real Average Value of Comm. Land per Hectare at 1989–1990 Prices	Total Value of Commercial Land in Major Urban Centres (g × h) ($m)
	a	b	c	d	e	f	g	h	j
1966–1967	7,545,924	14	538,995	377,296	398,479	150,345	161,698	341,168	55,166
1967–1968	7,766,265	14	554,733	388,313	398,479	154,735	166,420	341,168	56,777
1968–1969	7,993,040	14	570,931	399,652	398,479	159,253	171,279	341,168	58,435
1969–1970	8,226,437	14	587,603	411,322	398,479	163,903	176,281	341,168	60,141
1970–1971	8,466,649	14	604,761	423,332	398,479	168,689	181,428	341,168	61,897
1971–1972	8,753,668	14	625,262	437,683	398,479	174,408	187,579	341,168	63,996
1972–1973	9,050,417	14	646,458	452,521	398,479	180,320	193,938	341,168	66,165
1973–1974	9,357,227	14	668,373	467,861	398,479	186,433	200,512	341,168	68,408
1974–1975	9,674,437	14	691,031	483,722	398,479	192,753	207,309	341,168	70,727
1975–1976	10,002,400	14	714,457	500,120	398,479	199,287	214,337	341,168	73,125
1976–1977	10,112,426	14	722,316	505,621	398,479	201,479	216,695	341,168	73,929
1977–1978	10,223,663	14	730,262	511,183	398,479	203,696	219,078	341,168	74,743
1978–1979	10,336,123	14	738,295	516,806	398,479	205,936	221,488	341,168	75,565
1979–1980	10,449,821	14	746,416	522,491	398,479	208,202	223,925	341,168	76,396
1980–1981	10,565,900	14	754,707	528,295	398,479	210,514	226,412	341,168	77,245
1981–1982	10,696,917	14	764,066	534,846	398,479	213,125	229,220	341,168	78,202
1982–1983	10,829,559	14	773,540	541,478	398,479	215,768	232,062	341,168	79,172
1983–1984	10,964,200	14	783,157	548,210	398,479	218,450	234,947	341,168	80,156
1984–1985	11,084,899	14	791,779	554,245	398,479	220,855	237,534	341,168	81,039
1985–1986	11,324,853	14	808,918	566,243	398,479	225,636	242,675	341,168	82,793
1986–1987	11,505,569	14	821,826	575,278	398,479	229,236	246,548	341,168	84,114
1987–1988	11,704,512	14	836,037	585,226	398,479	233,200	250,811	341,168	85,569
1988–1989	11,899,565	14	849,969	594,978	398,479	237,086	254,991	341,168	86,995
1989–1990	12,077,391	14	862,671	603,870	398,479	240,629	258,801	341,168	88,295
1990–1991	12,213,873	14	872,420	610,694	398,479	243,349	261,726	341,168	89,292
1991–1992	12,334,572	14	881,041	616,729	398,479	245,753	264,312	341,168	90,175
1992–1993	12,435,600	14	888,257	621,780	398,479	247,766	266,477	341,168	90,913
1993–1994	12,536,849	14	895,489	626,842	398,479	249,784	268,647	341,168	91,654
1994–1995	12,638,923	14	902,780	631,946	398,479	251,817	270,834	341,168	92,400

Table A3.b Items L and M — Real Value of Residential and Commercial Land (Continued)

Year	Population of Minor Urban Centres k	Population Density (Persons per Hectare) m	Land Area of Minor Urban Centres (k/m) (Hectares) n	Resid. Land Area of Minor Urban Centres $(n \times 0.825)$ (Hectares) p	Real Ave. Value of Resid. Land per Hectare at 1989-1990 Prices q	Total Value of Resid. Land in Minor Urban Centres $(p \times q)$ ($m) r	Comm. Land Area of Minor Urban Centres $(n \times 0.175)$ (Hectares) s	Real Average Value of Comm. Land per Hectare at 1989-1990 Prices t	Total Value of Commercial Land in Major Urban Centres $(s \times t)$ ($m) u
1966-1967	2,294,442	14	163,889	135,208	398,479	53,878	28,681	341,168	9,785
1967-1968	2,321,295	14	165,807	136,791	398,479	54,508	29,016	341,168	9,899
1968-1969	2,369,195	14	169,228	139,613	398,479	55,633	29,615	341,168	10,104
1969-1970	2,404,513	14	171,751	141,695	398,479	56,462	30,056	341,168	10,254
1970-1971	2,718,703	14	194,193	160,209	398,479	63,840	33,984	341,168	11,594
1971-1972	2,634,556	14	188,183	155,251	398,479	61,864	32,932	341,168	11,235
1972-1973	2,523,368	14	180,241	148,698	398,479	59,253	31,542	341,168	10,761
1973-1974	2,417,107	14	172,651	142,437	398,479	56,758	30,214	341,168	10,308
1974-1975	2,259,650	14	161,404	133,158	398,479	53,061	28,246	341,168	9,637
1975-1976	2,065,980	14	147,570	121,745	398,479	48,513	25,825	341,168	8,811
1976-1977	2,092,694	14	149,478	123,319	398,479	49,140	26,159	341,168	8,925
1977-1978	2,110,718	14	150,766	124,382	398,479	49,563	26,384	341,168	9,001
1978-1979	2,133,121	14	152,366	125,702	398,479	50,090	26,664	341,168	9,097
1979-1980	2,158,489	14	154,178	127,197	398,479	50,685	26,981	341,168	9,205
1980-1981	2,223,111	14	158,794	131,005	398,479	52,203	27,789	341,168	9,481
1981-1982	2,315,771	14	165,412	136,465	398,479	54,378	28,947	341,168	9,876
1982-1983	2,347,705	14	167,693	138,347	398,479	55,128	29,346	341,168	10,012
1983-1984	2,371,424	14	169,387	139,745	398,479	55,685	29,643	341,168	10,113
1984-1985	2,413,841	14	172,417	142,244	398,479	56,681	30,173	341,168	10,294
1985-1986	2,354,519	14	168,180	138,748	398,479	55,288	29,431	341,168	10,041
1986-1987	2,383,887	14	170,278	140,479	398,479	55,978	29,799	341,168	10,166
1987-1988	2,413,816	14	172,415	142,243	398,479	56,681	30,173	341,168	10,294
1988-1989	2,459,591	14	175,685	144,940	398,479	57,756	30,745	341,168	10,489
1989-1990	2,471,377	14	176,527	145,635	398,479	58,032	30,892	341,168	10,539
1990-1991	2,529,379	14	180,670	149,053	398,479	59,394	31,617	341,168	10,787
1991-1992	2,583,545	14	184,539	152,245	398,479	60,666	32,294	341,168	11,018
1992-1993	2,624,968	14	187,498	154,686	398,479	61,639	32,812	341,168	11,194
1993-1994	2,678,965	14	191,355	157,868	398,479	62,907	33,487	341,168	11,425
1994-1995	2,761,139	14	197,224	162,710	398,479	64,837	34,514	341,168	11,775

Table A3.b Items L and M — Real Value of Residential and Commercial Land (Continued)

Year	Australian Population (Thousands) v	% of Aust. Population in Rural Areas w	Rural Population x	Pop. Density y	Land Area in Rural Areas (x/y) z	Resid. Land Area in Rural Areas $(z \times 0.95)$ (Hectares) aa	Real Ave. Value of Resid. Land per Hectare at 1989–1990 Prices bb	Total Value of Resid. Land in Rural Areas $(aa \times bb)$ ($m) cc	Comm. Land Area in Rural Areas $(z \times 0.05)$ (Hectares) dd	Real Ave. Value of Comm. Land per Hectare at 1989–1990 Prices ee	Total Value of Commercial Land in Rural Areas $(dd \times ee)$ ($m) ff
1966–1967	11,799	16.6	1,958,634	14	139,902	132,907	398,479	52,961	6,995	341,168	2,387
1967–1968	12,009	16.0	1,921,440	14	137,246	130,383	398,479	51,955	6,862	341,168	2,341
1968–1969	12,263	15.5	1,900,765	14	135,769	128,980	398,479	51,396	6,788	341,168	2,316
1969–1970	12,507	15.0	1,876,050	14	134,004	127,303	398,479	50,728	6,700	341,168	2,286
1970–1971	13,067	14.4	1,881,648	14	134,403	127,683	398,479	50,879	6,720	341,168	2,293
1971–1972	13,304	14.4	1,915,776	14	136,841	129,999	398,479	51,802	6,842	341,168	2,334
1972–1973	13,505	14.3	1,931,215	14	137,944	131,047	398,479	52,219	6,897	341,168	2,353
1973–1974	13,723	14.2	1,948,666	14	139,190	132,231	398,479	52,691	6,960	341,168	2,374
1974–1975	13,893	14.1	1,958,913	14	139,922	132,926	398,479	52,968	6,996	341,168	2,387
1975–1976	14,003	14.0	1,960,420	14	140,030	133,029	398,479	53,009	7,002	341,168	2,389
1976–1977	14,192	14.0	1,986,880	14	141,920	134,824	398,479	53,725	7,096	341,168	2,421
1977–1978	14,359	14.1	2,024,619	14	144,616	137,385	398,479	54,745	7,231	341,168	2,467
1978–1979	14,516	14.1	2,046,756	14	146,197	138,887	398,479	55,344	7,310	341,168	2,494
1979–1980	14,695	14.2	2,086,690	14	149,049	141,597	398,479	56,423	7,452	341,168	2,543
1980–1981	14,923	14.3	2,133,989	14	152,428	144,806	398,479	57,702	7,621	341,168	2,600
1981–1982	15,184	14.3	2,171,312	14	155,094	147,339	398,479	58,712	7,755	341,168	2,646
1982–1983	15,394	14.4	2,216,736	14	158,338	150,421	398,479	59,940	7,917	341,168	2,701
1983–1984	15,579	14.4	2,243,376	14	160,241	152,229	398,479	60,660	8,012	341,168	2,733
1984–1985	15,788	14.5	2,289,260	14	163,519	155,343	398,479	61,901	8,176	341,168	2,789
1985–1986	16,018	14.6	2,338,628	14	167,045	158,693	398,479	63,236	8,352	341,168	2,850
1986–1987	16,264	14.6	2,374,544	14	169,610	161,130	398,479	64,207	8,481	341,168	2,893
1987–1988	16,532	14.6	2,413,672	14	172,405	163,785	398,479	65,265	8,620	341,168	2,941
1988–1989	16,814	14.6	2,454,844	14	175,346	166,579	398,479	66,378	8,767	341,168	2,991
1989–1990	17,065	14.7	2,508,555	14	179,183	170,223	398,479	67,830	8,959	341,168	3,057
1990–1991	17,284	14.7	2,540,748	14	181,482	172,408	398,479	68,701	9,074	341,168	3,096
1991–1992	17,489	14.7	2,570,883	14	183,635	174,453	398,479	69,516	9,182	341,168	3,133
1992–1993	17,656	14.7	2,595,432	14	185,388	176,119	398,479	70,180	9,269	341,168	3,162
1993–1994	17,838	14.7	2,622,186	14	187,299	177,934	398,479	70,903	9,365	341,168	3,195
1994–1995	18,054	14.7	2,653,938	14	189,567	180,089	398,479	71,762	9,478	341,168	3,234

Table A3.b Items L and M — Real Value of Residential and
Commercial Land (Continued)

Year	Total Real Value of All Residential Land $(f+r+cc)$ ($m at 1989–1990 Prices) L	Total Real Value of All Commercial Land $(j+u+ff)$ ($m at 1989–1990 Prices) M
1966–1967	257,183	67,338
1967–1968	261,198	69,018
1968–1969	266,282	70,855
1969–1970	271,093	72,682
1970–1971	283,408	75,784
1971–1972	288,074	77,565
1972–1973	291,793	79,280
1973–1974	295,882	81,091
1974–1975	298,782	82,751
1975–1976	300,809	84,324
1976–1977	304,344	85,275
1977–1978	308,004	86,211
1978–1979	311,369	87,156
1979–1980	315,310	88,144
1980–1981	320,419	89,325
1981–1982	326,215	90,724
1982–1983	330,836	91,885
1983–1984	334,796	93,003
1984–1985	339,437	94,122
1985–1986	344,160	95,684
1986–1987	349,421	97,174
1987–1988	355,146	98,804
1988–1989	361,220	100,475
1989–1990	366,700	101,900
1990–1991	371,444	103,175
1991–1992	375,935	104,325
1992–1993	379,585	105,270
1993–1994	383,593	106,273
1994–1995	388,415	107,409

the area of residential and commercial land in each region over the study period.

Land values for residential and commercial land in Australia have been estimated for the years 1983–1984 to 1991–1992 by the Australian Valuation Office (Australian Bureau of Statistics, Catalogue No. 5241.0, page 28). Because 1989–1990 constant price values are required, the 1989–1990 nominal values for residential and commercial land were divided by the total land area for both in order to derive the average 1989–1990 per hectare value of each category. Both were then multiplied by the estimated area of residential and commercial land in each year in order to ascertain the real value of residential and commercial land over the study period.

Incidentally, the average 1989–1990 value of residential land came to $398,479 per hectare, or around $41,500 per quarter-acre block. Commercial land had an approximate average 1989–1990 value of $341,168 per hectare, or an equivalent value of around $35,500 per quarter-acre block. The data sources include those listed in Item 5, Appendix 2, and the Australian Bureau of Statistics, Catalogue No. 5241.0, Table 2.6, page 28.

Items N, O, P and Q — *Labour (capitalised value)* (Table A3.c).

To calculate the capitalised value of labour, the following present value (capitalisation) formula was used (Chiang, 1974, page 463):

$$K_L = \frac{W}{r+d}\left(1 \pm e^{\pm rt}\right) \qquad \text{(A3.1)}$$

where

- K_L equals the capitalised value of each unit of labour.
- W equals the average annual wage earned by labour.
- r equals the rate of discount.
- d equals the depreciation rate of worker knowledge and skill.
- t equals the average wage earning time period of all labour.

By assuming a discount rate of 5 percent, a depreciation rate of 10 percent, and an average wage earning time period for labour of 25 years, the preceding formula became

$$K_L = \frac{W}{0.15}\left(1 \pm e^{\pm 1.25}\right) \qquad \text{(A3.2)}$$

$$\approx 5\,W \qquad \text{(A3.3)}$$

This means the capitalised value of all forms of labour is approximately five times its annual gross wage or opportunity cost value.

a. *Paid labour force.*

The capitalised value of the paid labour force is equal to 5 times the number of paid/employed labour multiplied by the average real yearly wage (52 times the average weekly wage).

b. *Unemployed labour.*

The pool of unemployed labour must be included in the capital account if only because it is necessary to gain an understanding of the net psychic income derived from the stock of all potential labour, not just working labour.

Table A3.c   Items N, O, P, and Q — Capitalised Value of Labour

	Labour Force								Unpaid Labour				
Year	Number of Paid Employed (Thousands) a	Average Weekly Wage ($) b	Average Yearly Wage (b × 52) ($) c	Cap. Value of Paid LF [5 × (a × c)] ($b) N	Number of U/ed d	Ave. Jun. Female Weekly Wage ($) e	Average Yearly Wage (e × 52) ($) f	Cap. Value of U/ed Labour [5 × (d × f)] ($m) O	Yearly Value of Household Labour ($m) g	Cap. Value of Household Labour (5 × g) ($m) P	Yearly Value of Vol. Labour ($m) h	Cap. Value of Vol. Labour (5 × h) ($m) Q	Total Cap. Value of Labour (P + Q + R + S) ($m at 1989–1990 Prices) j
1966–1967	4,933	359.6	18,699	461,216	87	156.3	8,127.6	3,536	119,111	595,555	1,197	5,985	1,066,291
1967–1968	5,056	367.4	19,105	482,969	81	160.0	8,320.0	3,370	133,504	667,520	1,353	6,765	1,160,624
1968–1969	5,183	378.1	19,661	509,520	79	159.0	8,268.0	3,266	137,150	685,750	1,403	7,015	1,205,551
1969–1970	5,396	379.4	19,729	532,283	78	169.2	8,798.4	3,431	138,300	691,500	1,430	7,150	1,234,364
1970–1971	5,516	392.0	20,384	562,191	93	175.8	9,141.6	4,251	150,297	751,485	1,571	7,855	1,325,782
1971–1972	5,610	406.4	21,133	592,775	144	186.3	9,687.6	6,975	154,411	772,055	1,638	8,190	1,379,995
1972–1973	5,783	411.1	21,377	618,122	106	192.8	10,025.6	5,314	163,423	817,115	1,759	8,795	1,449,345
1973–1974	5,855	418.9	21,783	637,697	141	202.9	10,550.8	7,438	195,650	978,250	2,139	10,695	1,634,081
1974–1975	5,841	453.7	23,592	689,016	278	235.7	12,256.4	17,036	184,879	924,395	2,055	10,275	1,640,722
1975–1976	5,898	449.2	23,358	688,839	293	252.3	13,119.6	19,220	193,485	967,425	2,207	11,035	1,686,519
1976–1977	5,995	464.3	24,144	723,704	359	265.1	13,785.2	24,744	197,055	985,275	2,303	11,515	1,745,239
1977–1978	6,005	475.8	24,742	742,867	398	267.1	13,889.2	27,640	201,544	1,007,720	2,414	12,070	1,790,296
1978–1979	6,079	469.9	24,435	742,696	378	261.6	13,603.2	25,710	199,508	997,540	2,452	12,260	1,778,206
1979–1980	6,281	459.4	23,889	750,228	395	254.0	13,208.0	26,086	199,381	996,905	2,514	12,570	1,785,789
1980–1981	6,394	471.0	24,492	783,009	381	258.5	13,442.0	25,607	205,461	1,027,305	2,661	13,305	1,849,226
1981–1982	6,379	451.6	23,483	748,997	461	248.8	12,937.6	29,821	207,278	1,036,390	2,759	13,795	1,829,003
1982–1983	6,241	464.6	24,159	753,888	687	268.8	13,977.6	48,013	208,997	1,044,985	2,918	14,590	1,861,476
1983–1984	6,466	466.0	24,232	783,421	604	263.2	13,686.4	41,333	211,829	1,059,145	3,105	15,525	1,899,423
1984–1985	6,676	475.1	24,705	824,660	573	264.2	13,738.4	39,361	214,577	1,072,885	3,301	16,505	1,953,410
1985–1986	6,919	471.6	24,523	848,380	598	258.9	13,462.8	40,254	214,859	1,074,295	3,464	17,320	1,980,249
1986–1987	7,092	471.0	24,492	868,486	602	257.1	13,369.2	40,241	229,602	1,148,010	3,883	19,415	2,076,153
1987–1988	7,353	462.9	24,071	884,963	539	248.5	12,922.0	34,825	228,422	1,142,110	4,044	20,220	2,082,118
1988–1989	7,715	455.6	23,691	913,888	468	241.0	12,532.0	29,325	227,285	1,136,425	4,214	21,070	2,100,708
1989–1990	7,808	457.2	23,774	928,153	585	251.5	13,078.0	38,253	228,809	1,144,045	4,445	22,225	2,132,676
1990–1991	7,629	475.8	24,742	943,768	799	265.6	13,811.2	55,176	232,672	1,163,360	4,771	23,855	2,186,159
1991–1992	7,618	477.4	24,825	945,577	898	261.9	13,618.8	61,148	238,871	1,194,355	5,104	25,520	2,226,600
1992–1993	7,621	473.8	24,638	938,816	916	266.4	13,852.4	63,446	243,207	1,216,035	5,453	27,265	2,245,562
1993–1994	7,886	484.7	25,204	993,809	798	268.7	13,972.4	55,750	248,357	1,241,785	5,847	29,235	2,320,579
1994–1995	8,218	494.1	25,693	1,055,734	722	265.0	13,780.0	49,746	249,720	1,248,600	6,174	30,870	2,384,949

In the calculation of the capitalised value of unemployed labour, it was assumed that the weekly gross opportunity cost value was equivalent to the average junior female weekly wage — the lowest of all average weekly wages. This assumption was made because unemployed labour is invariably less productive than employed labour. For example, evidence suggests that prolonged periods of unemployment lead to a loss and, in some cases, the obsolescence of previously acquired skills.

To calculate the capitalised value of unemployed labour, the total number of unemployed Australians was multiplied by the average junior female weekly wage. It was next multiplied by 52 to obtain its yearly value. Finally, it was finally multiplied by a value of five as per Eq. (A3.3).

c. *Household labour.*

The annual real value of household labour has been calculated in the compilation of the uncancelled benefits account (Appendix 1, Item H). The estimated real value of household labour for each year over the study period was multiplied by five to obtain an equivalent measure of its capitalised value.

d. *Volunteer labour.*

Like household labour, volunteer labour has been calculated in Appendix 1 (Item I). To determine the capitalised value of volunteer labour, the estimated real value of volunteer labour was also multiplied by a value of five. The data sources are Foster (1996, Table 4.3, page 180; Table 4.17, page 207; Table 4.18, page 209; Table 4.19, page 220) and Chiang (1974)

Item R — *Stock of household dwellings.*

The stock of household dwellings was determined by subtracting the stock of private sector business dwellings from the total stock of private sector dwellings. The data sources are Foster (1996, Table 5.23, page 274), the Commonwealth Treasury of Australia, (1996, Table 1(b), page 50), and Davy (1996), Table 2, page 13).

Item S — Stock of consumer durables.

The data source is the Commonwealth Treasury of Australia, (1996, Table 1(b), page 50).

Item A1 — *Total human-made capital stock.*

Summing all items gives the total stock of service-yielding producer and consumer goods (human-made capital).

Item A2 — *Australian population.*

Indicated in the thousands, the data source is Foster (1996, Table 4.1, page 177).

Item A3 — *Average per capita ownership of human-made capital.*

The average per capita ownership of human-made capital equals Item A1 divided by Item A2.

appendix four

The natural capital account for Australia, 1966–1967 to 1994–1995

A4.1 Introduction

Appendix 4 reveals the methods used to calculate monetary values for the items appearing in the natural capital stock account. Tables revealing how the items have been calculated are located in this appendix. All items are measured in millions of Australian dollars at 1989–1990 prices. As in the previous three appendices, use has been made of the relevant price indexes to calculate constant real price measures of natural capital. The natural capital stock account for Australia appears in Table 14.6 in Chapter 14 and includes the following major natural capital assets.

   a. *Nonrenewable resources.*

   • Subsoil assets (mineral resources).
   • Agricultural land.

   b. *Renewable resources.*

   • Timber resource stocks.
   • Wetlands, mangroves, and saltmarshes.
   • Fishery stocks.
   • Livestock — fixed natural capital assets.
   • Water storage resources.

A4.2 Nonrenewable resources

Item 1 — *Agricultural land* (Table A4.a).

The annual area of agricultural land was multiplied by the 1989–1990 agricultural land price. This provides the real price estimate of agricultural land

Table A4.a   Item 1 — Real Value of Agricultural Land

Year	Total Area of Agricultural Land (Millions of Hectares) *a*	Real Average per Hectare Value of Agric. Land at 1989–1990 Prices ($) *b*	Total Real Value of Agricultural Land ($m at 1989–1990 Prices) 1
1966–1967	487.0	139	67,693
1967–1968	489.6	139	68,054
1968–1969	490.6	139	68,193
1969–1970	494.7	139	68,763
1970–1971	497.7	139	69,180
1971–1972	499.5	139	69,431
1972–1973	499.8	139	69,472
1973–1974	500.5	139	69,570
1974–1975	499.6	139	69,444
1975–1976	500.7	139	69,597
1976–1977	491.5	139	68,319
1977–1978	489.4	139	68,027
1978–1979	493.2	139	68,555
1979–1980	495.6	139	68,888
1980–1981	495.4	139	68,861
1981–1982	490.8	139	68,221
1982–1983	483.8	139	67,248
1983–1984	488.6	139	67,915
1984–1985	489.2	139	67,999
1985–1986	468.3	139	65,094
1986–1987	471.0	139	65,469
1987–1988	472.0	139	65,608
1988–1989	466.9	139	64,899
1989–1990	464.3	139	64,538
1990–1991	462.8	139	64,329
1991–1992	466.0	139	64,774
1992–1993	460.1	139	63,954
1993–1994	469.1	139	65,205
1994–1995	463.3	139	64,399

for each year over the study period. The data sources are the Australian Bureau of Statistics, Catalogue No. 4606.0, Table 2.10, page 23; the Australian Bureau of Statistics, Catalogue No. 5241.0, Table 2.7, page 28; the Australian Bureau of Statistics, Catalogue No. 7113.0 (various articles); and the Australian Bureau of Statistics, Catalogue No. 7102.0 (various articles).

Items 2, 3, 4 and 5 — *Subsoil assets.*

The real 1989–1990 prices of Australia's sub-soil assets was calculated using a gross price present value (PV) method. To do such, the following formula was used:

$$GPV = \sum_{t=1}^{T_A} \left(\frac{G_t \cdot Q_t}{(1+r)^t} \right) \tag{A4.1}$$

where:

- *GPV* equals the gross present value of the resource stock.
- T_A equals the average mine life over study period.
- G_t equals the gross resource price at 1989–1990.
- Q_t equals the average quantity of resource mined over study period.
- r equals the discount rate, assumed to be 5 percent.

Because G_t and Q_t effectively are constants and can be taken outside the summation sign, Eq. (A4.1) becomes

$$GPV = (G_t \cdot Q_t) \cdot \left(\frac{1 - (1 + r)^{-t_A}}{r} \right) \qquad (A4.2)$$

Once the present value of the resource stock was derived for 1989–1990, it was divided by its physical quantity as of 1989–1990 to obtain a per kilotonne value of each resource in 1989–1990 prices. The real value of each subsoil asset was then calculated by multiplying the 1989–1990 value by the amount of the stock at the end of each year in the study period. The gross price method was preferred to the net price method for reasons explained in the calculation of the user cost of nonrenewable resources.

The quantities we used to calculate the real 1989–1990 value of Australia's mineral resources differs from those used by the ABS in their balance sheet estimates (ABS, Catalogue No. 5241.0). The ABS used the quantity of economically demonstrated resources (EDR) as at the end of each financial year which varies considerably from year to year as resource prices and extraction costs fluctuate, and as mining technology improves. The ABS adopted this approach because their study aimed to estimate Australia's net financial worth between 1989 and 1992. This is not, however, our objective. Our objective is to estimate the stock of mineral and other natural capital resource assets. Monetary estimates are used simply as a convenient means to gaining a uniform and homogeneous measure of the natural capital stock. The quantities of individual mineral resource assets for each year over the study period was calculated by using the quantity of economically demonstrated resources as at 1989–1990 as the reference quantity, and by adding the quantity mined during the preceding years, and subtracting the quantity mined thereafter. The 1989–1990 economically demonstrated stock of resources, the 1989–1990 gross prices, the average mine lives as of 1989–1990, and the 1989–1990 gross present value prices for each of Australia's major minerals are as follows.

Item 2 — *metallic minerals* (Table A4.b).

Included are

- Bauxite: 5,600 megatonnes, $107 per tonne, 227.27 years, and $9.83 per tonne.

Table A4.b   Item 2: Real Value of Metallic Minerals

Year	Bauxite				Copper				Iron Ore			
	Bauxite Mined (Kt)	Identified Stock of Bauxite (Mt)	Per Tonne GPV of Bauxite ($)	Total Real Value of Bauxite Stocks ($m)	Copper Mined (Kt)	Identified Stock of Copper (Kt)	Per Tonne GPV of Copper ($)	Total Real Value of Copper Stocks ($m)	Iron Ore Mined (Kt)	Identified Stock of Iron Ore (Mt)	Per Tonne Value of GPV Ore ($)	Total Real Value of Iron Ore Stocks ($m)
1966–1967	4,244	6,134.0	9.83	60,297	111	6,704.9	1,924.05	12,901	11,071	16,527.5	2.85	47,103
1967–1968	4,955	6,129.1	9.83	60,249	92	6,704.8	1,924.05	12,900	17,314	16,510.2	2.85	47,054
1968–1969	7,921	6,121.1	9.83	60,171	109	6,704.7	1,924.05	12,900	26,632	16,483.6	2.85	46,978
1969–1970	9,256	6,111.9	9.83	60,080	131	6,704.6	1,924.05	12,900	38,576	16,445.0	2.85	46,868
1970–1971	12,733	6,099.1	9.83	59,955	158	6,704.4	1,924.05	12,900	51,188	16,393.8	2.85	46,722
1971–1972	13,697	6,085.4	9.83	59,820	177	6,704.3	1,924.05	12,899	62,063	16,331.7	2.85	46,545
1972–1973	14,702	6,070.7	9.83	59,675	186	6,704.1	1,924.05	12,899	64,401	16,267.3	2.85	46,362
1973–1974	18,545	6,052.2	9.83	59,493	220	6,703.9	1,924.05	12,899	84,828	16,182.5	2.85	46,120
1974–1975	22,205	6,030.0	9.83	59,275	251	6,703.6	1,924.05	12,898	96,950	16,085.6	2.85	45,844
1975–1976	19,755	6,010.2	9.83	59,081	219	6,703.4	1,924.05	12,898	97,651	15,987.9	2.85	45,566
1976–1977	22,806	5,987.4	9.83	58,856	218	6,703.2	1,924.05	12,897	93,255	15,894.7	2.85	45,300
1977–1978	24,642	5,962.8	9.83	58,614	222	6,703.0	1,924.05	12,897	95,932	15,798.7	2.85	45,026
1978–1979	25,541	5,937.3	9.83	58,363	222	6,702.7	1,924.05	12,896	83,134	15,715.6	2.85	44,789
1979–1980	27,629	5,909.6	9.83	58,092	238	6,702.5	1,924.05	12,896	91,717	15,623.9	2.85	44,528
1980–1981	27,179	5,882.4	9.83	57,824	242	6,702.2	1,924.05	12,895	95,534	15,528.3	2.85	44,256
1981–1982	25,441	5,857.0	9.83	57,574	231	6,702.0	1,924.05	12,895	84,661	15,443.7	2.85	44,014
1982–1983	23,625	5,833.4	9.83	57,342	245	6,701.8	1,924.05	12,895	87,694	15,356.0	2.85	43,765
1983–1984	24,372	5,809.0	9.83	57,103	262	6,701.5	1,924.05	12,894	71,040	15,284.9	2.85	43,562
1984–1985	31,537	5,777.5	9.83	56,793	236	6,701.3	1,924.05	12,894	94,406	15,190.5	2.85	43,293
1985–1986	31,839	5,745.6	9.83	56,480	260	6,701.0	1,924.05	12,893	92,900	15,097.6	2.85	43,028
1986–1987	33,168	5,712.5	9.83	56,153	248	6,700.8	1,924.05	12,893	94,015	15,003.6	2.85	42,760
1987–1988	35,142	5,677.3	9.83	55,808	233	6,700.5	1,924.05	12,892	101,748	14,901.9	2.85	42,470
1988–1989	37,335	5,640.0	9.83	55,441	238	6,700.3	1,924.05	12,892	96,064	14,805.8	2.85	42,197
1989–1990	39,983	5,600.0	9.83	55,048	296	6,700.0	1,924.05	12,891	105,810	14,700.0	2.85	41,895
1990–1991	41,831	5,558.2	9.83	54,637	330	6,699.7	1,924.05	12,891	110,509	14,589.5	2.85	41,580
1991–1992	39,855	5,518.3	9.83	54,245	320	6,699.4	1,924.05	12,890	117,673	14,471.8	2.85	41,245
1992–1993	41,320	5,477.0	9.83	53,839	378	6,699.0	1,924.05	12,889	112,115	14,359.7	2.85	40,925
1993–1994	42,159	5,434.8	9.83	53,424	420	6,698.6	1,924.05	12,888	120,534	14,239.2	2.85	40,582
1994–1995	42,655	5,392.2	9.83	53,005	410	6,698.1	1,924.05	12,888	128,493	14,110.7	2.85	40,215

Table A4.b — Item 2: Real Value of Metallic Minerals (Continued)

Year	Lead Mined (Kt)	Identified Stock of Lead (Kt)	Per Tonne GPV of Lead ($)	Total Real Value of Lead Stocks ($m)	Manganese Mined (Kt)	Identified Stock of Manganese (Kt)	Per Tonne GPV of Manganese ($)	Total Real Value of Manganese Stocks ($m)	Nickel Mined (Kt)	Identified Stock of Nickel (Kt)	Per Tonne GPV of Nickel ($)	Total Real Value of Nickel Stocks ($m)
1966-1967	371	20,570.0	702.70	14,455	318	144,721.0	36.65	5,304	2.0	3,001.3	4,457.69	13,379
1967-1968	382	20,188.0	702.70	14,186	558	144,163.0	36.65	5,284	2.6	3,001.3	4,457.69	13,379
1968-1969	389	19,799.0	702.70	13,913	744	143,419.0	36.65	5,256	4.6	3,001.3	4,457.69	13,379
1969-1970	453	19,346.0	702.70	13,594	903	142,516.0	36.65	5,223	11.2	3,001.3	4,457.69	13,379
1970-1971	457	18,889.0	702.70	13,273	751	141,765.0	36.65	5,196	29.8	3,001.3	4,457.69	13,379
1971-1972	404	18,485.0	702.70	12,989	1,051	140,714.0	36.65	5,157	35.5	3,001.2	4,457.69	13,379
1972-1973	396	18,089.0	702.70	12,711	1,165	139,549.0	36.65	5,114	35.5	3,001.2	4,457.69	13,379
1973-1974	403	17,686.0	702.70	12,428	1,522	138,027.0	36.65	5,059	40.1	3,001.1	4,457.69	13,378
1974-1975	375	17,311.0	702.70	12,164	1,522	136,505.0	36.65	5,003	45.9	3,001.1	4,457.69	13,378
1975-1976	408	16,903.0	702.70	11,878	1,555	134,950.0	36.65	4,946	75.8	3,001.0	4,457.69	13,378
1976-1977	397	16,506.0	702.70	11,599	2,154	132,796.0	36.65	4,867	82.5	3,000.9	4,457.69	13,377
1977-1978	432	16,074.0	702.70	11,295	1,389	131,407.0	36.65	4,816	85.9	3,000.8	4,457.69	13,377
1978-1979	400	15,674.0	702.70	11,014	1,257	130,150.0	36.65	4,770	82.4	3,000.7	4,457.69	13,377
1979-1980	422	15,252.0	702.70	10,718	1,724	128,426.0	36.65	4,707	69.7	3,000.7	4,457.69	13,376
1980-1981	398	14,854.0	702.70	10,438	2,020	126,406.0	36.65	4,633	74.3	3,000.6	4,457.69	13,376
1981-1982	388	14,466.0	702.70	10,165	1,449	124,957.0	36.65	4,580	74.3	3,000.5	4,457.69	13,376
1982-1983	455	14,011.0	702.70	9,846	1,127	123,830.0	36.65	4,538	87.6	3,000.4	4,457.69	13,375
1983-1984	481	13,530.0	702.70	9,508	1,370	122,460.0	36.65	4,488	76.6	3,000.4	4,457.69	13,375
1984-1985	441	13,089.0	702.70	9,198	1,849	120,611.0	36.65	4,420	76.9	3,000.3	4,457.69	13,375
1985-1986	498	12,591.0	702.70	8,848	2,003	118,608.0	36.65	4,347	85.8	3,000.2	4,457.69	13,374
1986-1987	445	12,146.0	702.70	8,535	1,649	116,959.0	36.65	4,287	76.7	3,000.1	4,457.69	13,374
1987-1988	489	11,657.0	702.70	8,191	1,854	115,105.0	36.65	4,219	74.6	3,000.1	4,457.69	13,374
1988-1989	462	11,195.0	702.70	7,867	1,986	113,119.0	36.65	4,146	58	3,000.1	4,457.69	13,373
1989-1990	495	10,700.0	702.70	7,519	2,119	111,000.0	36.65	4,068	63.1	3,000.0	4,457.69	13,373
1990-1991	565	10,135.0	702.70	7,122	1,932	109,068.0	36.65	3,997	76.8	2,999.9	4,457.69	13,373
1991-1992	579	9,556.0	702.70	6,715	1,470	107,598.0	36.65	3,943	67.6	2,999.9	4,457.69	13,372
1992-1993	580	8,976.0	702.70	6,307	1,285	106,313.0	36.65	3,896	57.9	2,999.8	4,457.69	13,372
1993-1994	510	8,466.0	702.70	5,949	2,092	104,221.0	36.65	3,820	65.4	2,999.7	4,457.69	13,372
1994-1995	505	7,961.0	702.70	5,594	1,985	102,236.0	36.65	3,747	75.9	2,999.7	4,457.69	13,372

Table A4.b — Item 2: Real Value of Metallic Minerals (Continued)

Year	Uranium				Zinc				Total Real Value of Metallic Mineral Stocks ($m at 1989–1990 Prices)
	Uranium Mined (Tonne)	Identified Stock of Uranium (Kt)	Per Kilogram GPV of Uranium ($)	Total Real Value of Uranium ($m)	Zinc Mined (Kt)	Identified Stock of Zinc (Kt)	Per Tonne GPV of Zinc ($)	Total Real Value of Zinc Stocks ($m)	
1966–1967	300	522.0	7.22	3,769	375	30,956.0	1,404.33	43,472	200,680
1967–1968	300	521.7	7.22	3,767	407	30,549.0	1,404.33	42,901	199,719
1968–1969	300	521.4	7.22	3,765	416	30,133.0	1,404.33	42,317	198,678
1969–1970	300	521.1	7.22	3,762	510	29,623.0	1,404.33	41,600	197,407
1970–1971	300	520.8	7.22	3,760	487	29,136.0	1,404.33	40,917	196,101
1971–1972	300	520.5	7.22	3,758	453	28,683.0	1,404.33	40,280	194,828
1972–1973	300	520.2	7.22	3,756	507	28,176.0	1,404.33	39,568	193,465
1973–1974	300	519.9	7.22	3,754	480	27,696.0	1,404.33	38,894	192,025
1974–1975	300	519.6	7.22	3,752	464	27,232.0	1,404.33	38,243	190,557
1975–1976	484	519.1	7.22	3,748	510	26,722.0	1,404.33	37,527	189,020
1976–1977	480	518.6	7.22	3,745	462	26,260.0	1,404.33	36,878	187,519
1977–1978	696	517.9	7.22	3,740	492	25,768.0	1,404.33	36,187	185,952
1978–1979	951	517.0	7.22	3,733	473	25,295.0	1,404.33	35,523	184,465
1979–1980	1,837	515.2	7.22	3,719	529	24,766.0	1,404.33	34,780	182,815
1980–1981	3,944	511.2	7.22	3,691	495	24,271.0	1,404.33	34,084	181,198
1981–1982	5,968	505.2	7.22	3,648	518	23,753.0	1,404.33	33,357	179,609
1982–1983	4,334	500.9	7.22	3,617	665	23,088.0	1,404.33	32,423	177,800
1983–1984	5,836	495.1	7.22	3,574	699	22,389.0	1,404.33	31,442	175,945
1984–1985	4,327	490.7	7.22	3,543	677	21,712.0	1,404.33	30,491	174,006
1985–1986	4,450	486.3	7.22	3,511	759	20,953.0	1,404.33	29,425	171,906
1986–1987	4,505	481.8	7.22	3,479	712	20,241.0	1,404.33	28,425	169,905
1987–1988	4,193	477.6	7.22	3,448	778	19,463.0	1,404.33	27,332	167,735
1988–1989	4,506	473.1	7.22	3,416	760	18,703.0	1,404.33	26,265	165,596
1989–1990	4,089	469.0	7.22	3,386	803	17,900.0	1,404.33	25,138	163,318
1990–1991	4,389	464.6	7.22	3,354	933	16,967.0	1,404.33	23,827	160,781
1991–1992	4,349	460.3	7.22	3,323	1023	15,944.0	1,404.33	22,391	158,124
1992–1993	2,704	457.6	7.22	3,304	1019	14,925.0	1,404.33	20,960	155,492
1993–1994	2,751	454.8	7.22	3,284	990	13,935.0	1,404.33	19,569	152,888
1994–1995	2,631	452.2	7.22	3,265	955	12,980.0	1,404.33	18,228	150,314

- Copper: 6,700 kilotonnes, $3,626 per tonne, 28.16 years, and $1,924.05 per tonne.
- Iron ore: 14,700 megatonnes, $25 per tonne, 185.07 years, and $2.85 per tonne.
- Lead: 10,700 megatonnes, $1,054 per tonne, 32.67 years, and $702.70 per tonne.
- Manganese ore: 111,000 megatonnes, $140 per tonne, 85.31 years, and $36.65 per tonne.
- Nickel: 3,000 kilotonnes, $12,468 per tonne, 51.34 years, and $4,457.69 per tonne.
- Uranium: 469 kilotonnes, $70 per kilogram, 206.14 years, and $7.22 per kilogram.
- Zinc: 17,900 kilotonnes, $2,378 per tonne, 36.96 years, and $1,404.33 per tonne.

Item 3 — *coal* (Table A4.c).

Included are

- Black coal: 58,000 megatonnes, $58 per tonne, 428.98 years, and $2.76 per tonne.
- Brown coal: 41,700 megatonnes, $58 per tonne, 1,254.93 years, and $0.93 per tonne.

Item 4 — *oil and gas* (Table A4.d).

Included are

- Petroleum (recoverable) — crude oil: 264,000 megalitres, $147 per kilolitre, 21.29 years, and $165.86 per litre.
- Petroleum — natural gas: 853,000 million m³, $59 per thousand m³, 76,416.49 years, and $15 per thousand m³.
- Petroleum — condensate: 78,000 megalitres, $147 per kilolitre, 59.4 years, and $54.34 per kilolitre.
- LPG: 106,000 megalitres, $76 per thousand m³, 50.24 years, and $35.02 per thousand m³.

Item 5 — *nonmetallic minerals* (Tables A4.e).

Included are

- Mineral sands — ilmenite: 80,700 kilotonnes, $83 per tonne, 69.13 years, and $23.19 per tonne.
- Mineral sands — rutile: 11,600 kilotonnes, $768 per tonne, 52.25 years, and $325.43 per tonne.
- Mineral sands — zircon: 18,000 kilotonnes, $658 per tonne, 54.3 years, and $267.86 per tonne.

Table A4.c   Item 3: Real Value of Coal Stocks

Year	Black Coal Mined (Kt)	Black Coal — Identified Stock of Black Coal (Mt)	Black Coal — Per Tonne GPV of Black Coal ($)	Black Coal — Total Real Value of Black Coal Stocks ($m)	Brown Coal Mined (Mt)	Brown Coal — Identified Stock of Brown Coal (Mt)	Brown Coal — Per Tonne GPV of Brown Coal ($)	Brown Coal — Total Real Value of Brown Coal Stocks ($m)	Total Real Value of Metallic Mineral Stocks ($m at 1989–1990 Prices)
1966–1967	35,950	53,490.1	2.76	147,633	15	42,408.0	0.93	39,439	187,072
1967–1968	39,060	53,451.1	2.76	147,525	15	42,393.0	0.93	39,425	186,950
1968–1969	44,860	53,406.2	2.76	147,401	15	42,378.0	0.93	39,412	186,813
1969–1970	50,410	53,355.8	2.76	147,262	15	42,363.0	0.93	39,398	186,660
1970–1971	52,970	53,302.8	2.76	147,116	19	42,344.0	0.93	39,380	186,496
1971–1972	58,860	53,244.0	2.76	146,953	20	42,324.0	0.93	39,361	186,315
1972–1973	66,190	53,177.8	2.76	146,771	21	42,303.0	0.93	39,342	186,112
1973–1974	66,580	53,111.2	2.76	146,587	23	42,280.0	0.93	39,320	185,907
1974–1975	72,210	53,039.0	2.76	146,388	24	42,256.0	0.93	39,298	185,686
1975–1976	77,130	52,961.9	2.76	146,175	30	42,226.0	0.93	39,270	185,445
1976–1977	85,960	52,875.9	2.76	145,937	32	42,194.0	0.93	39,240	185,178
1977–1978	87,790	52,788.1	2.76	145,695	32	42,162.0	0.93	39,211	184,906
1978–1979	91,900	52,696.2	2.76	145,442	32	42,130.0	0.93	39,181	184,622
1979–1980	90,200	52,606.0	2.76	145,193	33	42,097.0	0.93	39,150	184,343
1980–1981	106,200	52,499.8	2.76	144,899	32	42,065.0	0.93	39,120	184,020
1981–1982	110,200	52,389.6	2.76	144,595	35	42,030.0	0.93	39,088	183,683
1982–1983	120,340	52,269.3	2.76	144,263	40	41,990.0	0.93	39,051	183,314
1983–1984	129,440	52,139.8	2.76	143,906	30	41,960.0	0.93	39,023	182,929
1984–1985	145,140	51,994.7	2.76	143,505	40	41,920.0	0.93	38,986	182,491
1985–1986	163,620	51,831.1	2.76	143,054	40	41,880.0	0.93	38,948	182,002
1986–1987	182,430	51,648.6	2.76	142,550	43	41,837.0	0.93	38,908	181,459
1987–1988	167,690	51,481.0	2.76	142,087	40	41,797.0	0.93	38,871	180,959
1988–1989	184,050	51,296.9	2.76	141,579	50	41,747.0	0.93	38,825	180,404
1989–1990	196,900	51,100.0	2.76	141,036	47	41,700.0	0.93	38,781	179,817
1990–1991	204,680	50,895.3	2.76	140,471	49	41,651.0	0.93	38,735	179,207
1991–1992	218,250	50,677.1	2.76	139,869	51	41,600.0	0.93	38,688	178,557
1992–1993	222,510	50,454.6	2.76	139,255	48	41,552.0	0.93	38,643	177,898
1993–1994	221,500	50,233.1	2.76	138,643	50	41,502.0	0.93	38,597	177,240
1994–1995	237,230	49,995.8	2.76	137,988	50	41,452.0	0.93	38,550	176,539

Table A4.d Item 4: Real Value of Oil and Gas Stocks

Year	Petroleum (Crude Oil)				Petroleum (Condensate)				Petroleum (Natural Gas)			
	Petroleum (Crude Oil) Extracted (M/Litres)	Identified Stock of Petroleum (Crude Oil) (M/Litres)	Per Kilolitre GPV of Petroleum (Crude Oil) ($)	Total Real Value of Crude Oil Stocks ($m)	Petroleum (Condensate) Extracted (M/Litres)	Identified Stock of Petroleum (Condensate) (M/Litres)	Per Kilolitre GPV of Condensate ($)	Total Real Value of Petroleum (Condensate) ($m)	Petroleum (Natural Gas) Extracted (Million m³)	Identified Stock of Petroleum (Natural Gas) (Million m³)	Per Thousand m³ GPV of Petroleum (Natural Gas) ($)	Total Real Value of Petroleum (Natural Gas) ($m)
1966–1967	10,000	787,015.0	165.86	130,534	200	103,438.0	54.34	5,621	0.75	853,197.6	15.00	12,798
1967–1968	11,000	776,015.0	165.86	128,710	250	103,188.0	54.34	5,607	1.00	853,196.6	15.00	12,798
1968–1969	12,000	764,015.0	165.86	126,720	300	102,888.0	54.34	5,591	1.25	853,195.4	15.00	12,798
1969–1970	13,000	751,015.0	165.86	124,563	350	102,538.0	54.34	5,572	1.50	853,193.9	15.00	12,798
1970–1971	14,936	736,079.0	165.86	122,086	400	102,138.0	54.34	5,550	1.96	853,191.9	15.00	12,798
1971–1972	19,038	717,041.0	165.86	118,928	450	101,688.0	54.34	5,526	2.63	853,189.3	15.00	12,798
1972–1973	20,668	696,373.0	165.86	115,500	500	101,188.0	54.34	5,499	3.71	853,185.6	15.00	12,798
1973–1974	23,193	673,180.0	165.86	111,654	550	100,638.0	54.34	5,469	4.40	853,181.2	15.00	12,798
1974–1975	23,134	650,046.0	165.86	107,817	600	100,038.0	54.34	5,436	4.82	853,176.3	15.00	12,798
1975–1976	23,827	626,219.0	165.86	103,865	650	99,388.0	54.34	5,401	5.38	853,171.0	15.00	12,798
1976–1977	24,598	601,621.0	165.86	99,785	700	98,688.0	54.34	5,363	6.40	853,164.6	15.00	12,797
1977–1978	25,369	576,252.0	165.86	95,577	750	97,938.0	54.34	5,322	7.05	853,157.5	15.00	12,797
1978–1979	24,896	551,356.0	165.86	91,448	800	97,138.0	54.34	5,278	7.86	853,149.7	15.00	12,797
1979–1980	23,711	527,645.0	165.86	87,515	850	96,288.0	54.34	5,232	9.09	853,140.6	15.00	12,797
1980–1981	23,093	504,552.0	165.86	83,685	900	95,388.0	54.34	5,183	10.52	853,130.0	15.00	12,797
1981–1982	22,386	482,166.0	165.86	79,972	950	94,438.0	54.34	5,132	11.65	853,118.4	15.00	12,797
1982–1983	21,085	461,081.0	165.86	76,475	984	93,454.0	54.34	5,078	11.75	853,106.6	15.00	12,797
1983–1984	25,732	435,349.0	165.86	72,207	1,096	92,358.0	54.34	5,019	12.31	853,094.3	15.00	12,796
1984–1985	29,241	406,108.0	165.86	67,357	1,715	90,643.0	54.34	4,926	13.17	853,081.2	15.00	12,796
1985–1986	29,782	376,326.0	165.86	62,417	1,952	88,691.0	54.34	4,819	14.50	853,066.7	15.00	12,796
1986–1987	29,457	346,869.0	165.86	57,532	2,047	86,644.0	54.34	4,708	14.90	853,051.8	15.00	12,796
1987–1988	28,551	318,318.0	165.86	52,796	2,713	83,931.0	54.34	4,561	15.48	853,036.3	15.00	12,796
1988–1989	25,574	292,744.0	165.86	48,555	2,681	81,250.0	54.34	4,415	15.96	853,020.3	15.00	12,795
1989–1990	28,744	264,000.0	165.86	43,787	3,250	78,000.0	54.34	4,239	20.32	853,000.0	15.00	12,795
1990–1991	28,661	235,339.0	165.86	39,033	3,294	74,706.0	54.34	4,060	21.24	852,978.8	15.00	12,795
1991–1992	27,780	207,559.0	165.86	34,426	3,529	71,177.0	54.34	3,868	23.39	852,955.4	15.00	12,795
1992–1993	27,037	180,522.0	165.86	29,941	3,666	67,511.0	54.34	3,669	24.57	852,930.8	15.00	12,794
1993–1994	25,216	155,306.0	165.86	25,759	3,730	63,781.0	54.34	3,466	26.72	852,904.1	15.00	12,794
1994–1995	26,767	128,539.0	165.86	21,319	4,395	59,386.0	54.34	3,227	29.47	852,874.6	15.00	12,793

Table A4.d Item 4: Real Value of Oil and Gas Stocks (Continued)

Year	LPG Extracted (Naturally Occurring) (M/litres)	Identified Stock of LPG (M/litres)	Per Thousand m³ GPV of LPG ($)	Total Real Value of LPG ($m)	Total Real Value of Oil and Gas Stocks ($m at 1989–1990 Prices)
1966–1967	500	164,777.0	35.02	5,770	154,724
1967–1968	500	164,277.0	35.02	5,753	152,868
1968–1969	500	163,777.0	35.02	5,735	150,844
1969–1970	500	163,277.0	35.02	5,718	148,651
1970–1971	752	162,525.0	35.02	5,692	146,126
1971–1972	1,228	161,297.0	35.02	5,649	142,901
1972–1973	1,796	159,501.0	35.02	5,586	139,382
1973–1974	2,030	157,471.0	35.02	5,515	135,435
1974–1975	2,172	155,299.0	35.02	5,439	131,489
1975–1976	2,232	153,067.0	35.02	5,360	127,423
1976–1977	2,530	150,537.0	35.02	5,272	123,217
1977–1978	2,918	147,619.0	35.02	5,170	118,866
1978–1979	3,171	144,448.0	35.02	5,059	114,582
1979–1980	3,112	141,336.0	35.02	4,950	110,494
1980–1981	2,984	138,352.0	35.02	4,845	106,510
1981–1982	3,033	135,319.0	35.02	4,739	102,639
1982–1983	2,909	132,410.0	35.02	4,637	98,987
1983–1984	3,132	129,278.0	35.02	4,527	94,549
1984–1985	3,864	125,414.0	35.02	4,392	89,471
1985–1986	4,016	121,398.0	35.02	4,251	84,284
1986–1987	3,927	117,471.0	35.02	4,114	79,150
1987–1988	3,923	113,548.0	35.02	3,976	74,129
1988–1989	3,763	109,785.0	35.02	3,845	69,610
1989–1990	3,785	106,000.0	35.02	3,712	64,533
1990–1991	3,547	102,453.0	35.02	3,588	59,475
1991–1992	3,589	98,864.0	35.02	3,462	54,550
1992–1993	3,778	95,086.0	35.02	3,330	49,734
1993–1994	3,701	91,385.0	35.02	3,200	45,219
1994–1995	3,609	87,776.0	35.02	3,074	40,414

Table A4.e   Item 5: Real Value of Nonmetallic Minerals

Year	Ilmenite (Mineral Sand)				Rutile			
	Ilmenite Mined (Kt)	Identified Stock of Ilmenite (Kt)	Per Tonne GPV of Ilmenite ($)	Total Real Value of Ilmenite ($m)	Rutile Mined (Kt)	Identified Stock of Rutile (Kt)	Per Tonne GPV of Rutile ($)	Total Real Value of Rutile ($m)
1966–1967	522	80,724.9	23.19	1,872	251	18,042.0	325.43	5,871
1967–1968	554	80,724.3	23.19	1,872	270	17,772.0	325.43	5,784
1968–1969	562	80,723.8	23.19	1,872	293	17,479.0	325.43	5,688
1969–1970	721	80,723.1	23.19	1,872	326	17,153.0	325.43	5,582
1970–1971	886	80,722.2	23.19	1,872	371	16,782.0	325.43	5,461
1971–1972	814	80,721.4	23.19	1,872	375	16,407.0	325.43	5,339
1972–1973	707	80,720.7	23.19	1,872	313	16,094.0	325.43	5,237
1973–1974	720	80,719.9	23.19	1,872	335	15,759.0	325.43	5,128
1974–1975	817	80,719.1	23.19	1,872	319	15,440.0	325.43	5,025
1975–1976	991	80,718.1	23.19	1,872	348	15,092.0	325.43	4,911
1976–1977	959	80,717.2	23.19	1,872	390	14,702.0	325.43	4,784
1977–1978	1,033	80,716.1	23.19	1,872	325	14,377.0	325.43	4,679
1978–1979	1,255	80,714.9	23.19	1,872	257	14,120.0	325.43	4,595
1979–1980	1,181	80,713.7	23.19	1,872	275	13,845.0	325.43	4,506
1980–1981	1,385	80,712.3	23.19	1,872	312	13,533.0	325.43	4,404
1981–1982	1,321	80,711.0	23.19	1,872	231	13,302.0	325.43	4,329
1982–1983	1,149	80,709.8	23.19	1,872	221	13,081.0	325.43	4,257
1983–1984	893	80,708.9	23.19	1,872	163	12,918.0	325.43	4,204
1984–1985	1,493	80,707.5	23.19	1,872	170	12,748.0	325.43	4,149
1985–1986	1,419	80,706.0	23.19	1,872	212	12,536.0	325.43	4,080
1986–1987	1,238	80,704.8	23.19	1,872	216	12,320.0	325.43	4,009
1987–1988	1,498	80,703.3	23.19	1,872	246	12,074.0	325.43	3,929
1988–1989	1,610	80,701.7	23.19	1,871	231	11,843.0	325.43	3,854
1989–1990	1,690	80,700.0	23.19	1,871	243	11,600.0	325.43	3,775
1990–1991	1,616	80,698.4	23.19	1,871	244	11,356.0	325.43	3,696
1991–1992	1,466	80,696.9	23.19	1,871	192	11,164.0	325.43	3,633
1992–1993	1,787	80,695.1	23.19	1,871	183	10,981.0	325.43	3,574
1993–1994	1,795	80,693.3	23.19	1,871	187	10,794.0	325.43	3,513
1994–1995	1,777	80,691.6	23.19	1,871	231	10,563.0	325.43	3,438

The data sources are the Australian Bureau of Statistics, Catalogue No. 5241.0, Table 3.3, pages 48–51 and Table 3.4, pages 57–59; the Australian Bureau of Statistics, Catalogue No. 8405.0 (various articles); the Australian Bureau of Agricultural and Resource Economics, *Commodity Statistical Bulletin*, (various articles); and the Bureau of Resource Sciences, (1994).

A4.3 Renewable resources

Items 6, 7, 8 and 9 — *Timber stocks* (Table A4.f).

As mentioned in Appendix 2, Australian timber resource stocks are classified into four major categories. These are

- Native forest timber.
- Woodland forest timber.
- Plantation timber — broadleaved.
- Plantation timber — coniferous.

Table A4.e   Item 5: Real Value of Nonmetallic Minerals (Continued)

Year	Zircon Mined (Kt)	Identified Stock of Zircon (Kt)	Per Tonne GPV of Zircon ($)	Total Real Value of Zircon ($m)	Total Real Value of Nonmetallic Minerals ($m at 1989–1990 Prices)
1966–1967	232	27,510.0	275.30	7,574	15,317
1967–1968	290	27,220.0	275.30	7,494	15,149
1968–1969	292	26,928.0	275.30	7,413	14,973
1969–1970	363	26,565.0	275.30	7,313	14,767
1970–1971	389	26,176.0	275.30	7,206	14,540
1971–1972	406	25,770.0	275.30	7,094	14,306
1972–1973	357	25,413.0	275.30	6,996	14,106
1973–1974	375	25,038.0	275.30	6,893	13,893
1974–1975	368	24,670.0	275.30	6,792	13,688
1975–1976	382	24,288.0	275.30	6,686	13,470
1976–1977	420	23,868.0	275.30	6,571	13,227
1977–1978	398	23,470.0	275.30	6,461	13,012
1978–1979	392	23,078.0	275.30	6,353	12,820
1979–1980	445	22,633.0	275.30	6,231	12,608
1980–1981	492	22,141.0	275.30	6,095	12,371
1981–1982	434	21,707.0	275.30	5,976	12,176
1982–1983	462	21,245.0	275.30	5,849	11,977
1983–1984	382	20,863.0	275.30	5,744	11,819
1984–1985	458	20,405.0	275.30	5,617	11,638
1985–1986	501	19,904.0	275.30	5,480	11,431
1986–1987	452	19,452.0	275.30	5,355	11,236
1987–1988	457	18,995.0	275.30	5,229	11,030
1988–1989	484	18,511.0	275.30	5,096	10,822
1989–1990	511	18,000.0	275.30	4,955	10,602
1990–1991	443	17,557.0	275.30	4,833	10,400
1991–1992	288	17,269.0	275.30	4,754	10,259
1992–1993	355	16,914.0	275.30	4,656	10,101
1993–1994	413	16,501.0	275.30	4,543	9,927
1994–1995	511	15,990.0	275.30	4,402	9,711

Based on a study of timber resources in each of the preceding categories by the ABS, the real 1989–1990 net present values of timber resources are $39,091 per average km^2 for native forest timber, $171,039 per average km^2 for plantation timber — broadleaved, and $556,976 per average km^2 for plantation timber — coniferous. It was assumed that the value of woodland forest timber is half as much as native forest timber.

To calculate the total real value of Australia's timber resource stocks (at 1989–1990 prices), the previous values were multiplied by the total area of each the four timber categories. The data sources are the Australian Bureau of Statistics, Catalogue No. 5241.0, Table 4.2, page 72; Table 4.3, page 73; Table 4.4, page 73; Table 4.5, page 74; Table 4.7, page 79; Table 4.8, page 80; Table 4.9, page 81; and Table 4.9, page 82; the Australian Bureau of Statistics, Catalogue No. 4601.0, Table 6.1.12, page 137; the Department of Primary Industries, *Australian Forest Resources*, (various articles); and the Australian Bureau of Agricultural and Resource Economics, *Commodity Statistical Bulletin* (various articles).

Table A4.f Items 6, 7, 8, and 9 — Real Value of Timber Stocks

Year	Native Timber			Woodland Timber			Plantation — Broadleaved			Plantation — Coniferous			Total Value of Timber Stocks (6 + 7 + 8 + 9) ($m at 1989–1990 prices) 9A
	Area of Native Forest (km²) *a*	Value of Timber per km² ($) *b*	Total Value of Native Timber ($m) 6	Area of Woodland (km²) *c*	Value of Timber per km² ($) *d*	Total Value of Woodland Timber ($m) 7	Area of Broadleaved Forest (km²) *e*	Value of Timber per km² ($) *f*	Total Value of Broadleaved Timber ($m) 8	Area of Coniferous Forest (km²) *g*	Value of Timber per km² ($) *h*	Total Value of Coniferous Timber ($m) 9	
1966–1967	403,410	39,091	15,770	710,000	39,091	27,755	222	171,039	38	4,900	556,976	2,729	46,291
1967–1968	393,410	39,091	15,379	705,000	39,091	27,559	242	171,039	41	5,050	556,976	2,813	45,792
1968–1969	383,410	39,091	14,988	700,000	39,091	27,364	262	171,039	45	5,200	556,976	2,896	45,293
1969–1970	373,410	39,091	14,597	695,000	39,091	27,168	282	171,039	48	5,350	556,976	2,980	44,793
1970–1971	390,620	39,091	15,270	690,000	39,091	26,973	302	171,039	52	5,500	556,976	3,063	45,358
1971–1972	407,830	39,091	15,942	685,000	39,091	26,777	322	171,039	55	5,650	556,976	3,147	45,922
1972–1973	425,040	39,091	16,615	680,000	39,091	26,582	342	171,039	58	5,800	556,976	3,230	46,486
1973–1974	426,810	39,091	16,684	675,000	39,091	26,386	362	171,039	62	5,950	556,976	3,314	46,447
1974–1975	428,580	39,091	16,754	670,000	39,091	26,191	382	171,039	65	6,100	556,976	3,398	46,407
1975–1976	430,370	39,091	16,824	665,000	39,091	25,996	402	171,039	69	6,250	556,976	3,481	46,369
1976–1977	431,000	39,091	16,848	660,000	39,091	25,800	422	171,039	72	6,400	556,976	3,565	46,285
1977–1978	438,300	39,091	17,134	655,000	39,091	25,605	442	171,039	76	6,550	556,976	3,648	46,462
1978–1979	419,932	39,091	16,416	650,000	39,091	25,409	458	171,039	78	6,816	556,976	3,796	45,699
1979–1980	408,840	39,091	15,982	650,000	39,091	25,409	479	171,039	82	7,183	556,976	4,001	45,474
1980–1981	408,670	39,091	15,975	650,000	39,091	25,409	502	171,039	86	7,411	556,976	4,128	45,598
1981–1982	408,590	39,091	15,972	650,000	39,091	25,409	423	171,039	72	7,773	556,976	4,329	45,783
1982–1983	408,410	39,091	15,965	650,000	39,091	25,409	430	171,039	74	7,745	556,976	4,314	45,762
1983–1984	409,380	39,091	16,003	650,000	39,091	25,409	420	171,039	72	7,927	556,976	4,415	45,899
1984–1985	413,100	39,091	16,148	640,160	39,091	25,024	390	171,039	67	8,030	556,976	4,473	45,712
1985–1986	412,820	39,091	16,138	640,060	39,091	25,021	410	171,039	70	8,360	556,976	4,656	45,885
1986–1987	408,410	39,091	15,965	639,960	39,091	25,017	470	171,039	80	8,580	556,976	4,779	45,841
1987–1988	408,280	39,091	15,960	639,860	39,091	25,013	520	171,039	89	8,900	556,976	4,957	46,019
1988–1989	409,720	39,091	16,016	634,660	39,091	24,809	600	171,039	103	9,140	556,976	5,091	46,019
1989–1990	408,190	39,091	15,957	634,560	39,091	24,806	960	171,039	164	9,260	556,976	5,158	46,084
1990–1991	408,180	39,091	15,956	634,460	39,091	24,802	1,060	171,039	181	9,400	556,976	5,236	46,175
1991–1992	410,570	39,091	16,050	634,360	39,091	24,798	1,170	171,039	200	9,560	556,976	5,325	46,372
1992–1993	407,190	39,091	15,917	634,260	39,091	24,794	1,310	171,039	224	9,570	556,976	5,330	46,266
1993–1994	407,190	39,091	15,917	634,260	39,091	24,794	1,460	171,039	250	9,590	556,976	5,341	46,302
1994–1995	407,090	39,091	15,914	634,160	39,091	24,790	1,550	171,039	265	9,640	556,976	5,369	46,338

Item 10 — *Wetlands, mangroves, and saltmarshes* (Table A4.g).

To calculate the real 1989–1990 value of wetlands, mangroves, and salt-marshes, use was made of previous estimates of the cumulative area of each ecosystem type lost over the study period (see Item 8 in Appendix 2). By working back from the estimated area of remaining wetlands, mangroves, and saltmarshes as of 1994–1995, the area lost each year was added to the following year's total to calculate the area of all three categories over the study period.

The estimated value of a hectare of wetlands, mangroves, and saltmarsh was

- Wetlands — $3,000 per hectare.
- Mangroves — $10,000 per hectare.
- Saltmarsh — $6,500 per hectare.

The data sources are also discussed in Item 8, Appendix 2, and include the Australian Bureau of Statistics, Catalogue No. 4601.0, Table 12.2.2.9, page 356, Paijmans et al. (1995), State of the Environment Advisory Council, (1996, Table 8.6 and 8.7, page 8–22), and Galloway et al. (1984)

Item 11 — *Fishery stocks* (Table A4.h).

There are seven major fishery categories which make up the total stock of fishery resources. These include prawns, rock lobster, abalone, scallops, oys-ters, fish (excluding tuna), and tuna.

As mentioned in Appendix 2, there is no accurate measure of Australia's fishery stocks nor, therefore, any knowledge of the actual net change in fishery stocks. To estimate fishery stock levels, the method used to calculate the net change in each of the fishery categories was used again. To recall, where a fishery stock category has been fully exploited, it was assumed the stock had declined by an amount equivalent to 25 percent of the annual production tonnage. Where a category has been overexploited, the stock was assumed to have fallen by an amount equivalent to 50 percent of the annual production tonnage. As for the stock of oysters, which is currently increasing, it was assumed to have declined by 25 percent of the annual production tonnage between 1966–1967 and 1974–1975 (prior to extensive aquaculture), to have remained unchanged between 1974–1975 and 1984–1985; and to have increased at the rate of 12.5 percent of the annual production tonnage since 1984–1985.

The starting stock level of each fish category as at 1965–1966 was assumed to be 33.33 times greater than the harvest rate in 1966–1967. In other words, it was assumed the harvest rate in 1966–1967 was 3 percent of the total stock as of the end of 1965-66. Based on this assumption, the stock of each category as of the end of the 1965–1966 financial year was

---

Table A4.g Item 10 — Real Value of Wetlands, Mangroves, and Saltmarshes

	Wetlands				Mangroves			
Year	Annual Loss and/or Degradation of Wetlands (Hectares) a	Total Area of Insignificantly Affected Wetlands (Hectares) b	Average Value of a Hectare of Wetlands ($) c	Total Value of Wetlands (b × c) ($m) d	Annual Loss of Mangroves (Hectares) e	Total Area of Mangroves (Hectares) f	Average Value of a Hectare of Mangroves ($) g	Total Value of Mangroves (f × g) ($m) h
Up to 1966-67		11,980,000				1,191,136		
1966–1967	109,400	11,870,600	3,000	35,612	2,096	1,189,040	10,000	11,890
1967–1968	109,400	11,761,200	3,000	35,284	2,096	1,186,944	10,000	11,869
1968–1969	109,400	11,651,800	3,000	34,955	2,096	1,184,848	10,000	11,848
1969–1970	109,400	11,542,400	3,000	34,627	2,096	1,182,752	10,000	11,828
1970–1971	109,400	11,433,000	3,000	34,299	2,096	1,180,656	10,000	11,807
1971–1972	109,400	11,323,600	3,000	33,971	2,096	1,178,560	10,000	11,786
1972–1973	109,400	11,214,200	3,000	33,643	2,096	1,176,464	10,000	11,765
1973–1974	109,400	11,104,800	3,000	33,314	2,096	1,174,368	10,000	11,744
1974–1975	109,400	10,995,400	3,000	32,986	2,096	1,172,272	10,000	11,723
1975–1976	109,400	10,886,000	3,000	32,658	2,096	1,170,176	10,000	11,702
1976–1977	109,400	10,776,600	3,000	32,330	2,096	1,168,080	10,000	11,681
1977–1978	109,400	10,667,200	3,000	32,002	2,096	1,165,984	10,000	11,660
1978–1979	109,400	10,557,800	3,000	31,673	2,096	1,163,888	10,000	11,639
1979–1980	109,400	10,448,400	3,000	31,345	2,096	1,161,792	10,000	11,618
1980–1981	109,400	10,339,000	3,000	31,017	2,096	1,159,696	10,000	11,597
1981–1982	109,400	10,229,600	3,000	30,689	2,096	1,157,600	10,000	11,576
1982–1983	54,700	10,174,900	3,000	30,525	2,834	1,154,766	10,000	11,548
1983–1984	54,700	10,120,200	3,000	30,361	2,834	1,151,932	10,000	11,519
1984–1985	54,700	10,065,500	3,000	30,197	2,834	1,149,098	10,000	11,491
1985–1986	54,700	10,010,800	3,000	30,032	2,834	1,146,264	10,000	11,463
1986–1987	54,700	9,956,100	3,000	29,868	2,834	1,143,430	10,000	11,434
1987–1988	54,700	9,901,400	3,000	29,704	2,834	1,140,596	10,000	11,406
1988–1989	54,700	9,846,700	3,000	29,540	2,834	1,137,762	10,000	11,378
1989–1990	54,700	9,792,000	3,000	29,376	2,834	1,134,928	10,000	11,349
1990–1991	54,700	9,737,300	3,000	29,212	2,834	1,132,094	10,000	11,321
1991–1992	54,700	9,682,600	3,000	29,048	2,834	1,129,260	10,000	11,293
1992–1993	54,700	9,627,900	3,000	28,884	2,834	1,126,426	10,000	11,264
1993–1994	54,700	9,573,200	3,000	28,720	2,834	1,123,592	10,000	11,236
1994–1995	54,700	9,518,500	3,000	28,556	2,834	1,120,758	10,000	11,208

Table A4.g   Item 10 — Real Value of Wetlands, Mangroves, and Saltmarshes (Continued)

		Saltmarshes			
Year	Annual Loss of Saltmarsh (Hectares) j	Total Area of Saltmarshes (Hectares) k	Average Value of a Hectare of Saltmarshes ($) m	Total Value of Mangroves $(k \times m)$ ($m) n	Total Real Value of Wetlands, Mangroves, and Saltmarshes $(d + h + n)$ ($m at 1989–1990 Prices) 10
Up to 1966–1967		2,430,890			
1966–1967	5,438	2,425,452	6,500	15,765	63,268
1967–1968	5,438	2,420,014	6,500	15,730	62,883
1968–1969	5,438	2,414,576	6,500	15,695	62,499
1969–1970	5,438	2,409,138	6,500	15,659	62,114
1970–1971	5,438	2,403,700	6,500	15,624	61,730
1971–1972	5,438	2,398,262	6,500	15,589	61,345
1972–1973	5,438	2,392,824	6,500	15,553	60,961
1973–1974	5,438	2,387,386	6,500	15,518	60,576
1974–1975	5,438	2,381,948	6,500	15,483	60,192
1975–1976	5,438	2,376,510	6,500	15,447	59,807
1976–1977	5,438	2,371,072	6,500	15,412	59,423
1977–1978	5,438	2,365,634	6,500	15,377	59,038
1978–1979	5,438	2,360,196	6,500	15,341	58,654
1979–1980	5,438	2,354,758	6,500	15,306	58,269
1980–1981	5,438	2,349,320	6,500	15,271	57,885
1981–1982	5,438	2,343,882	6,500	15,235	57,500
1982–1983	5,841	2,338,041	6,500	15,197	57,270
1983–1984	5,841	2,332,200	6,500	15,159	57,039
1984–1985	5,841	2,326,359	6,500	15,121	56,809
1985–1986	5,841	2,320,518	6,500	15,083	56,578
1986–1987	5,841	2,314,677	6,500	15,045	56,348
1987–1988	5,841	2,308,836	6,500	15,007	56,118
1988–1989	5,841	2,302,995	6,500	14,969	55,887
1989–1990	5,841	2,297,154	6,500	14,932	55,657
1990–1991	5,841	2,291,313	6,500	14,894	55,426
1991–1992	5,841	2,285,472	6,500	14,856	55,296
1992–1993	5,841	2,279,631	6,500	14,818	54,966
1993–1994	5,841	2,273,790	6,500	14,780	54,735
1994–1995	5,841	2,267,949	6,500	14,742	54,505

- Prawns — 356.4 kilotonnes.
- Rock lobster — 297.0 kilotonnes.
- Abalone — 231.0 kilotonnes.
- Scallops — 200.0 kilotonnes.
- Oysters — 200.0 kilotonnes.
- Fish (excluding tuna) — 1214.4 kilotonnes.
- Tuna — 200.0 kilotonnes.

The change in each category of fishery stocks was subtracted/added to the previous year's level to estimate their total stock levels over the study period.

Based on ABS estimates, the real 1989–1990 prices of each category of fish stock are $9,717 per tonne of prawns, $15,586 per tonne of rock lobster, $17,902 per tonne of abalone, $4,048 per tonne of scallops, $5,258 per tonne of oysters, $2,297 per tonne of fish, and $8,013 per tonne of tuna.

Table A4.h Item 11 — Real Value of Fishery Stocks

Year	Prawns (Fully Exploited)					Rock Lobster (Fully Exploited)				
	Annual Harvest (kt) a	Net Decline in Prawn Stocks ($a \times 0.25$) (kt) b	Prawn Stocks (kt) c	Value of Prawns per Tonne ($) d	Total Value of Prawn Stocks ($m at 1989–1990 Prices) e	Annual Harvest (kt) f	Net Decline in Lobster Stocks ($f \times 0.25$) (kt) g	Rock Lobster Stocks (kt) h	Value of Rock Lobster per Tonne ($) j	Total Value of Rock Lobster ($m at 1989–1990 Prices) k
1965-66			356.4					297.0		
1966–1967	10.8	2.7	353.7	9,717	3,437	9.0	2.3	294.8	15,586	4,594
1967–1968	12.8	3.2	350.5	9,717	3,406	10.0	2.5	292.3	15,586	4,555
1968–1969	14.8	3.7	346.8	9,717	3,370	11.0	2.8	289.5	15,586	4,512
1969–1970	16.8	4.2	342.6	9,717	3,329	12.0	3.0	286.5	15,586	4,465
1970–1971	18.8	4.7	337.9	9,717	3,283	13.0	3.3	283.3	15,586	4,415
1971–1972	17.5	4.4	333.5	9,717	3,241	13.1	3.3	280.0	15,586	4,364
1972–1973	16.8	4.2	329.3	9,717	3,200	13.0	3.3	276.7	15,586	4,313
1973–1974	24.5	6.1	323.2	9,717	3,141	11.8	3.0	273.8	15,586	4,267
1974–1975	16.3	4.1	319.1	9,717	3,101	12.3	3.1	270.7	15,586	4,219
1975–1976	19.5	4.9	314.3	9,717	3,054	12.9	3.2	267.5	15,586	4,169
1976–1977	23.1	5.8	308.5	9,717	2,997	12.9	3.2	264.3	15,586	4,119
1977–1978	19.3	4.8	303.7	9,717	2,951	14.5	3.6	260.6	15,586	4,062
1978–1979	21.7	5.4	298.2	9,717	2,898	15.4	3.9	256.8	15,586	4,002
1979–1980	22.0	5.5	292.7	9,717	2,844	14.5	3.6	253.2	15,586	3,946
1980–1981	26.9	6.7	286.0	9,717	2,779	15.6	3.9	249.3	15,586	3,885
1981–1982	21.8	5.5	280.6	9,717	2,726	16.0	4.0	245.3	15,586	3,822
1982–1983	22.2	5.6	275.0	9,717	2,672	18.0	4.5	240.8	15,586	3,752
1983–1984	22.7	5.7	269.3	9,717	2,617	16.0	4.0	236.8	15,586	3,690
1984–1985	20.5	5.1	264.2	9,717	2,567	14.1	3.5	233.2	15,586	3,635
1985–1986	18.5	4.6	259.6	9,717	2,522	11.8	3.0	230.3	15,586	3,589
1986–1987	20.8	5.2	254.4	9,717	2,472	12.3	3.1	227.2	15,586	3,541
1987–1988	22.6	5.7	248.7	9,717	2,417	15.9	4.0	223.2	15,586	3,479
1988–1989	28.4	7.1	241.6	9,717	2,348	17.7	4.4	218.8	15,586	3,410
1989–1990	23.3	5.8	235.8	9,717	2,291	15.7	3.9	214.9	15,586	3,349
1990–1991	29.1	7.3	228.5	9,717	2,221	14.7	3.7	211.2	15,586	3,292
1991–1992	24.5	6.1	222.4	9,717	2,161	18.5	4.6	206.6	15,586	3,220
1992–1993	24.8	6.2	216.2	9,717	2,101	18.4	4.6	202.0	15,586	3,148
1993–1994	22.8	5.7	210.5	9,717	2,045	16.8	4.2	197.8	15,586	3,083
1994–1995	25.3	6.3	204.2	9,717	1,984	16.3	4.1	193.7	15,586	3,019

Table A4.h Item 11 — Real Value of Fishery Stocks (Continued)

Year	Abelone (Fully Exploited)					Scallops (Overexploited)				
	Annual Harvest (kt) m	Net Decline in Abelone Stocks ($m \times 0.25$) (kt) n	Abelone Stocks (kt) p	Value of Abelone per Tonne ($) q	Total Value of Abelone Stocks ($m at 1989–1990 Prices) r	Annual Harvest (kt) s	Net Decline in Scallops Stocks ($s \times 0.5$) (kt) t	Scallop Stocks (kt) u	Value of Scallops per Tonne ($) v	Total Value of Scallop Stocks ($m at 1989–1990 Prices) w
1965–1966			231.0					200.0		
1966–1967	7.0	1.8	229.3	17,902	4,104	6.0	3.0	197.0	4,048	797
1967–1968	7.0	1.8	227.5	17,902	4,073	7.0	3.5	193.5	4,048	783
1968–1969	7.0	1.8	225.8	17,902	4,041	8.0	4.0	189.5	4,048	767
1969–1970	7.0	1.8	224.0	17,902	4,010	9.0	4.5	185.0	4,048	749
1970–1971	7.7	1.9	222.1	17,902	3,976	9.3	4.7	180.4	4,048	730
1971–1972	8.0	2.0	220.1	17,902	3,940	10.1	5.1	175.3	4,048	710
1972–1973	6.4	1.6	218.5	17,902	3,911	17.0	8.5	166.8	4,048	675
1973–1974	6.0	1.5	217.0	17,902	3,884	12.4	6.2	160.6	4,048	650
1974–1975	5.0	1.3	215.7	17,902	3,862	6.1	3.1	157.6	4,048	638
1975–1976	5.3	1.3	214.4	17,902	3,838	4.6	2.3	155.3	4,048	628
1976–1977	6.9	1.7	212.7	17,902	3,807	4.4	2.2	153.1	4,048	620
1977–1978	5.1	1.3	211.4	17,902	3,784	9.1	4.6	148.5	4,048	601
1978–1979	6.2	1.6	209.9	17,902	3,757	10.5	5.3	143.3	4,048	580
1979–1980	6.4	1.6	208.3	17,902	3,728	16.4	8.2	135.1	4,048	547
1980–1981	6.9	1.7	206.5	17,902	3,697	9.8	4.9	130.2	4,048	527
1981–1982	7.7	1.9	204.6	17,902	3,663	15.4	7.7	122.5	4,048	496
1982–1983	6.1	1.5	203.1	17,902	3,635	30.5	15.3	107.2	4,048	434
1983–1984	8.5	2.1	201.0	17,902	3,597	29.7	14.9	92.4	4,048	374
1984–1985	7.7	1.9	199.0	17,902	3,563	19.2	9.6	82.8	4,048	335
1985–1986	7.0	1.8	197.3	17,902	3,532	16.2	8.1	74.7	4,048	302
1986–1987	6.7	1.7	195.6	17,902	3,502	11.7	5.9	68.8	4,048	279
1987–1988	6.8	1.7	193.9	17,902	3,471	6.0	3.0	65.8	4,048	266
1988–1989	5.5	1.4	192.5	17,902	3,447	5.4	2.7	63.1	4,048	255
1989–1990	5.1	1.3	191.3	17,902	3,424	6.3	3.2	60.0	4,048	243
1990–1991	5.2	1.3	190.0	17,902	3,400	14.5	7.3	52.7	4,048	213
1991–1992	5.0	1.3	188.7	17,902	3,378	29.8	14.9	37.8	4,048	153
1992–1993	4.7	1.2	187.5	17,902	3,357	33.6	16.8	21.0	4,048	85
1993–1994	4.7	1.2	186.4	17,902	3,336	24.6	12.3	8.7	4,048	35
1994–1995	5.1	1.3	185.1	17,902	3,313	12.7	6.4	2.4	4,048	10

Table A4.h   Item 11 — Real Value of Fishery Stocks (Continued)

Year	Oysters (Increasing)					Fish - Excluding Tuna (Fully Exploited)				
	Annual Harvest (kt) x	Net Decline in Oyster Stocks (kt) y	Oyster Stocks (kt) z	Value of Oysters per Tonne ($) aa	Total Value of Oyster Stocks ($m at 1989–1990 prices) bb	Annual Harvest (kt) cc	Net Decline in Fish Stocks (cc × 0.25) (kt) dd	Fish Stocks (kt) ee	Value of Fish per Tonne ($) ff	Total Value of Fish Stocks ($m at 1989–1990 Prices) gg
1965-1966			200.0					1214.4		
1966-1967	6.0	1.5	198.5	5,258	1,044	36.8	9.2	1205.2	2,297	2,768
1967-1968	7.0	1.8	196.8	5,258	1,035	38.8	9.7	1195.5	2,297	2,746
1968-1969	8.0	2.0	194.8	5,258	1,024	40.8	10.2	1185.3	2,297	2,723
1969-1970	9.0	2.3	192.5	5,258	1,012	42.8	10.7	1174.6	2,297	2,698
1970-1971	9.8	2.5	190.1	5,258	999	44.8	11.2	1163.4	2,297	2,672
1971-1972	10.4	2.6	187.5	5,258	986	46.8	11.7	1151.7	2,297	2,645
1972-1973	9.2	2.3	185.2	5,258	974	45.8	11.5	1140.3	2,297	2,619
1973-1974	10.5	2.6	182.5	5,258	960	56.0	14.0	1126.3	2,297	2,587
1974-1975	8.9	2.2	180.3	5,258	948	46.3	11.6	1114.7	2,297	2,560
1975-1976	10.3	0.0	180.3	5,258	948	44.3	11.1	1103.6	2,297	2,535
1976-1977	10.8	0.0	180.3	5,258	948	48.5	12.1	1091.5	2,297	2,507
1977-1978	9.8	0.0	180.3	5,258	948	50.8	12.7	1078.8	2,297	2,478
1978-1979	6.7	0.0	180.3	5,258	948	50.2	12.6	1066.2	2,297	2,449
1979-1980	8.4	0.0	180.3	5,258	948	43.6	10.9	1055.3	2,297	2,424
1980-1981	8.3	0.0	180.3	5,258	948	58.0	14.5	1040.8	2,297	2,391
1981-1982	7.8	0.0	180.3	5,258	948	57.9	14.5	1026.4	2,297	2,358
1982-1983	7.8	0.0	180.3	5,258	948	58.5	14.6	1011.7	2,297	2,324
1983-1984	7.0	0.0	180.3	5,258	948	64.0	16.0	995.7	2,297	2,287
1984-1985	7.5	0.0	180.3	5,258	948	73.0	18.3	977.5	2,297	2,245
1985-1986	6.8	-0.9	181.2	5,258	953	76.0	19.0	958.5	2,297	2,202
1986-1987	6.9	-0.9	182.1	5,258	957	85.0	21.3	937.2	2,297	2,153
1987-1988	7.5	-0.9	183.0	5,258	962	94.0	23.5	913.7	2,297	2,099
1988-1989	8.0	-1.0	184.0	5,258	967	97.6	24.4	889.3	2,297	2,043
1989-1990	7.5	-0.8	184.8	5,258	972	120.9	30.2	859.1	2,297	1,973
1990-1991	8.1	-1.2	186.0	5,258	978	145.4	36.4	822.7	2,297	1,890
1991-1992	8.2	-1.1	187.1	5,258	984	139.8	35.0	787.8	2,297	1,810
1992-1993	8.6	-1.1	188.2	5,258	990	135.0	33.8	754.0	2,297	1,732
1993-1994	8.8	-1.1	189.3	5,258	995	123.6	30.9	723.1	2,297	1,661
1994-1995	9.3	-1.2	190.5	5,258	1,002	129.3	32.3	690.8	2,297	1,587

Table A4.h Item 11 — Real Value of Fishery Stocks (Continued)

Year	Annual Harvest (kt) hh	Net Decline in Tuna Stocks (hh × 0.5) (kt) jj	Tuna Stocks (kt) kk	Value of Tuna per Tonne ($) mm	Total Value of Tuna Stocks ($m at 1989–1990 Prices) nn	Total Value of Fishery Stocks (e + k + r + w + bb + gg + nn) ($m at 1989–1990 Prices) 11
1965–1966			200.0			
1966–1967	6.0	3.0	197.0	8,013	1,579	18,323
1967–1968	6.0	3.0	194.0	8,013	1,555	18,152
1968–1969	6.0	3.0	191.0	8,013	1,530	17,968
1969–1970	6.0	3.0	188.0	8,013	1,506	17,770
1970–1971	6.8	3.4	184.6	8,013	1,479	17,555
1971–1972	10.2	5.1	179.5	8,013	1,438	17,323
1972–1973	13.6	6.8	172.7	8,013	1,384	17,076
1973–1974	9.7	4.9	167.9	8,013	1,345	16,834
1974–1975	11.1	5.6	162.3	8,013	1,301	16,629
1975–1976	10.7	5.4	157.0	8,013	1,258	16,430
1976–1977	10.1	5.1	151.9	8,013	1,217	16,215
1977–1978	12.3	6.2	145.8	8,013	1,168	15,992
1978–1979	11.3	5.7	140.1	8,013	1,123	15,756
1979–1980	13.6	6.8	133.3	8,013	1,068	15,505
1980–1981	18.2	9.1	124.2	8,013	995	15,222
1981–1982	21.7	10.9	113.4	8,013	908	14,921
1982–1983	21.3	10.7	102.7	8,013	823	14,589
1983–1984	15.8	7.9	94.8	8,013	760	14,273
1984–1985	13.0	6.5	88.3	8,013	708	14,001
1985–1986	13.5	6.8	81.6	8,013	653	13,753
1986–1987	11.8	5.9	75.6	8,013	606	13,510
1987–1988	10.5	5.3	70.4	8,013	564	13,259
1988–1989	8.6	4.3	66.1	8,013	530	13,000
1989–1990	8.2	4.1	62.0	8,013	497	12,749
1990–1991	11.7	5.9	56.1	8,013	450	12,444
1991–1992	13.2	6.6	49.5	8,013	397	12,102
1992–1993	10.2	5.1	44.4	8,013	356	11,769
1993–1994	7.7	3.9	40.6	8,013	325	11,481
1994–1995	7.9	4.0	36.6	8,013	294	11,208

These values were multiplied by the total stock of each fishery category to determine their total value. The total of each category was then summed to ascertain an estimate of the total real value of all fishery stocks. The data sources include those discussed in Item 7, Appendix 2, the State of the Environment Advisory Council, (1996, Figure 8.4, page 8–14 and Table 8.11, page 8–29), and the Australian Bureau of Agricultural and Resource Economics, *Commodity Statistical Bulletin* (various articles).

Item 12 — *Livestock (fixed natural capital assets)* (Table A4.j).

As mentioned in Appendix 3, livestock estimates are restricted to beef and dairy cattle as well as sheep and lambs. Livestock regarded as fixed natural capital assets are those used repeatedly or continuously over periods of more than one year to produce other goods (such as wool, milk, etc.). The real 1989–1990 (capital asset) prices of beef and dairy cattle, sheep and lambs were derived from a recent study by the ABS on livestock. To calculate the total value of such livestock assets, the quantities of beef and dairy cattle, sheep, and lambs were multiplied by the per head asset value of each. The data sources are the Australian Bureau of Statistics, Catalogue No. 7221.0 and 7102.0 (various articles), the Australian Bureau of Agricultural and Resource Economics, *Commodity Statistical Bulletin* (various articles), and the Australian Bureau of Statistics, Catalogue No. 5241.0, Table 5.2 and Table 5.3, page 107, and Table 5.4 and 5.5, page 108.

Item 13 — *Stored surface water resources* (Table A4.k).

To calculate the real value of Australia's stored surface water resources, use was made of an ANCOLD register of the cumulative constructed storage capacity of Australia's largest dams. The price of water in storage was assumed to be an average $0.50 per kilolitre in 1989–1990 prices. This is much higher than the price paid for water by all users anywhere in Australia. The higher price recognises that the price of water fails to be anywhere near the private marginal cost of provision, let alone its social marginal cost. The data sources are the State of the Environment Advisory Council, (1996, Figure 7.5, page 7–8) and ANCOLD (1990).

Item 14 — *Nonrenewable natural capital.*

The sum of Items 1 to 5.

Item 15 — *Renewable natural capital.*

The sum of Items 6 to 13.

Item 16 — *Total natural capital.*

Table A4.j Item 12 - Livestock (Fixed Natural Capital Assets)

Year	Beef cattle (Thousands) a	Per Head Value of Beef Cattle ($) b	Total Value of Beef Cattle ($) c	Dairy Cattle (Thousands) d	Per Head Value of Dairy Cattle ($) e	Total Value of dairy cattle ($) f	Sheep (Thousands) g	Per Head Value of Sheep ($) h	Total Value of Sheep ($) j	Lambs (Thousands) k	Per Head Value of Lambs ($) m	Total Value of Lambs ($) n	Value of Livestock (Fixed Natural Capital Assets) (c+f+j+n) ($m at 1989–1990 Prices) 12
1966–1967	8,736	731.59	6,391	4,620	543.92	2,513	111,168	9.93	1,104	31,652	71.00	2,247	12,255
1967–1968	9,427	731.59	6,897	4,490	543.92	2,442	115,920	9.93	1,151	33,005	71.00	2,343	12,833
1968–1969	10,419	731.59	7,622	4,340	543.92	2,361	117,360	9.93	1,165	33,415	71.00	2,372	13,521
1969–1970	11,475	731.59	8,395	4,230	543.92	2,301	121,680	9.93	1,208	34,645	71.00	2,460	14,364
1970–1971	12,966	731.59	9,486	4,110	543.92	2,236	125,496	9.93	1,246	35,732	71.00	2,537	15,504
1971–1972	14,950	731.59	10,937	4,010	543.92	2,181	123,624	9.93	1,228	35,199	71.00	2,499	16,845
1972–1973	16,077	731.59	11,762	3,980	543.92	2,165	112,392	9.93	1,116	32,001	71.00	2,272	17,315
1973–1974	17,331	731.59	12,679	3,760	543.92	2,045	96,552	9.93	959	27,491	71.00	1,952	17,635
1974–1975	18,605	731.59	13,611	3,730	543.92	2,029	100,440	9.93	997	28,598	71.00	2,030	18,668
1975–1976	19,091	731.59	13,967	3,600	543.92	1,958	104,976	9.93	1,042	29,889	71.00	2,122	19,089
1976–1977	18,093	731.59	13,237	3,270	543.92	1,779	103,968	9.93	1,032	29,541	71.00	2,097	18,145
1977–1978	16,813	731.59	12,300	3,060	543.92	1,664	94,536	9.93	939	26,917	71.00	1,911	16,814
1978–1979	15,520	731.59	11,354	2,870	543.92	1,561	91,800	9.93	912	26,134	71.00	1,856	15,682
1979–1980	14,957	731.59	10,942	2,830	543.92	1,539	93,960	9.93	933	26,753	71.00	1,899	15,314
1980–1981	14,317	731.59	10,474	2,800	543.92	1,523	95,544	9.93	949	27,204	71.00	1,931	14,877
1981–1982	13,933	731.59	10,193	2,780	543.92	1,512	94,608	9.93	939	26,937	71.00	1,913	14,557
1982–1983	12,621	731.59	9,233	2,760	543.92	1,501	97,200	9.93	965	27,675	71.00	1,965	13,665
1983–1984	12,390	731.59	9,064	2,800	543.92	1,523	93,960	9.93	933	26,753	71.00	1,899	13,420
1984–1985	12,787	731.59	9,355	2,810	543.92	1,528	97,272	9.93	966	27,696	71.00	1,966	13,816
1985–1986	12,230	731.59	8,947	2,710	543.92	1,474	104,904	9.93	1,042	29,869	71.00	2,121	13,584
1986–1987	12,339	731.59	9,027	2,640	543.92	1,436	108,288	9.93	1,075	30,832	71.00	2,189	13,727
1987–1988	12,333	731.59	9,023	2,540	543.92	1,382	110,304	9.93	1,095	31,406	71.00	2,230	13,729
1988–1989	12,730	731.59	9,313	2,510	543.92	1,365	109,728	9.93	1,090	31,242	71.00	2,218	13,986
1989–1990	13,229	731.59	9,678	2,490	543.92	1,354	116,352	9.93	1,261	33,128	71.00	2,352	14,646
1990–1991	13,587	731.59	9,940	2,430	543.92	1,322	122,616	9.93	1,218	34,912	71.00	2,479	14,958
1991–1992	13,728	731.59	10,043	2,430	543.92	1,322	117,504	9.93	1,167	33,456	71.00	2,375	14,907
1992–1993	13,606	731.59	9,954	2,440	543.92	1,327	106,704	9.93	1,060	30,381	71.00	2,157	14,498
1993–1994	13,485	731.59	9,865	2,450	543.92	1,333	104,400	9.93	1,037	29,725	71.00	2,110	14,345
1994–1995	13,363	731.59	9,776	2,460	543.92	1,338	100,800	9.93	1,001	28,700	71.00	2,038	14,153

Table A4.k   Item 13 — Real Value of Stored Surface Water Resources

Year	Cumulative Storage Capacity (Gigalitres) *a*	Assumed Real 1989–1990 Price for Water ($0.50 per Kilolitre) *b*	Total Real Value of Stored Surface Water ($a \times b$) ($) 13
1959–1960	19,500		
1966–1967	29,650	0.50	14,825
1967–1968	31,100	0.50	15,550
1968–1969	32,550	0.50	16,275
1969–1970	34,000	0.50	17,000
1970–1971	37,450	0.50	18,725
1971–1972	40,900	0.50	20,450
1972–1973	44,350	0.50	22,175
1973–1974	47,800	0.50	23,900
1974–1975	51,250	0.50	25,625
1975–1976	54,700	0.50	27,350
1976–1977	58,150	0.50	29,075
1977–1978	61,600	0.50	30,800
1978–1979	65,050	0.50	32,525
1979–1980	68,500	0.50	34,250
1980–1981	69,500	0.50	34,750
1981–1982	70,500	0.50	35,250
1982–1983	71,500	0.50	35,750
1983–1984	72,500	0.50	36,250
1984–1985	73,500	0.50	36,750
1985–1986	74,500	0.50	37,250
1986–1987	75,500	0.50	37,750
1987–1988	76,500	0.50	38,250
1988–1989	77,500	0.50	38,750
1989–1990	78,500	0.50	39,250
1990–1991	79,333	0.50	39,667
1991–1992	80,167	0.50	40,083
1992–1993	81,000	0.50	40,500
1993–1994	81,833	0.50	40,917
1994–1995	82,667	0.50	41,333

The sum of Items 14 and 15.

Item 17 — *Australian population.*

Indicated in the thousands. The data source is Foster (1996), Table 4.1, page 177).

Item 18 — *Average per capita ownership of natural capital.*

Item 16 divided by Item 17.

Acronyms

A	Waste assimilative capacity of the ecosphere
A-M	Atomistic-mechanistic
E	Technical efficiency
GDP	Gross Domestic Product
h	Total low entropy harvested
h_R	Renewable low entropy harvested
h_{NR}	Nonrenewable low entropy harvested
MUB	Marginal uncancelled benefits
MUC	Marginal uncancelled costs
N-RKn	Nonrenewable natural capital
r	Recycling
RKn	Renewable natural capital
RKn/A ratio	Renewable natural capital/waste assimilative capacity ratio
RKn/s_R ratio	Renewable natural capital/low entropy surplus ratio
SD	Sustainable development
SNB	Sustainable net benefits
s_R	Renewable low entropy surplus
s_{NR}	Nonrenewable low entropy surplus
UB	Uncancelled benefits
UC	Uncancelled costs
EEE	Ecological economic efficiency
UNDP	United Nations Development Programme
W	Waste
WCED	World Commission on Environment and Development
WTO	World Trade Organisation

References

Alchain, A. and Demsetz, H., Production, information costs, and economic organisations, *Am. Econ. Rev.*, 62, 777, 1972.

Allen, R., *How to Save the World*, Kogan Page, London, 1980.

Allen, P., Evolution, innovation, and economics, in *Technical Change and Economic Theory*, Dosi, G., Freeman, C., Nelson, R., Silverberg, G., and Soete, L., Eds., Pinter Publishers, London, 1988.

Alonzo Smith, G., The teleological view of wealth: a historical perspective, in *Economics, Ecology, and Ethics: Essays Toward a Steady-State Economy*, Daly, H., Ed., W. H. Freeman, San Francisco, 1980, 215.

ANCOLD, *Register of Large Dams in Australia*, National Committee on Large Dams, c/o Hydro Electric Commission, Hobart, 1990.

d'Arge, R., Sustenance and sustainability: how can we preserve and consume without major conflict?, in *Investing in Natural Capital*, Jansson, A., Hammer, M., Folke, C., and Costanza, R., Eds., Island Press, Washington, D.C., 1994, 113.

Arthur, W., Competing technologies, increasing returns, and lock-in by historical events, *The Econ. J.*, 99, 116, 1989.

Atkins, A., *The Economic and Social Costs of Road Accidents in Australia: With Preliminary Cost Estimates for Australia, 1978*, report prepared for the Office of Road Safety, AGPS, Canberra, 1981.

Australian Bureau of Agricultural and Resource Economics, *Commodity Statistical Bulletin*, AGPS, Canberra, various articles.

Australian Bureau of Statistics, *Agriculture, Australia*, Catalogue No. 7113.0, AGPS, Canberra.

Australian Bureau of Statistics, *Australian Agriculture and the Environment*, Catalogue No. 4606.0, AGPS, Canberra.

Australian Bureau of Statistics, *Australian National Accounts: National Income, Expenditure and Product*, Catalogue No. 5204.0, AGPS, Canberra.

Australian Bureau of Statistics, *Australian Yearbook*, Catalogue No. 1301.0, AGPS, Canberra.

Australian Bureau of Statistics, *Australians and the Environment*, Catalogue No. 4601.0, AGPS, Canberra.

Australian Bureau of Statistics, *Census of Population and Housing*, Catalogue No. 2822.0, AGPS, Canberra.

Australian Bureau of Statistics, *Characteristics of Australian Farms*, Catalogue No. 7102.0, AGPS, Canberra.

Australian Bureau of Statistics, *Cost of Environmental Protection, Australia: Selected Industries, 1991–1992*, Catalogue No. 4603.0, AGPS, Canberra.

Australian Bureau of Statistics, *Education and Training in Australia*, Catalogue No. 4224.0, AGPS, Canberra, various articles.

Australian Bureau of Statistics, *Employment, Underemployment, and Unemployment, 1966–1983 in Australia*, Catalogue No. 6246.0, AGPS, Canberra.

Australian Bureau of Statistics, *Energy Accounts for Australia, 1993–1994*, Catalogue No. 4604.0, AGPS, Canberra.

Australian Bureau of Statistics, *How Australians Use Their Time*, Catalogue No. 4153.0, AGPS, Canberra, 1993.

Australian Bureau of Statistics, *Income Distribution in Australia*, Catalogue No. 6523.0, AGPS, Canberra, various articles.

Australian Bureau of Statistics, *Livestock and Livestock Products*, Catalogue No. 7221.0, AGPS, Canberra, various articles.

Australian Bureau of Statistics, *Mining Production, Australia*, Catalogue No. 8405.0, AGPS, Canberra, various articles.

Australian Bureau of Statistics, *National Balance Sheets for Australia: Issues and Experimental Estimates, 1989 to 1992*, Occasional Paper, Catalogue No. 5241.0, AGPS, Canberra, 1995.

Australian Bureau of Statistics, *National Crime Statistics*, Catalogue No. 4510.0, AGPS, Canberra.

Australian Bureau of Statistics, *Schools, Australia*, Catalogue No. 4221.0, AGPS, Canberra, various articles.

Australian Bureau of Statistics, *Survey of Motor Vehicle Usage*, Catalogue No. 9208.0, AGPS, Canberra, various articles.

Australian Bureau of Statistics, *Voluntary Work in Australia*, Catalogue No. 4441.0, AGPS, Canberra, various articles.

Ayres, R., *Resources, Environment and Economics*, John Wiley & Sons, New York, 1978.

Ayres, R. and Miller, S., The role of technological change, *J. Environ. Econ. Manage.*, 7, 353, 1980.

Bailey, E., Contestability and the design of regulatory and antitrust policy, *Am. Econ. Assoc. Pap. Proc.*, 71(2), 178, 1981.

Barney, G., *The Global 2000 Report to the President of the United States*, Penguin Books, Harmondsworth, Middlesex, 1980.

Barry, B., Justice between generations, in *Law, Morality and Society: Essays in Honour of H. L. A. Hart*, Hacker, P. and Raz, J., Eds., Clarendon Press, Oxford, 1977, 268.

Baumol, W., Panzar, J., and Willig, R., *Contestable Markets and the Theory of Industry Structure*, Harcourt Brace Jovanovich, San Diego, 1982.

Beattie, A., Ed., *Biodiversity: Australia's Living Wealth*, Reed, Sydney, 1995.

Beckerman, W., Economic growth and the environment: whose growth? whose environment?, *World Develop.*, 20(4), 481, 1992.

van den Bergh, J. and van der Straaten, J., The significance of sustainable development for ideas, tools, and policy, in *Toward Sustainable Development: Concepts, Methods, and Policy*, van den Bergh, J. and van der Straaten, J., Eds., Island Press, Washington, D.C., 1994, 1.

van den Bergh, J. and Verbruggen, H., Spatial sustainability, trade, and indicators: an evaluation of the ecological footprint, *Ecol. Econ.*, 29(1), 61, 1999.

Berkes, F. and Folke, C., Investing in cultural capital for sustainable use of natural capital, in *Investing in Natural Capital: The Ecol. Econ. Approach to Sustainability*, Jansson, A., Hammer, M., Folke, C., and Costanza, R., Eds., Island Press, Washington, D.C., 1994, 128.

Berlin, I., Two concepts of liberty, in *Political Philosophy*, Quinton, A., Ed., Oxford University Press, London, 1967, 141.

Berry, W., *The Gift of Good Land: Further Essays Cultural and Agricultural*, North Point, San Francisco, 1981.

Biodiversity Unit, *Native Vegetation Clearance, Habitat Loss, and Biodiversity Decline: An Overview of Recent Native Vegetation Clearance in Australia and its Implications for Biodiversity*, Department of Environment, Sports, and Territories, Biodiversity Series Paper No. 6, AGPS, Canberra, 1995.

Bishop, R., Economic efficiency, sustainability, and biodiversity, *Ambio*, May, 69, 1993.

Bishop, M. and Thompson, D., Regulatory reform and productivity growth in the UK's public utilities, *Appl. Econ.*, 24, 1181, 1992.

Blandy, R. and Brummitt, W., *Labour Productivity and Living Standards*, Allen & Unwin, Sydney, 1990.

Blum, H., *Times Arrow and Evolution*, 3rd ed., Harper Torchbook, Princeton, 1962.

Boulding, K., *The Image*, University of Michigan Press, Ann Arbor, MI, 1956.

Boulding, K., *The Meaning of the Twentieth Century*, Harper & Row, New York, 1964.

Boulding, K., The economics of the coming spaceship Earth, in *Environmental Quality in a Growing Economy*, Jarrett, H., Ed., Johns Hopkins University Press, Baltimore, MD, 1966, 3.

Boulding, K., *A Primer on Social Dynamics*, The Free Press, New York, 1970a.

Boulding, K., *Economics as a Science*, McGraw-Hill, New York, 1970b.

Boulding, K., *Evolutionary Economics*, Sage Publications, Beverly Hills, CA, 1981.

Boulding, K., *The World as a Total System*, Sage Publications, London, 1985.

Boulding, K., *Three Faces of Power*, Sage Publications, Newbury Park, 1989.

Boulding, K., *Towards a New Economics*, Edward Elgar, Aldershot, 1990.

Boulding, K., What do we want to sustain?: environmentalism and human evaluations, in *Ecological Economics: the Science and Management of Sustainability*, Costanza, R., Ed., Columbia University Press, New York, 1991, 22.

Bredvold, L. and Ross, R., *The Philosophy of Edmund Burke: a Selection from His Speeches and Writings*, University of Michigan Press, Ann Arbor, MI, 1960.

Brown, A., Should Telstra Be Privatised?, School of Economics Working Paper No. 8, Griffith University, 1966.

Brown, T., The concept of value in resource allocation, *Land Econ.*, 60(3), 231, 1984.

Buitenkamp, M., Venner, H., and Wams, T., Eds., *Action Plan Sustainable Netherlands*, Dutch Friends of the Earth, Amsterdam, 1993.

Bunke, H., Economics, affluence, and existentialism, *Q. Rev. Econ. Bus.*, 4, 9, 1964.

Bureau of Resource Economics, *Energy Demand and Supply in Australia, 1960–1961 to 1984–1985*, AGPS, Canberra, 1987.

Bureau of Resource Sciences, *Australia's Identified Mineral Resources*, AGPS, Canberra, 1994.

Bureau of Transport and Communications, *Social Cost of Transport Accidents in Australia*, AGPS, Canberra, 1979.

Burmeister, E., *Capital Theory and Dynamics*, Cambridge University Press, London, 1980.

Capra, F., *The Turning Point*, Fontana, London, 1982.

Castaneda, B., An index of sustainable economic welfare (ISEW) for Chile, *Ecol. Econ.*, 28(2), 231, 1999.

Catton, W., *Overshoot*, University of Illinios Press, Urbana, 1980.

Clare, R. and Johnston, K., Profitability and productivity of government business enterprises, Economic Planning Advisory Council Research Paper No. 2, 1992.

Clark, C., *Mathematical Bioeconomics*, John Wiley & Sons, New York, 1976.

Clark, N. and Juma, N., Evolutionary theories in economic thought, in *Technical Change and Economic Theory*, Dosi, G., Freeman, C., Nelson, R., Silverberg, G., and Soete, L., Eds., Pinter Publishers, London, 1988, 197.

Clark, N., Organisation and information in the evolution of economic systems, in *Evolutionary Theories of Economic and Technological Change*, Saviotti, P. and Metcalfe, J., Eds., Harwood Academic Publishers, Reading, 1991, 88.

Clark, W. and Munn, R., *Sustainable Development of the Biosphere*, Cambridge University Press, Cambridge, 1986.

Coase, R., The nature of the firm, *Econometrica*, 4, 386,1937.

Coase, R., The problem of social cost, *J. Law Econ.*, October 1960, 1, 1960.

Cobb, J., Ecology, ethics and theology, in *Toward a Steady State Economy*, Daly, H., Ed., W. H. Freeman, San Francisco, 1973, 307.

Cobb, C. and Cobb, J., *The Green National Product*, University Press of America, New York, 1994.

Cobb, C., Halstead, T., and Rowe, J., If the GDP is up, why is America down?, *Atlantic Mon.*, October, 59, 1995.

Colvin, P., Ontological and epistemological commitments and social relations in the sciences, in *The Social Production of Scientific Knowledge*, Mendelsohn, E., Weingart, P., and Whitley, R., Eds., D. Reidel Publishing, Boston, 1977, 103.

Common, M., *Natural Resource Accounting and Sustainability*, Centre for Resource and Environmental Studies Conference Paper, Australian National University, 1990.

Common, M., What is ecological economics?, *Ecol. Econ. Conf. Pa.*, Inaugural ANZEE Conference, Coffs Harbour, 1923 November 1995, 1.

Common, M. and Perrings, C., Towards an ecological economics of sustainability, *Ecol. Econ.*, 6, 7, 1992.

Commonwealth Treasury of Australia, *Economic Roundup,*, AGPS, Canberra, 1996.

Coombs, H., *The Quality of Life and Its Assessment*, University of Tasmania Occasional Paper No. 11, 1977.

Costanza, R., Social traps and environmental policy, *BioScience*, 37(6), 407, 1987.

Costanza, R., Ed., *Ecological Economics: the Science and Management of Sustainability*, Columbia University Press, New York, 1991.

Costanza, R., Toward an operational definition of ecosystem health, in *Ecosystem Health: New Goals for Environmental Management*, Costanza, R., Norton, B., and Haskell, B., Eds., Island Press, Washington, D.C., 1992, 239.

Costanza, R., Three general policies to achieve sustainability, in *Investing in Natural Capital*, Jansson, A., Hammer, M., Folke, C., and Costanza, R., Eds., Island Press, Washington, D.C., 1994, 392.

Costanza, R. and Daly, H., Toward an ecological economics, *Ecol. Model.*, 38, 1, 1987.

Costanza, R. and Perrings, C., A flexible assurance bonding system for improved environmental management, *Ecol. Econ.*, 2, 57, 1990.

Costanza, R., Norton, B., and Haskell, B., Eds., *Ecosystem Health: New Goals for Environmental Management*, Island Press, Washington, D.C., 1992.

Costanza, R., Wainger, L., Folke, C., and Maler, K.-G., Modeling complex ecological economic systems: toward an evolutionary, dynamic understanding of people and nature, *BioScience*, 43, 545, 1993.

Costanza, R., d'Arge, R., de Groot, R., Farber, S., Grasso, M., Hannon, B., Limburg, K., Naeem, S., O'Neill, R., Parvelo, J., Raskin, R., Sutton, P., and van den Belt, M., The value of the world's ecosystem services and natural capital, *Nature*, 15, 253, 1997a, 253.

Costanza, R., Cumberland, J., Daly, H., Goodland, R., and Norgaard, R., *An Introduction to Ecol. Econ.*, St. Lucie Press, Boca Raton, FL, 1997b.

Crowe, B., The tragedy of the commons revisited, *Science*, 28, 1103-1969.

Daily, G. and Ehrlich, P., Population, sustainability, and Earth's carrying capacity, *BioScience*, 42(10), 761, 1992.

Daly, H., On economics as a life science, *J. Pol. Economy*, 76 (1), 392-405.

Daly, H., Ed. (1973), *Towards a Steady State Econ.*, W. H. Freeman, San Francisco.

Daly, H. (1979), Entropy, growth, and the political economy of scarcity, in *Scarcity and Growth Reconsidered*, Smith, V. K., Ed., Johns Hopkins University Press, Baltimore, MD, 1979, 67–94.

Daly, H., Ed., *Economics, Ecology, and Ethics: Essays toward a Steady-State Economy*, W. H. Freeman, San Francisco, CA, 1980.

Daly, H., The economic growth debate: what some economists have learned but many have not, *J. Environ. Econ. Manage.*, 14, 323, 1987.

Daly, H., Toward some operational principles of sustainable development, *Ecol. Econ.*, 2(1), 1, 1990.

Daly, H. (1991a), Ecological economics and sustainable development, in *Ecological Physical Chemistry*, Rossi, C. and Tiezzi, E., Eds., Proc. Inter. Workshop, 8–12 November 1990, Sienna, Italy, Elsevier Science Publishers, Amsterdam, 1991a, 185.

Daly, H., *Steady-State Economics: Second Edition with New Essays*, Island Press, Washington D.C., 1991b.

Daly, H., Allocation, distribution, and scale: towards an economics that is efficient, just, and sustainable, *Ecol. Econ.*, 6, 185, 1992.

Daly, H., The perils of free trade, *Sci. Am.*, 269, 24, 1993.

Daly, H., Are efficiency, equity, and scale independent — a reply, *Ecol. Econ.*, 10 (2), 1994.

Daly, H., *Beyond Growth: the Economics of Sustainable Development*, Beacon Press, Boston, 1966.

Daly, H., *Ecological Economics and the Ecology of Economics*, Edward Elgar, Aldershot, 1999.

Daly, H. and Cobb, J., *For the Common Good: Redirecting the Economy Toward Community, the Environment, and a Sustainable Future*, Beacon Press, Boston, 1989.

Daly, H. and Goodland, R., An ecological-economic assessment of deregulation of international commerce under GATT, *Ecol. Econ.*, 9, 73, 1994.

David, P., Clio and the economics of QWERTY, *Am. Econ. Rev.*, 75(2), 332, 1985.

Davy, G., *Some Notes on Personal Share Ownership in Australia*, Griffith University, mimeo, 1996.

Dean, J., Trade and the environment: a survey of the literature, in *International Trade and the Environment*, Low, P., Ed., World Bank, Washington, D.C., 1992, 15.

Demsetz, H., Why regulate utilities?, *J. Law Econ.*, 11, 55, 1968.

Department of Arts, Heritage and Environment, *State of the Environment in Australia 1986*, AGPS, Canberra, 1987.

Department of Employment, Education, and Training, *Selected Higher Education Student Statistics*, AGPS, Canberra, 1995 various articles.

Department of Primary Industries, *Australian Forest Resources*, AGPS, Canberra, various articles.

DeSerpa, A., *Microeconomic Theory: Issues and Applications*, 2nd ed., Allyn and Bacon, Needham Heights, 1988.

Diefenbacher, H., The index of sustainable economic welfare in Germany, in *The Green National Product*, Cobb, C. and Cobb, J., Eds., University Press of America, New York, 1994.

Dilnot, R., Distribution and composition of personal sector wealth, *Aust. Econ. Rev.*, XX, 33, 1990.

Dosi, G., Freeman, C., Nelson, R., Silverberg, G., and Soete, L., Eds., *Technical Change and Economic Theory*, Pinter Publishers, London, 1988.

Dosi, G. and Metcalfe, J., On some notions of irreversibility in economics, in *Evolutionary Theories of Economic and Technological Change*, Saviotti, P. and Metcalfe, J., Eds., Harwood Academic Publishers, Reading, 1991, 133.

Downs, A., *An Economic Theory of Democracy*, Harper & Row, New York, 1957.

Downs, A., *Inside Bureaucracy*, Little Brown, Boston, 1967.

Dragun, A., Externalities, property rights, and power, *J. Econ. Iss.*, XVII(3), 667, 1983.

Dury, G., *An Introduction to Environmental Systems*, Heinemann, Exeter, 1981.

Easterlin, R., Does economic growth improve the human lot? some empirical evidence, in *Nations and Households in Economic Growth*, David, P. and Weber, R., Eds., Academic Press, New York, 1974.

Economic Planning Advisory Council, *Income Distribution in Australia: Recent Trends and Research*, Commission Paper No. 7, 46, 1995.

Ehrlich, P., Biodiversity and ecosystem function: need we know more?, in *Biodiversity and Ecosystem Function*, Schulze, E.-D. and Mooney, H., Eds., Springer-Verlag, Berlin, 1993, VII.

Ehrlich, P., Ehrlich, A., and Holdren, J., Availability, entropy, and the laws of thermodynamics, in *Economics, Ecology, and Ethics: Essays toward a Steady-State Economy*, Daly, H., Ed., W. H. Freeman, San Francisco, 1980, 44.

Ehrlich, P. and Ehrlich, A., *The Population Explosion*, Simon & Schuster, New York, 1990.

Ekins, P., An indicator framework for economic progress, *Development*, 3/4, 92, 1990.

Ekins, P., A four-capital model of wealth creation, in *Real-Life Economics: Understanding Wealth Creation*, Ekins, P. and Max-Neef, M., Eds., Routledge, London, 1992, 147.

Ekins, P.,, Limits to growth and sustainable development: grappling with ecological realities, *Ecol. Econ.*, 8, 269, 1993.

Ekins, P., The environmental sustainability of economic processes: a framework for analysis, in *Toward Sustainable Development: Concepts, Methods, and Policy*, van den Bergh, J. and van der Straaten, J., Eds., Island Press, Washington, D.C., 1994, 25.

Ekins, P. and Max-Neef, M., Eds., *Real-Life Economics: Understanding Wealth Creation*, Routledge, London, 1992.

Ekins, P., Folke, C., and Costanza, R., Trade, environment, and development: the issues in perspective, *Ecol. Econ.*, 9, 1, 1994.

El Serafy, S., The proper calculation of income from depletable natural resources, in *Environmental Accounting for Sustainable Development*, Ahmad, Y., El Serafy, S., and Lutz, E., Eds., World Bank, Washington, D.C., 1989, 10.

El Serafy, S., Weak and strong sustainability: natural resources and national account-ing — part I, *Environ. Tax. Acc.*, 1(1), 27, 1996.

EPAC (Economic Planning and Advisory Commission), Microeconomic reform pay-ing off for government business enterprises and the economy generally, media release July 1995, 14 September 1995.

Estrin, S., Profit-sharing, motivation and company performance: a survey, Depart-ment of Economics Pamphlet, London School of Economics, 1986.

Faber, M. and Proops, J., *Evolution, Time, Production and the Environment*, Springer, Heidelberg, 1990.

Faber, M., Manstetten, R., and Proops, J., Toward an open future: ignorance, novelty, and evolution, in *Ecosystem Health: New Goals for Environmental Management*, Costanza, R., Norton, B., and Haskell, B., Eds., Island Press, Washington, D.C., 1992, 72.

Fabricius, C., *Guide to Environmental Legislation in Australia and New Zealand: a Sum-mary and Brief Description of Environment Protection and Related Legislation of the Federal, State, and Territory Governments of Australia, and of New Zealand*, 4th ed., Dept. of the Environment, Sport and Territories, Report No. 29, AGPS, Canberra, 1994.

Feige, E. and Blau, D., The economics of natural resource scarcity and implications for development policy and international co-operation, in *Resources and Devel-opment*, Dorner, P. and El-Shafie, M., Eds., Croom Helm, London, 1980, 109.

Field, B., *Environmental Economics: an Introduction*, 2nd ed., McGraw-Hill, New York. 1998.

Fischer, S., Dornbusch, R., and Schmalensee, R., *Economics*, 2nd ed., McGraw-Hill, New York, 1988.

Fisher, I., *Nature of Capital and Income*, A. M. Kelly, New York, 1906.

Fisher, I., *100% Money*, A. M. Kelly, New York, 1935.

Folke, C., Hammer, M., Costanza, R., and Jansson, A., Investing in natural capital — why, what, and how, in *Investing in Natural Capital*, Jansson, A., Hammer, M., Folke, C., and Costanza, R., Eds., Island Press, Washington D.C., 1994, 1.

Foster, R., *Australian Economic Statistics: 1949–1950 to 1994–1995*, Reserve Bank of Australia Occasional Paper No. 8, 1996.

Freeman, A., *Air and Water Pollution Control: a Benefit-Cost Assessment*, John Wiley & Sons, New York, 1982.

Friedman, M., *Essays in Positive Economics*, University of Chicago Press, Chicago, 1953.

Galbraith, J., *Economics, Peace, and Laughter*, Harmondsworth, Middlesex, 1970.

Galligan, B. and Lynch, G., *Integrating Conservation and Development: Australia's Re-source Assessment Commission and the Testing Case of Coronation Hill*, Federalism Research Paper No. 14, Australian National University, Canberra, 1992.

Galloway, R., Story, R., Cooper, R., and Yapp, G., *Coastal Land of Australia*, CSIRO Division of Water and Land Resources, Natural Resources Series No. 1, AGPS, Canberra, 1984.

Garrod, B., Are economic globalization and sustainable development compatible? Business strategy and the role of the multinational enterprise, *Int. J. Sustain. Dev.*, 1(1), 43, 1998.

Georgescu-Roegen, N., *The Entropy Law and the Economic Process*, Harvard University Press, Cambridge, MA, 1971.

Georgescu-Roegen, N., The entropy law and the economic problem, in *Toward a Steady State Economy*, Daly, H., Ed., W. H. Freeman, San Francisco, CA, 1973, 37.

Georgescu-Roegen, N., Comments on the papers by Daly and Stiglitz, in *Scarcity and Growth Reconsidered*, Smith, V. K., Ed., Johns Hopkins University Press, Baltimore, MD, 95, 1979.

Giarini, O., The modern economy as a service economy: the production of utilisation value, in *Real-Life Economics: Understanding Wealth Creation*, Ekins, P. and Max-Neef, M., Eds., Routledge, London, 1992, 136.

Gilder, G., *Wealth and Poverty*, Basic, New York, 1981.

Goldschmidt-Clermont, L., Measuring households' non-monetary production, in *Real-Life Economics: Understanding Wealth Creation*, Ekins, P. and Max-Neef, M., Eds., Routledge, London, 1992, 265.

Goldsmith, E., World ecological areas programme: the problem, *The Ecologist*, 10(1), 1, 1980.

Goodland, R. and Ledec, G., Neoclassical economics and the principles of sustainable development, *Ecol. Model.*, 38, 19, 1987.

Gowdy, J., The social context of natural capital: the social limits to sustainable development, *Int. J. Soc. Econ.*, 21(8), 43, 1994a.

Gowdy, J., *Coevolutionary Economics: the Economy, Society, and the Environment*, Kluwer Academic Publishers, Boston, 1994b.

Gowdy, J. and McDaniel, C., One world, one experiment: addressing the biodiversity-economics conflict, *Ecol. Econ.*, 15, 181, 1995.

Graetz, R., Wilson, M., and Campbell, S., *Landcover Disturbance over the Australian Continent: a Contemporary Assessment*, Department of Environment, Sports, and Territories, Biodiversity Series Paper No. 7, 1995.

Grossman, G. and Krueger, A., Economic growth and the environment, *Q. J. Econ.*, 112, 353, 1995.

Hamilton, K., Green adjustments to GDP, *Res. Pol.*, 20(3), 155, 1994.

Hamilton, K, Pearce, D., Atkinson, G., Gomez-Lobo, A., and Young, C., *The Policy Implications of Natural Resource Accounting*, Centre for Social and Economic Research on the Global Environment, London, 1993.

Hanley, N. and Moffatt, I., Faichney, R., and Wilson, M., Measuring sustainability: a time series of alternative indicators for Scotland, *Ecol. Econ.*, 28(1), 55, 1999.

Hardin, G., The tragedy of the commons, *Science*, 162, 1243, 1968.

Hartwick, J., Intergenerational equity and the investing of rents from exhaustible resources, *Am. Econ. Rev.*, 65(5), 972, 1977.

Hartwick, J., Substitution among exhaustible resources and intergenerational equity, *Rev. Econ. Stud.*, 45, 347, 1978.

Haskell, B., Norton, B., and Costanza, R., Introduction: what is ecosystem health and why should we worry about it?, in *Ecosystem Health: New Goals for Environmental Management*, Costanza, R., Norton, B., and Haskell, B., Eds., Island Press, Washington, D.C., 1992, 3.

Haught, J., *The Promise of Nature: Ecology and Cosmic Purpose*, Paulist Press, Mahwah, NJ, 1993.

Hawtrey, R., The need for faith, *Econ. J.*, 56, 351, 1946.

Hayek, F., The use of knowledge in society, *Am. Econ. Rev.*, 35(4), 519, 1945.

Heer, D., Marketable licences for babies: Boulding's proposal revisited, *Soc. Biol.*, Spring, 1, 1975.

Henderson, L., *The Fitness of the Environment: an Inquiry into the Biological Significance of the Properties of Matter*, Beacon Press, Boston, 1913.

Hicks, J., *Value and Capital*, 2nd ed., Clarendon Press, London, 1946.

Hinterberger, F., Biological, cultural, and economic evolution and the economy-ecology Rrelationship, in *Toward Sustainable Development: Concepts, Methods, and Policy*, van den Bergh, J. and van der Straaten, J., Eds., Island Press, Washington, D.C., 1994, 57.

Hirsch, F., *The Social Limits to Growth*, Routledge & Kegan Paul, London, 1976.

Hirschman, A., Rival interpretations of market society: civilising, destructive, or feeble?, *J. Econ. Lit.*, XX, 1463, 1982.

Hobson, J., *Wealth and Life: a Study in Values*, MacMillan, London, 1929.

Hodge, I., *Environmental Economics*, MacMillan, London, 1995.

Hodgson, G., *Economics and Institutions*, Polity Press, Cambridge, 1988.

Hodgson, G., Evolution and intention in economic theory in *Evolutionary Theories of Economic and Technological Change*, Saviotti, P. and Metcalfe, J., Eds., Harwood Academic Publishers, Reading, 1991, 108.

Holland, J. and Miller, J., Artificial adaptive agents in economic theory, *Am. Econ. Rev.*, 81, 365, 1991.

Holmberg, J. and Sandbrook, R., Sustainable development: what is to be done?, in *Policies for a Small Planet*, Holmberg, J., Ed., Earthscan Publications, London, 1992, 19.

Horkheimer, M., *Eclipse of Reason*, Oxford University Press, New York, 1947.

Howarth, R. and Norgaard, R., Intergenerational resource rights, efficiency, and social optimality, *Land Econ.*, 66(1), 1, 1990.

Hundloe, T., Sustainable development: environmental limits and the limits to economics, Industry Commission Occasional Paper, 1991.

Hutchinson, G., Efficiency gains through privatisation of UK industries, in *Privatisation and Economic Efficiency*, Ott, A. and Hartley, K., Eds., Edward Elgar, Aldershot, 1991, 87.

Iggulden, J., *Silent Lies*, Evandale, Bellingen, 1996.

Ise, J., The theory of value as applied to natural resources, *Am. Econ. Rev.*, 15, 284, 1925.

Jackson, J., McIver, R., and McConnell, C., *Economics*, 4th ed., McGraw-Hill, Sydney, 1994.

Jackson, T. and Marks, N., Consumption, sustainable welfare, and human needs with reference to UK expenditure patterns between 1954 and 1994, *Ecol. Econ.*, 28(3), 421, 1999.

Jackson, T. and Stymne, S., *Sustainable Economic Welfare in Sweden: a Pilot Index 1950–1992*, Stockholm Environment Institute, The New Economics Foundation, 1996.

Jacobs, M., What is socio-ecological economics?, *Ecol. Econ. Bull.*, 1(2), 14, 1996.

Jacobsen, V., Property rights, prices and sustainability, Working Papers in Economics, Department of Economics, The University of Waikato, Hamilton, No. 91/5, September 1991.

Jaffe, A., Peterson, S., Portney, P., and Stavins, R., Environmental regulation and the competitiveness of US manufacturing: what does the evidence tell us?, *J. Econ. Lit.*, 33(1), 132, 1995.

Jansson, A., Hammer, M., Folke, C., and Costanza, R., Eds., *Investing in Natural Capital*, Island Press, Washington, D.C., 1994.

Johnson, L., *A Morally Deep World*, Cambridge University Press, Cambridge, 1991.

Kafka, P., Conditions of creation: is there hope in spite of the entropy law?, paper presented at the First International Conference of the European Association for Bioeconomic Studies, Rome, 1991.

Kapp, W., *The Public Cost of Private Enterprise*, Schocken Books, New York, 1951.

Kluckholm, C., *Signification and Significance: a Study of the Relation of Signs and Values*, Harvard University Press, Cambridge, 1964.

Knight, F., *Risk, Uncertainty, and Profit*, Houghton Mifflin, Boston, 1921.

Knight, F., *The Economic Organisation*, Harper & Row, London, 1933.

Kryger, T., Trends in Unemployment and Underemployment, Department of the Parliamentary Library, Parliamentary Research Service, Background Paper No. 36, AGPS, Canberra, 1933.

Laszlo, E., *The Systems View of the World*, G. Braziller, New York, 1972.

Lauderdale, J., *An Inquiry into the Nature and Origin of Public Wealth and into the Means and Causes of Its Increase*, 2nd ed., Constable, Edinburgh, 1819.

Lawn, P., In defence of the strong sustainability approach to national income accounting, *Environ. Tax. Acc.*, 3(1), 29, 1998.

Lawn, P., Grounding the ecol. Econ. Paradigm with ten core principles, *International Journal of Agricultural Resources, Governance and Ecology* (forthcoming).

Lawn, P., On Georgescu-Roegen's contribution to ecological economics, *Ecol. Econ.*, 29(1), 5, 1999b.

Lawn, P. and Sanders, R., Has Australia surpassed its optimal macroeconomic scale? Finding out with the aid of benefit and cost accounts and a sustainable net benefit index, *Ecol. Econ.*, 28(2), 213, 1999.

Lele, S., Sustainable development: a critical review, *World Dev.*, 19(6), 607, 1991.

Leipert, C., From gross to adjusted national product, in *The Living Economy: a New Economics in the Making*, Ekins, P., Ed., Routledge & Kegan Paul, London, 1986, 132.

Leonard, H., *Pollution and the Struggle for the World Product: Multinational Corporations, Environment, and International Comparative Advantage*, Cambridge University Press, Cambridge, 1988.

Leontief, W., *The Structure of American Economy, 1991–1939*, Oxford University Press, New York, 1941.

Lintott, J., Environmental accounting: useful to whom and for what?, *Ecol. Econ.*, 16(3), 179, 1996.

Lipsey, R. and Lancaster, K., The general theory of second best, *Rev. Econ. Stud.*, 24, 11, 1956.

Lone, O., Environmental and resource accounting, in *Real-Life Economics: Understanding Wealth Creation*, Ekins, P. and Max-Neef, M., Eds., Routledge, London, 1992, 239.

Lovelock, J., *Gaia: a New Look at Life on Earth*, Oxford University Press, Oxford, 1979.

Lovelock, J., *Ages of Gaia: a Biography of Our Living Planet*, Norton & Company, New York, 1988.

Lowi, T., The public character of private markets, Cornell University mimeo, draft July 10, 1985.

Lowi, T., Risks and rights in the history of American governments, in *Risk*, Burger, J., Ed., Michigan Press, Ann Arbor, MI, 1993, 17.

Marshall, A., *Principles of Economics*, 8th ed., Macmillan, London, 1925.

Maslow, A., *Motivation and Personality*, Harper & Row, New York, 1954.

Max-Neef, M., Economic growth and quality of life, *Ecol. Econ.*, 15(2), 115, 1995.

McDonald, J., Legal framework and critical issues, in *International Trade, Investment, and Environment*, Buckley, R. and Wild, C., Eds., Proc. 1993 Fenner Conference on the Environment, 27–29 July 1993, Australian Academy of Science, Canberra, Griffith University Publishers, 1994, 65.

McLeod, J. and Chaffee, S., The construction of social reality, in *The Social Influence Processes*, Tedeschi, J., Ed., Aldine-Atherton, Chicago, 1972.

Meadows, D., Herman Daly's farewell address to the World Bank, *ISEE Newsl.*, 5(4), 13, 1994.

Meadows, D. H., Meadows, D. L., Randers, J., and Behrens, W., III, Eds., *The Limits to Growth*, Universe Books, New York, 1972.

Megginson, W., Nash, R., and Van Randenborgh, M., The financial and operating performance of newly privatised firms: an international empirical analysis, *J. Financ.*, 49(2), 403, 1994.

Miles, I., Social indicators for real-life economics, in *Real-Life Economics: Understanding Wealth Creation*, Ekins, P. and Max-Neef, M., Eds., Routledge, London, 1992, 265.

Mishan, E., *The Costs of Economic Growth*, Staples Press, London, 1967.

Mishan, E., The growth of affluence and the decline of welfare, in *Economics, Ecology, and Ethics: Essays toward a Steady-State Economy*, Daly, H., Ed., W. H. Freeman, San Francisco, 1980, 267.

Mitlin, D., Sustainable development: a guide to the literature, *Environ. Urban.*, 4(1), 111, 1992.

Mokyr, J., Is economic change optimal?, *Am. Econ. Hist. Rev.*, XXXII(1), 3, 1992.

Moss, M., Ed., *The Measurement of Economic and Social Performance*, National Bureau of Economic Research, New York, 1973.

Murray, C., *Losing Ground: American Social Policy, 1950–1980*, Basic Books, New York, 1984.

National Greenhouse Gas Inventory Committee, *Workbook for Waste*, Workbook No. 8, Revision No. 1, AGPS, Canberra, 1996.

Nelson, R. and Winter, S., *An Evolutionary Theory of Economic Change*, Harvard University Press, Cambridge, 1982.

Nicolis, G. and Prigogine, I., *Self-Organisation in Non-Equilibrium Systems*, John Wiley & Sons, New York, 1977.

Niskanen, W., *Bureaucracy and Representative Government*, Aldine Atherton, Chicago, 1971.

Nordhaus, W., Too slow or not too slow: the economics of the greenhouse effect, *Econ. J.*, 101, 920, 1991.

Nordhaus, W. and Tobin, J., Is economic growth obsolete?, in *Economic Growth*, The National Bureau of Economic Research, Fiftieth Anniversary Colloquium, Columbia University Press, New York, 1972.

Norgaard, R., Coevolutionary development potential, *Land Econ.*, 60, 160, 1984.

Norgaard, R., Environmental economics: an evolutionary critique and a plea for pluralism, *J. Environ. Econ. Manage.*, 12, 382, 1985.

Norgaard, R., Thermodynamic and economic concepts as related to resource use policies: synthesis, *Land Econ.*, 64, 325, 1986.

Norgaard, R., Sustainable development: a co-evolutionary view, *Futures*, December, 606, 1988.

Norgaard, R., The case for methodological pluralism, *Ecol. Econ.*, 1, 37, 1989.

Norgaard, R., Economic indicators of resource scarcity: a critical essay, *J. Environ. Econ. Manage.*, 19, 19, 1990.

Norgaard, R., Coevolution of economy, society and environment, in *Real-Life Economics: Understanding Wealth Creation*, Ekins, P. and Max-Neef, M., Eds., Routledge, London, 1992, 76.

Norgaard, R., *Development Betrayed: the End of Progress and a Coevolutionary Revisioning of the Future*, Routledge, New York, 1994.

Norton, B., On the inherent danger of undervaluing species, in *The Preservation of Species*, Norton, B., Ed., Princeton University Press, Princeton, 1986, 110.

Noss, R. and Cooperrider, A., *Saving Nature's Legacy — Protecting and Restoring Biodiversity*, Island Press, Washington, D.C., 1994

O'Connor, M., Non-market codependencies and the conditions of market performance, Working Papers in Economics No. 63, Department of Economics, The University of Auckland, 1989a.

O'Connor, M., Fables of the bees: equilibrium and coevolutionary conceptions of the problem of externality, Working Papers in Economics No. 64, Department of Economics, The University of Auckland, 1989b.

O'Connor, M., Entropy, structure, and organisational change, *Ecol. Econ.*, 3, 95, 1991a.

O'Connor, M., Passing on the gift: an input-output analysis of the dynamics of sustainable activity, Working Papers in Economics No. 85, Department of Economics, The University of Auckland, 1991b.

O'Connor, M., Entropic irreversibility, involuntary exchange, global instability and uncontrolled economy-environment coevolution, Working Papers in Economics No. 89, Department of Economics, The University of Auckland, 1991c.

OECD, *Fighting Noise: Strengthening Noise Abatement Policies*, OECD, Paris, 1986.

OECD, Wetlands in a changing world, mimeo, 28 July 1990.

Okun, A., *Equality and Efficiency: the Big Trade-Off*, Brookings Institute, Washington, D.C., 1975.

Ostrom, E., *Governing the Commons: the Evolution of Institutions for Collective Action*, Cambridge University Press, Cambridge, 1990.

Paijmans, K., Galloway, R., Faith, R., Fleming, P., Haantjens, H., Heyligers, P., Kalma, J., and Loffler, E., Aspects of Australian Wetlands, CSIRO Division of Water and Land Resources Technical Paper No. 44, 1985.

Pasek, J., Obligations to future generations: a philosophical note, *World Dev.*, 20(4), 513, 1992.

Paterson, J., A review of the cost of road accidents to road safety, Report No. NR/23 for the Department of Transport, AGPS, Canberra, 1973.

Pearce, D., Foundations of ecological economics, *Ecol. Model.*, 38, 9, 1987.

Pearce, D., Barbier, E., and Markandya, A., Sustainable development and cost-benefit analysis, LEEC Paper 88-03, London Environmental Economics Centre, London.

Pearce, D., Markandya, A., and Barbier, E., *Blueprint for a Green Economy*, Earthscan, London, 1989.

Pearce, D., Barbier, E., and Markandya, A., *Sustainable Development: Economics and Environment in the Third World*, Edward Elgar, Aldershot, 1990.

Pearce, D. and Turner, R., *Economics of Natural Resources and the Environment*, Harvester Wheatsheaf, London, 1990.

Pearce, D. and Atkinson, G., Capital theory and the measurement of sustainable development: an indicator of weak sustainability, *Ecol. Econ.*, 8, 103, 1993.

Pearce, D. and Warford, J., *World without End: Economics, Environment, and Sustainable Development*, Oxford University Press, New York, 1994.

Pearce, D., Hamilton, K., and Atkinson, G., Measuring sustainable development: progress on indicators, *Environ. Dev. Econ.*, 1(1), 85, 1996.

Perrings, C., Conservation of mass and instability in a dynamic economy-environment system, *J. Environ. Econ. Manage.*, 13, 199, 1986.

Perrings, C., *Economy and Environment: a Theoretical Essay on the Interdependence of Economic and Environmental Systems*, Cambridge University Press, Cambridge, 1987.

Pezzey, J., Economic analysis of sustainable growth and sustainable development, World Bank Environmental Working Paper No. 15, Washington, D.C., 1989.

Pigram, J., *Issues in the Management of Australia's Water Resources,* Longman Cheshire, Melbourne, 1986.

Pindyck, R. and Rubinfield, D., *Microeconomics,* 3rd ed., Prentice Hall, NJ, 1995.

Pitchford, J., *Australia's Foreign Debt: Myths and Realities,* Allen & Unwin, Sydney, 1990.

Pizzigati, S., *The Maximum Wage,* Apex Press, New York, 1992.

Platt, J., Social traps, *Am. Psychol.,* 28, 641, 1973.

Polanyi, K., *The Great Transformation,* Beacon Press, Boston, 1994.

Polanyi, K., Our obsolete market mentality, in *Primitive, Archaic, and Modern Economics,* Dalton, G., Ed., Archer Books, Garden City, NY, 1968, 59.

Power, T., *The Economic Pursuit of Quality,* M. E. Sharpe, Armonk, NY, 1988.

Prakash, A. and Gupta, A., Are efficiency, equity, and scale independent?, *Ecol. Econ.,* 10(2), 89, 1994.

Prigogine, I., *Introduction to Non-Equilibrium Thermodynamics,* John Wiley & Sons, New York, 1962.

Prigogine, I. and Stengers, I., *Order out of Chaos,* Heinemann, London, 1984.

Ratnayake, R. and Wydeveld, M., The multinational corporation and the environment: testing the pollution haven hypothesis, Department of Economics Working Paper Series No. 179, University of Auckland, 1998.

Redclift, M., The meaning of sustainable development, *Geoforum,* 23(3), 395, 1992.

Redefining Progress, Gross production vs genuine progress, excerpt from *The Genuine Progress Indicator: Summary of Data and Methodology,* San Francisco, 1995.

Rees, W., Revisiting carrying capacity: area-based indicators of sustainability, *Pop. Environ.,* 17, 195, 1996.

Rees, W. and Wackernagel, M., Ecological footprints and appropriated carrying capacity: measuring the natural capital requirements of the human economy, in *Investing in Natural Capital,* Jansson, A., Hammer, M., Folke, C., and Costanza, R., Eds., Island Press, Washington, D.C., 1994, 362.

Repetto, R., Magrath, W., Wells, M., Beer, C., and Rossini, F., *Wasting Assets: Natural Resources in the National Income Accounts,* World Resources Institute, Washington, D.C., 1989.

Robinson, J., *Economic Philosophy,* C. A.Watts & Co., London, 1962.

Rosenberg, K. and Oegema, T., *A Pilot ISEW for The Netherlands 1950–1992,* Instituut Voor Milieu — En Systeemanalyse, Amsterdam, 1995.

Rossi, C. and Tiezzi, E., Eds., *Ecological Physical Chemistry,* Proc. International Workshop, 8–12 November 1990, Sienna, Italy, Elsevier Science Publishers, Amsterdam, 1990.

Rothschild, M., *Bionomics: the Inevitability of Capitalism,* Futura Publications, London, 1990.

Ruttan, V., Constraints on the design of sustainable systems of agricultural production, *Ecol. Econ.,* 10(3), 209, 1994.

Rymes, T., Some theoretical problems in accounting for sustainable consumption, Carleton Economic Papers, 92-02, 1992.

Sachs, I., Sustainable development, decentralised bio-industrialisation and new rural–urban configurations: India, Brazil, mimeo, 1989.

Samuelson, P., Nordhaus, W., Richardson, S., Scott, G., and Wallace, R., *Economics,* 3rd Australian ed., McGraw-Hill, Sydney, 1992.

Saviotti, P. and Metcalfe, J., Eds., *Evolutionary Theories of Economic and Technological Change,* Harwood Academic Publishers, Reading, 1991.

Scarth, W., *Macroeconomics: an Introduction to Advanced Methods*, Harcourt Brace Jovanovich, Toronto, 1988.

Schumacher, E., The age of plenty: a Christian view, in *Toward a Steady State Economy*, Daly, H., Ed., W. H. Freeman, San Francisco, 1973, 126.

Schumpeter, J., *History of Economic Analysis*, Oxford University Press, New York, 1954.

Seldon, T. and Song, D., Environmental quality and development: is there a Kuznets curve for air pollution?, *J. Environ. Econ. Manage.*, 27, 147, 1964.

Sen, A., *On Ethics and Economics*, Basil Blackwell, Oxford, 1987.

Silverberg, G., Modelling economic dynamics and technical change: mathematical approaches to self-organisation and evolution, in *Technical Change and Economic Theory*, Dosi, G., Freeman, C., Nelson, R., Silverberg, G., and Soete, L., Eds., Pinter Publishers, London, 1988, 531.

Simon, J., *The Ultimate Resource*, Princeton University Press, NJ., 1981.

Simon, J. and Kahn, H., *The Resourceful Earth*, Basil Blackwell, Oxford, 1984.

Singer, R., Comment, in *The Measurement of Economic and Social Performance*, Moss, M., Ed., National Bureau of Economic Research, New York, 1973, 532.

Smart, J. and Williams, B., *Utilitarianism: For and Against*, Cambridge University Press, Cambridge, 1973.

Smith, V. K., Ed., *Scarcity and Growth Reconsidered*, Johns Hopkins University Press, Baltimore, MD, 1979.

Soddy, F., *Cartesian Economics: the Bearing of Physical Science upon State Stewardship*, Hendersons, London, 1922.

Soddy, F., *Wealth, Virtual Wealth and Debt*, George Allen and Unwin, London, 1926.

Solow, R., The economics of resources or the resources of economics, *Am. Econ. Rev.*, 64, 1, 1974.

Solow, R., Georgescu-Roegen versus Solow/Stiglitz, *Ecol. Econ.*, 22(3), 267, 1997.

State of the Environment Advisory Council, *State of the Environment in Australia 1996*, CSIRO Publishing, Collingwood, 1996.

Steadman, L. and Bryan, R., Cost of road accidents in Australia, Bureau of Transport and Communications Occasional Paper 91, AGPS, Canberra, 1991.

Stevens, C., Do environmental policies affect competitiveness?, *OECD Ob.*, 183, 22, 1993.

Stiglitz, J., A neoclassical analysis of the economics of natural resources, in *Scarcity and Growth Reconsidered*, Smith, V. K., Ed., Johns Hopkins University Press, Baltimore, MD, 1979, 36.

Stockhammer, E., Hochrieter, H., Obermayr, B., and Steiner, K., The ISEW (index of sustainable economic welfare) as an alternative to GDP in measuring economic welfare, Draft, 1995.

Subcommittee on Health and Environment, Ozone layer depletion, U.S. House of Representatives, Serial 100-7, 9 March 1987.

Swaney, J., Externality and community, *J. Econ. Issu.*, XV(3), 615, 1981.

Tawney, R., *The Acquisitive Society*, G. Bell & Sons, London, 1921.

Tawney, R., *Religion and the Rise of Capitalism*, Harmondsworth, Middlesex, 1926.

Tawney, R., Book review of *Wealth and Life* by J. A. Hobson, *Pol. Q.* I, 276, 1930.

Tinbergen, J., *On the Theory of Economic Policy*, North-Holland, Amsterdam, 1952.

Tisdell, C., Sustainable development: differing perspectives of ecologists and economists, and relevance to LDCs, *World Dev.*, 16(3), 373, 1988.

Troub, R., Kenneth Boulding: economics from a different perspective, *J. Econ. Issues*, XII(2), 501, 1978.

Tullock, G., *The Politics of Bureaucracy*, University of America Press, Lanham, 1965.

Turner, R. and Pearce, D., Sustainable economic development: economic and ethical principle, *Economics and Ecology: New Frontiers and Sustainable Development*, Barbier, E., Ed., Chapman & Hall, London, 1993, 177.

Ulanowicz, R., Contributory values of ecosystem resources, in *Ecological Economics: the Science and Management of Sustainability*, Costanza, R., Ed., Columbia University Press, New York, 1991, 190.

United Nations Development Programme (UNDP), *Human Development Report*, Oxford University Press, New York, 1994.

Usback, S. and James, R., *A Directory of Important Wetlands in Australia*, Australian Nature Conservation Agency, Canberra, 1993.

Veblen, T., *The Theory of the Leisure Class: an Economic Study of Institutions*, Macmillan, New York, 1899.

Victor, P., Indicators of sustainable development: some lessons from capital theory, *Ecol. Econ.*, 4, 191, 1991.

Vining, A. and Boardman, A., Ownership versus competition: efficiency in public enterprise, *Public Choice*, 73, 205, 1992.

Vitousek, P., Ehrlich, P., Ehrlich, A., and Matson, P., Human appropriation of the products of photosynthesis, *BioScience*, 36, 368, 1986.

Wackernagel, M. and Rees, W., *Our Ecological Footprint: Reducing Human Impact on the Earth*, New Society Publishers, Gabriola Island, 1996.

Wackernagel, M., Onisto, L., Bello, P., Callejas Linares, A., Susana Lopez Falfan, S., Mendez Garcia, J., Suarez Guerrero, A. I., and Suarez Guerrero, Ma. G., National natural capital accounting with the ecological footprint concept, *Ecol. Econ.*, 29(3), 375-1999.

Waddington, C., *Tools of Thought*, Paladin, St. Albans, 1977.

Walker, J., Estimates of the costs of crime in Australia, *Trends and Issues in Crime and Criminal Justice*, No. 39, Australian Institute of Criminology, Canberra, 1992.

Wallace, R. and Norton, B., Policy implications of Gaian theory, *Ecol. Econ.*, 6, 103, 1992.

WCN/IUCN, *The World Conservation Strategy: Living Resource Conservation for Sustainable Development*, Gland, Switzerland, 1980.

Weber, M., *The Protestant Ethic and the Spirit of Capitalism*, Unwin University Books, London, 1930.

Weiss, P., The living system: determinism stratified, in *Beyond Reductionism: New Perspectives in the Life Sciences*, Koestler, A. and Smythies, F., Eds., Hutchinson New York, 1969, 3.

Weisskopf, W., *Alienation and Economics*, E. P. Dutton & Co., New York, 1971.

Weisskopf, W., Economic growth versus existential balance, in *Towards a Steady State Economy*, Daly, H., Ed., W. H. Freeman, San Francisco, 1973, 240.

Weitzman, M., *The Share Economy*, Harvard University Press, Cambridge, MA, 1984.

Wheelock, J., The household in the total economy, in *Real-Life Economics: Understanding Wealth Creation*, Ekins, P. and Max-Neef, M., Eds., Routledge, London, 1972, 124.

Wilber, C., The new economic history re-examined: R. H. Tawney on the origins of capitalism, *Am. J. Econ. Sociol.*, 33, 249, 1974.

Williams, R., *A Sociological Interpretation of American Society*, 3rd ed., A. Knopf, New York, 1970.

Williamson, O., *Markets and Hierarchies: a Study in the Economics of Internal Organisation*, Free Press, New York, 1975.

World Bank, Environment, growth, and development, Development Committee Pamphlet No.14, Washington, D.C., 1987.

World Bank, *World Development Report*, Oxford University Press, Oxford, 1992.

World Commission on Environment and Development (WCED), *Our Common Future*, Oxford University Press, Oxford, 1987.

World Resources Institute (WRI), *Resource Flows: the Material Basis of Industrial Economies*, World Resources Institute, Washington, D.C., 1997.

Zolotas, X., *Economic Growth and Declining Social Welfare*, New York University Press, New York, 1981.

Index

Milton Keynes UK
Ingram Content Group UK Ltd.
UKHW021905071024
449327UK00021B/1627